THE TANK MUSEUM

DATA BOOK OF WHEELED VEHICLES

Army Transport

1939-1945

LONDON HER MAJESTY'S STATIONERY OFFICE

First published 1983

ISBN 0 11 290408 4

Produced in the UK for HMSO
Dd 736166 C22 3/83

INTRODUCTION

The British Army lorry has always been a confirmed favourite with transport enthusiasts and vehicle preservationists alike. The very service it is called upon to perform gives it a rugged, purposeful air which is not so evident in its more colourful civilian counterpart. The tasks it performs lack the agressive *élan* of the tank and armoured car but lorries are every bit as important as these fighting vehicles to the successful conduct of war and have even greater importance in peacetime. Allied to their individuality is their variety, for there is a military counterpart to almost every civilian type, albeit geared to a more specialized role.

The history of mechanized military transport is much older than the armoured fighting vehicle; the British Army has been testing mechanical road vehicles for haulage and load carrying since 1859, but it was not until 1914 that they came into their own. The First World War finally proved that the forbearing horse had had his day. As someone pointed out, lorries don't catch diseases, and never need feeding when they are not working. The concept of a purpose-built military vehicle as distinct from a modified commercial type dates from the years between the wars. Before this time, a Government-sponsored subsidy scheme ensured that large numbers of trucks were built to an acceptable common standard, available for call-up when hostilities occurred. In due course the divergent requirements of the War Office and the civilian operator reduced the effectiveness of this scheme and the military authorities were encouraged to develop many more specialized types.

When the plans were formulated they were arranged into a series of groups based on the load-carrying capability of the chassis. This was calculated with a generous margin of safety to take account of the various military accoutrements that the vehicle must carry, and the extra stresses induced by cross-country operation. These groupings ranged in payload value from 8 cwts to 10 tons but by far the most important were the 15 cwts, 30 cwts and 3 ton classes. The 8 cwt was never much more than a glorified pick-up truck, while vehicles above 3 tons were generally considered to be road bound and therefore readily adaptable from commercial heavy-haulage types. Thus it was the other three classes which were in the mainstream of military development. The 15 cwt was characterized by the front-end design which required the crewmen to sit either side of the engine in order to obtain maximum utilization of the chassis for load carrying. This led to the unique design typefied by the Morris-Commercial, Guy Ant and Bedford MWD models. In 1926 the Royal Army Service Corps, in conjunction with certain well-known manufacturers, perfected a form of articulating twin-axle rear bogie for six-wheel trucks which set new standards in military transport design. As a result both the 30 cwt and 3 ton classes were dominated by a range of forward-control lorries of relatively standardized design from firms such as Morris, Guy, AEC, Thornycroft, Albion and Crossley. A range of specialized body types was constructed for fitting to these chassis but the general pre-war practice was to operate mostly General Service (G.S.) wagons with a truck-type body, while the workshop, wireless and bridge-carrying bodies were stored seperately, to be fitted on mobilization.

The late 1930s witnessed a move towards the adoption of four-wheel drive in an effort to improve off-road performance. The outbreak of war in 1939 caught this process in its early stages and although the 4x4, as the Army termed it, soon began to dominate, the proven rigid six-wheelers remained in service for the duration of the conflict. Two particular aspects of military service that demand even more specialized products are the problems of artillery haulage and tank transportation. Between the wars the former practice was dominated by the tracked Dragon but once the four- and six-wheel-drive tractor reached maturity it soon eclipsed its predecessor. On the other hand the articulated tank transporter was a very slow starter. Had this type of vehicle been available in France in 1940 in adequate numbers, the effect of the Blitzkreig might have been somewhat diminished. Nothing is more certain to wear out a tank and its crew quicker than a long road journey on tracks, and the otherwise invulnerable Matildas could have been used far more judiciously. As it was, the famous Scammell was never available in the quantity required and Britain was obliged to turn to the United States to make up the shortfall with the equally legendary Diamond T.

When it got under way, transatlantic production was the ultimate key to success. The huge motor industry of Canada and the United States was readily harnessed to mass production. America produced phenomenal quantities of vehicles of all types to serve all the Allied nations, and in doing so introduced two famous words to the international vocabulary, JEEP and DUKW. Canada introduced a degree of standardization second to none with a range of vehicles which combined British design principles with American production techniques. The blunt nose of a Canadian Military Pattern truck probably figures somewhere in more pictures of British and Commonwealth operations than any other service vehicle.

This amazing variety of vehicles would have defeated the most conscientious transport officer and before confusion reigned he turned to his Data Book of Wheeled Vehicles. Here he could find the answers he sought – details of weight, size, carrying capacity and haulage ability – information without which his motorized charge would have been just so much useless metal. This reprint, a joint production of the Tank Museum and Her Majesty's Stationery Office, contains every item that appeared in the ultimate 1945 edition, and therefore covers almost every type of wheeled transport in the Allied inventory. In addition to primary examples mentioned above it includes amphibians, half-tracks, motorcycles and trailers with a smattering of light armoured vehicles where these employed conventional chassis.

The original edition consisted of loose-bound pages with gaps in the numbering sequence at the end of each section. These were to be filled with new entries as they were issued. The omission of a number of pages was an error in the original volume. This edition reproduces wherever possible two original pages per page with minimal change to the sequence. A new consecutive pagination has been added at the top of pages but the original page numbers have been reprinted at the foot in brackets as there are cross-references to these.

This book provides the researcher and the modeller with a wealth of information difficult to obtain from any other single source. It will be an invaluable addition to any transport enthusiast's library, the more so since the original is virtually unobtainable at any price.

DATA BOOK OF
WHEELED VEHICLES
FIFTH EDITION

This Fifth Edition is a reprint of the Fourth Edition, incorporating all amendments. No attempt has been made to include details of alterations to vehicles since 1945 nor have new vehicles produced since that date been included.

Issued by DGFV Ministry of Supply.

ARRANGEMENT OF BOOK

VEHICLE INFORMATION for all types except Tank Transporters, Trailers and motor cycles has been divided into three sections thus

1. CHASSIS INFORMATION

giving details of
Abridged chassis specification
Type of control (normal or forward)
Tyre equipment
Chassis dimensions, wheelbase, track turning circles and ground clearances.
Brief Performance figures.

WILL BE FOUND BY REFERRING TO

INDEX TO CHASSIS

BY MAKE

Index Pages Nos. IND.010 to IND.014

2. BODY INFORMATION

giving details of
Body design
Cab details
Special fitments e.g. winch, Generator, drawbar gear etc.

WILL BE FOUND BY REFERRING TO

INDEX TO VEHICLES

BY TYPE

Index Pages IND.015 onwards

ARRANGEMENT OF BOOK (contd.)

3. LATEST WEIGHT AND DIMENSIONS

including:-
Overall Dimensions
Inside Body Dimensions
Stripped Shipping Heights
Unladen Axle Weights
Laden Axle Weights.

WILL BE FOUND IN THE WEIGHT SHEETS

at the back of the book

Tank Transporters, Trailers and Motor Cycles

All relevant information for above types is included in the one section.

For Tank Transporters and Trailers see Index to Vehicles by Type.

For Motor Cycles see Index to Chassis by make.

SPECIAL INFORMATION

GENERAL DEFINITIONS OF TERMS AND ABBREVIATIONS

Br. = British
Ca. = Canadian
U.S. = U.S.A.

C.I. = Compression Ignition, Oil or Diesel engine

---

4 x 2 = Four wheel vehicle, rear wheel drive
4 x 4 - Four wheel vehicle, four wheel drive
6 x 4 = Six wheel vehicle, four rear wheel drive
6 x 6 = Six wheel vehicle, six wheel drive
4 x 2 - 2 = Semi-Trailer type, 4 x 2 motive unit coupled to a 2 wheel semi-trailer
4 x 2 - 4 = 4 x 2 motive unit coupled to a 4 wheeled semi-trailer

---

PERFORMANCE FIGURES

TRACTIVE EFFORT PER TON AT 100% is given for all tractors and the heavier types of vehicles.

PETROL CONSUMPTION FIGURES are based for average conditions, on metalled roads in the U.K. Dense traffic conditions, hilly country or poor roads will increase consumption considerably.

RADIUS OF ACTION indicates distance vehicle can travel between refuelling points based on m.p.g. figures as above.

TANK CAPACITY DATA

Shows total fuel carried in vehicle tanks including reserve, but excluding spare fuel carried in cans.

GROUND CLEARANCE DATA

Data is for minimum clearance laden. Belly clearances are also laden.

WEIGHT DATA

UNLADEN WEIGHTS are for the complete vehicle (chassis, cab and body) excluding all the following items:-

Wt. of fuel in tanks and spare petrol carriers.
Wt. of spare wheel and tyre.
Wt. of vehicle tools and chains.
Wt. of W.D. vehicle equipment, (picks, shovels etc.)
Wt. of driver and mate.

(N.B. Tank Transporter unladen weights include all these items.)

LADEN WEIGHTS are for the complete vehicle with full rated payload, fully equipped and ready for service (i.e. including all the equipment and personnel weights omitted from the unladen weight).

WEIGHTS OF STANDARD ITEMS

In the weight sheets at the back of the book, the column headed "Weights of Standard Items" shows the weight of fuel, oil and water, spare wheel and tyre, and vehicle tools and chains. The weight of these items is not included in the unladen weights.

STRIPPED SHIPPING HEIGHTS

STRIPPED SHIPPING HEIGHT Indicates the height that the vehicle can be reduced to when the superstructure is detached or cab folded down (in the case of vehicles with split or folding cabs).

Normally it is not necessary to reduce vehicle heights below 7' 0" to 6' 6" for shipping. Certain vehicles can however be cut down still further by more complicated disassembly. This is indicated as follows. The normal stripped height (entailing only simply disassembly such as removing superstructure and splitting cab top) is shown first and this is followed by a "+" sign with numerals indicating the number of inches the height can be further reduced by more complicated disassembly. E.G. 7' 0" + 16" indicates a normal stripped height of 7' 0", which can if necessary be reduced to 5' 8" by further disassembly.

MASTER INDEX
INDEX TO CHASSIS BY MAKE

Including semi-trailer types, but excluding full trailers which are shown on page 031

ABBREVIATIONS USED

Br. = British
C. = Canadian
U. = U.S.A.
4 x 2 = Four wheel rear wheel drive
4 x 4 = " " four " "
6 x 4 = Six " four rear wheel drive
6 x 6 = " " six wheel drive
4 x 2-2 = Semi-trailer type, 4 x 2 motive unit, coupled to a 2-whld. Semi-trailer.

---

x See pages IND.036 to 038 for alternative nomenclatures

x See pages IND.036-038 for alternative nomenclatures

This Index shows where the Body details are given for a particular type and make.

The relevant Chassis details will be found by referring to Index to Chassis by Make (Page IND. 010). Full Trailers and Motor Cycles are shown at the end.

This Index is arranged by load capacity; starting with Cars and working up to Tank Transporters and Tractors.

The Latest Weights and Dimensions are given in the Weight Sheets right at the back of the book (except in the case of Tank Transporters and Trailers where latest weights and dimensions are included in the text pages).

This Index shows all vehicles under their new Official W. D. Nomenclatures. *

*A few vehicles are however shown in the Index under two different sections: (a) where the new official nomenclature is not so widely known as an alternative name; (b) where the vehicle has a dual function. These few "unofficial" names are given in brackets to distinguish them from the official nomenclatures.

ABBREVIATIONS USED

Br. = British type
C. = Canadian type
U. = U.S.A. type

(* There is one minor variation from the new W. D. nomenclatures. This Data Book shews the semi-trailer drive notations as "4 x 2 - 2". In the new official nomenclatures a "plus" sign is used in place of the "minus" sign.)

CARS, UTILITIES AND LIGHT RECCY. CARS

AMBULANCES

	Page	Wt. Sh.
Ambulance, 2 stretcher 4 x 4		
C. Chevrolet	609	901
Br. Humber	13	901
Ambulance, 4-stretcher 4 x 2		
Br. Austin, Bedford	81	901
C. Ford	623	901
Ambulance, 4-stretcher 4 x 4		
U. Dodge	408	901

AMPHIBIANS

	Page	Wt. Sh.
Amphibian 5-cwt. 4 x 4 (Amphibious Jeep)		
U. Ford	692	902
Amphibian 10-cwt. Tracked G.S. (Amphibious Weasel, M.29C)		
U. Studebaker	690	902
Amphibian, 2½ ton 6 x 6 G.S.		
U. G.M.C. DUKW	694	902
Amphibian 4 ton 8 x 8 G.S.		
Br. Morris Terrapin	696	902

TRUCKS 8-CWT. TO 15-CWT.

	Page	Wt. Sh.
Truck, 8-cwt, 4 x 2 G.S.		
Br. Ford, Morris, Humber	11	902
C. Chevrolet, Ford	213	902
Truck, 8-cwt. 4 x 4 G.S.		
Br. Morris, Humber	11	902
C. Dodge	213	902
Truck, 8 cwt. 4 x 2 Fitted for Wireless		
Br. Morris, Humber	12	902
C. Ford	213	902
Truck, 8-cwt. 4 x 4 Fitted for Wireless		
Br. Morris, Humber	12	902

TRUCKS 8-CWT. TO 15-CWT. (contd.)

	Page	Wt. Sh.
Truck, 10-cwt, 4 x 4 G.S.		
U. Dodge Pickup	406	902
Truck, 10-cwt. Tracked, G.S. (WEASEL or M.29)		
U. Studebaker	689	903
Truck, 15-cwt. 4 x 2 G.S.		
Br. Bedford, Ford, Guy, Morris	18A	903
C. Chevrolet, Ford	216	903
C. Dodge	265	903
Truck, 15-cwt. 4 x 4 G.S.		
Br. Morris	611	903
Br. Guy	614	903
C. Chevrolet	613	903
C. Ford	615	903
Truck, 15-cwt. 4 x 2 A/A		
Br. Bedford, Ford	616	903
C. Chevrolet, Ford	616	903
Truck, 15-cwt. 4 x 4, Armoured Personnel		
C. G.M. Canada	683	903
U. White (M.3A1 Scout Car)	365	903
Truck, 15-cwt. 4 x 2, A/T		
Br. Bedford, Morris	20	904
Truck, 15-cwt. 4 x 2, Compressor		
Br. Morris	19	904
Truck, 15-cwt. 4 x 2, Light Warning		
Br. Ford, Guy	619	904
Truck, 15-cwt. 4 x 2, Office		
Br. Ford	617	904
Br. Morris	23	904
Truck, 15-cwt. 4 x 2 Recorder (P.C.C.)		
Br. Guy	619	904

	Page	Wt. Sh.
Truck, 15-cwt. 4 x 2 Water 200 gall.		
Br. Bedford, Morris	22	904
C. Dodge	265	904
Truck, 15-cwt. 4 x 2 Wireless (House)		
Br. Guy, Morris (latest type)	618	904
Br. Morris (Early type)	21	904
Truck, 15-cwt. 4 x 2 Fitted for Wireless		
Br. Bedford	115	904

15-CWT. HALF TRACKS

	Page	Wt. Sh.
Truck, 15-cwt. Half Tracked G.S.		
U. International (M.14 modified)	681	904
Truck, 15-cwt. Half Tracked, Command (M.14 modified)		
U. International	681	904
Truck, 15-cwt. Half Tracked, Personnel (M.5, M.9 and M.14) (modified)		
U. International	680	904

LORRIES - 30-CWT.

	Page	Wt. Sh.
Lorry, 30 cwt, 4 x 2 G.S.		
Br. Austin, Bedford, Commer, Ford	28	905
Lorry, 30 cwt. 4 x 4 G.S.		
Br. Ford	48A	905
C. Chevrolet, Ford	219	905
U. Dodge	245	905
U. G.M.C.	249	905
Lorry, 30-cwt. 4 x 4, G.S. w/winch		
U. Chevrolet	411	905

	Page	Wt. Sh.
Lorry, 30-cwt. 6 x 4, G.S.		
Br. Morris	31	905
Lorry, 30 cwt. 4 x 4 A/T Portee		
Br. Morris	85	905
C. Chevrolet	268	905
Lorry, 30 cwt. 4 x 2 Battery Slave		
Br. Bedford	621	905
Lorry, 30-cwt. 6 x 4 Breakdown, light		
Br. Morris	34	905
Lorry, 30 cwt. 6 x 4 Office		
Br. Morris	32	905
Lorry, 30-cwt. 4 x 4 Tipping w/winch		
U. Chevrolet	411	905
Lorry, 30-cwt. 6 x 4, Winch		
Br. Morris	31	905

LORRIES 1½-3 TON

	Page	Wt. Sh.
Lorry, 1½-3 ton 4 x 4, House Type Bodies as under		
U. G.M.C.		
Basic Body Details	433	906
Tool and Bench Truck	434	906
Spare Parts Truck	434	906
Automotive Repair	435	906
Machine Shop	435	906
Small Arms Repair	436	906
Tank Maintenance	436	906
Welding	437	906
Lorry, 2½ ton, 6 x 6 G.S.		
U. Studebaker, G.M.C.	416-7	906
Lorry, 2½-5 ton, 6 x 4, G.S.		
U. Studebaker, G.M.C.	415	906

	Page	Wt. Sh.
Lorry, 2½ ton 6 x 6, Petrol 600 gal.		
U. Studebaker, G.M.C.	420	906
Lorry, 2½ ton 6 x 6, Store (Binned)		
U. Studebaker	420	906

LORRIES 3 TON

	Page	Wt. Sh.
Lorry, 3 ton 4 x 2 G.S.		
Br. Austin, Bedford, Commer, Dennis, Leyland	43	906
Br. Thornycroft	117	906
C. Chevrolet, Ford	222	906
C. latest Canadian Ford	623	906
C. Dodge (Early Stake Body)	267	907
C. Dodge (Latest G.S. Body)	267	907
U. Chevrolet	271	907
U. Dodge WF.32	273	907
U. Dodge V.K. 62	247	907
U. Ford 6-cyl.	275	907
U. G.M.C. A.C. 504	253	907
Lorry, 3 ton 4 x 4 G.S.		
Br. Albion, Austin, Bedford, Karrier, Thornycroft	51	907
Br. Ford	48B	907
C. Chevrolet, Ford (latest type)	627	907
C. Chevrolet, Ford (early type)	226	907
Lorry, 3 ton 4 x 4, G.S. with Winch		
Br. Karrier	629	907
C. Chevrolet Winch Lorry	628	908
Lorry, 3 ton 6 x 4, G.S.		
Br. Albion, A.E.C., Austin, Crossley Guy, Leyland, Thornycroft	67	908
U. Chevrolet-Thornton	243	908
Lorry, 3 ton 6 x 6, G.S.		
U. G.M.C.	251	908

	Page	Wt. Sh.
Lorry, 3 ton 4 x 4, A/T Portee, 6 pdr.		
Br. Austin, Bedford	131	908
Lorry, 3 ton 4 x 2, Battery Slave		
Br. Bedford	621	908
Lorry, 3 ton 6 x 4, Breakdown Gantry		
Br. Crossley, Guy, Leyland	68	908
Br. Austin	638	908
U. Dodge W.K.60	235	908
Lorry, 3 ton 4 x 2, Breakdown Light (Holmes Wrecker)		
C. Chevrolet, Ford	230	909
Lorry, 3 ton 4 x 4, Breakdown Loght (Holmes Wrecker)		
C. Chevrolet, Ford	230	909
Lorry, 3 ton 6 x 4, Camera		
Br. Leyland	648	909
Lorry, 3 ton 4 x 4, Cipher Office		
Br. Bedford	640	909
Lorry, 3 ton 4 x 4, Command H.P.		
Br. Bedford	640	909
Lorry, 3 ton 4 x 4, Command L.P.		
Br. Bedford	640	909
Lorry, 3 ton 6 x 4, Crane Turntable (Coles Crane)		
Br. Austin	636	909
Br. Crossley, Leyland	78	909
Lorry, 3 ton 6 x 4 Dark Room		
Br. Leyland	648	909
Lorry, 3 ton 6 x 4, Derrick		
Br. Crossley, Guy, Leyland	71	909

	Page	Wt. Sh.
Lorry, 3 ton 6 x 4, F.B.E.		
Br. Albion	72	909
Lorry, 3 ton 4 x 4, Fire Tender		
Br. Bedford	639	909
Lorry, 3 ton 4 x 4 Kitchen		
Br. Bedford	634	909
Lorry, 3 ton 6 x 4 Longlanding Bay		
Br. Albion	74	910
Lorry, 3 ton 4 x 4 Machinery ("House Type Body")		
Br. Ford	632	910
Lorry, 3 ton 4 x 4 Machinery ("Flat Floor" Body)		
Br. Albion	630	910
Br. Ford	631	910
Lorry, 3 ton 6 x 4 Machinery (House Type Body)		
Br. Albion, Leyland, Thornycroft	69	910
Lorry, 3 ton 6 x 4 Machinery ("Flat Floor Body")		
Br. Albion, Crossley, Guy, Karrier Layland, Thornycroft	70	910
Lorry, 3 ton 4 x 2 Machinery R.A.S.C. (See under "Workshop")		
Lorry, 3 ton 4 x 4 Machinery, 24 k.w. R.E.		
Br. Ford	633	910
Lorry, 3 ton 6 x 4, Machinery, 24 k.w. R.E.		
Br. Leyland	633	910
Lorry, 3 ton 4 x 4 Mobile Terminal Carrier		
Br. Bedford	640	910

	Page	Wt. Sh.
Lorry, 3 ton 4 x 4 Mobile Ops. Room		
Br. Bedford	637	910
Lorry, 3 ton 4 x 2, Petrol 800 gall.		
Br. Bedford	119	910
C. Ford	624	911
U. Brockway	276	911
Lorry, 3 ton 6 x 4, Photo Mechanical		
Br. Leyland	647	911
Lorry, 3 ton 6 x 4, Plotter		
Br. Leyland	649	911
Lorry, 3 ton 6 x 4, Pontoon		
Br. Albion, Leyland	73	911
Lorry, 3 ton 6 x 4, Printing		
Br. Leyland	646	911
Lorry, 3 ton 6 x 4, Processing		
Br. Leyland	645	911
Lorry, 3 ton 4 x 4, Recorder A.A. Mk.I (Westex Recorder)		
Br. Austin	637	811
Lorry, 3 ton 4 x 2, Searchlight		
Br. Thornycroft, Tilling Stevens	45	911
Lorry, 3 ton 6 x 4, Searchlight		
Br. Guy, Leyland	76	911
U. G.M.C.	255	911
Lorry, 3 ton 6 x 4, Small Box Girder		
Br. Albion, A.E.C., Karrier	75	911
Lorry, 3 ton 4 x 2, Store (Binned)		
Br. Bedford	121	211
C. Chevrolet, Ford	231	912

LORRIES 3 TON (contd.)

	Page	Wt. Sh.
Lorry, 3 ton 4 x 4, Store (Binned)		
C. Chevrolet, Ford	231	912
Lorry, 3 ton 4 x 4, T.E.V.		
Br. Bedford	640	912
Lorry, 3 ton 4 x 2, Tipping		
Br. Dennis	44	912
Br. Thornycroft 3-way	117	912
C Ford	223	912
C. Dodge	625	912
Lorry, 3 ton 6 x 4, Trestle and Sliding Bay		
Br. Albion, Leyland	74	912
Lorry, 3 ton 4 x 4, Troop Carrying		
Br. Bedford	129	912
Lorry, 3 ton 4 x 2, Water		
Br. Bedford 350 gallon	118	912
Br. Bedford 500 gallon	644	912
C. Dodge 450 gallon	644	912
Lorry, 3 ton 4 x 4 Winch		
C. Chevrolet	628	912
Br. Karrier (G.S. w/winch)	629	912
Lorry, 3 ton 6 x 4, Wireless		
Br. Guy	77	912
Br. Leyland	77	913
Lorry, 3 ton 6 x 4, Wireless H.P.		
Br. Thornycroft	123	913
Lorry, 3 ton 4 x 4, Wireless (I)		
Br. Bedford	640	913
Lorry, 3 ton 4 x 4, Wireless (R)		
Br. Bedford	640	913

LORRIES 3 TON (contd.)

	Page	Wt. Sh.
Lorry, 3 ton 4 x 4, Wireless H.P.		
Br. Bedford	640	913
Lorry, 3 ton 4 x 2, Workshop R.A.S.C.		
Br. Bedford, Commer.	120	913
C. Chevrolet, Ford	233	913
Lorry, 3 ton 4 x 4, Workshop R.A.S.C.		
C. Chevrolet, Ford	233	913
Lorry, 3 ton 4 x 2, X-Ray		
Br. Bedford	635	913

LORRIES OVER 3 TONS

(Including Semi-Trailer Types)

	Page	Wt. Sh.
Lorry, 4 ton 4 x 4 G.S. w/winch		
U. F.W.D. H.A.R.	650	913
Lorry 4 ton 6 x 6 G.S. w/winch		
U. Diamond T.	425	913
Lorry, 5 ton 4 x 2, G.S.		
U. Mack	653	913
Lorry, 5 ton 6 x 4 G.S.		
U. Studebaker, G.M.C.	417	913
Lorry, 5 ton 4 x 4 G.S.		
U. F.W.D. S.U.	665	913
Lorry, 6 ton 4 x 2, G.S.		
Br. Dennis, E.R.F., Foden, Maudsley	56	914
Br. Dennis Mk. II	659	914
U. International K.R. 10	286	914
Lorry, 6 ton 4 x 2 -2 G.S.		
Br. Bedford-Scammell semi-trailer	656	914
U. Mack semi-trailer	655	914
U. Mack semi-trailer early type	285	914

	Page	Wt. Sh.
Lorry 6 ton 4 x 4-2 G.S.		
Br. Bedford QL & semi-trailer	125	914
C. Ford & semi-trailer	658	914
Lorry, 6 ton 6 x 4, Crane Turntable		
Br. Thornycroft	660	914
Lorry, 6 ton 4 x 2-2 Petrol 1750 gal.		
Br. Bedford-Scammell (S. trailer)	657	914
Lorry, 6 ton 4 x 2 Petrol, 1600 gal.		
U. Brockway	277	914
Lorry, 7 ton 4 x 2 G.S.		
U. International	287	914
Lorry, 7 ton 4 x 2-2 G.S.		
U. Studebaker (semi-trailer)	419	914
Lorry, 7 ton 4 x 2-2, Low loading		
U. International (semi-trailer)	663	914
Lorry, 10 ton 6 x 4, G.S.		
Br. Albion, (Early type)	60	914
Br. Foden, Leyland (Early types)	60	915
Br. Albion (Latest Type)	57A	915
Br. Foden (Latest Type)	127	915
Br. Leyland (Latest Type) Mk.II	667	915
U. Mack NR, White	668	916
Lorry, 10 ton 4 x 2-2 G.S.		
U. Brockway	284	915
U. International	664	915
Lorry, 10 ton 6 x 4, Auto Processing		
Br. Foden	645	915
Lorry, 10 ton 6 x 4, Enlarging		
Br. Foden	645	915
Lorry, 10 ton 6 x 4, Printing		
Br. Foden	646	915
Lorry, 10 ton 6 x 4, Photo Mechanical		
Br. Foden	647	915

TANK TRANSPORTERS (All types)

(For Wts. and Dimensions see Text Pages)

	Page
Trailer 7½ ton 6 wh. Light Recovery	507
Transporter, 13 ton 6 x 4	
U. Mack N.R.	511
Transporter, 15 ton 6 x 4-4 Carriage of cable Drums	
Br. Albion (20 ton model derated)	531
Transporter, 18 ton 6 x 4	
U. White 920, Mack EXBX	521/2
U. White-Ruxtall	522
Transporter, 20 ton 6 x 4-4	
Br. Albion (See also under 15 ton)	531
Br. Scammell 6 x 4-8	539
U. Federal (Reo)	535
Transporter, 30 ton 6 x 4-8, Recovery	
Br. Scammell	543
U. Shelvoke & Drewry Diamond T	556
Tractor 6 x 4 for 40 ton Trailer	
U. Diamond T	547
40 ton 12 dual whl. Trailers for above	
Br. Crane Mk.I Type	551
Br. Dyson Mk.II Type	552
U. Rogers, Winter Weiss, Fruehauf (all to same design)	549
45 Ton Tracked Recovery Trailer	
Br. Boulton & Paul	558

TRACTORS
ARTILLERY AND A.A. TRACTORS
BREAKDOWN TRACTORS
A.T. PORTEES -
SELF PROPELLED GUNS

	Page	Wt. Sh.
Tractor 4 x 4 F/A		
Br. Morris (Latest)	669	915
Br. Morris (Early type)	84	915
C. Chevrolet, Ford	240	915
Br. Guy	83	915
Tractor 4 x 4 Medium (M.A.T.)		
Br. A.E.C.	86	915
U. F.W.D.	237	915
Tractor 6 x 6 Medium (Truck 6 ton 6 x 6)		
U. Mack N.M.5	429	915
U. White 666	431	916
Tractor 6 x 4 Heavy (H.A.T.)		
Br. Scammell	89	916
Br. Albion	677	916
Tractor 6 x 6 Heavy (Truck 7½ ton 6 x 6)		
U. Mack N.O.	675	916
Tractor, 6 x 4, Light A.A. (Bofors)		
Br. Morris	35	916
Tractor, 4 x 4 Light A.A. (Bofors)		
Br. Bedford	50A	916
C. Ford 30-cwt. (Early type)	671	916
C. Ford 3 ton (Latest type)	672	916

TRACTORS, ARTILLERY AND A.A. TRACTORS, BREAKDOWN TRACTORS, A.T. PORTEES, SELF PROPELLED GUNS (contd.)

	Page	Wt.	Sh.
Tractor, 4 x 4, A/T. 17 Pdr.			
Br. Morris	670		916
C. Chevrolet	670		816
Tractor, 6 x 4, Breakdown Heavy			
Br. Scammell	91		916
U. Mack LMSW	283		916
Tractor, 6 x 6 Breakdown Heavy			
U. Diamond T. (4 ton 6 x 6)	426		916
U. Ward La France (M.I.)	440		916
Tractor, 6 x 4, for 40 ton Trailer			
U. Diamond T.	547		916
Tractor, 4 x 2 for 6 ton semi-trailer (See Bedford-Scammell)			
Br. Bedford	656		916
(Portee 2 pdr. A.T.)			
Br. Morris	85		917
C. Chevrolet	268		917
(Portee and Fire, 6 pdr. A.T.)			
Br. Austin, Bedford	131		917
Carrier Self-Propelled 40 mm. A.A. (S.P. BOFORS)			
Br. Morris	135		917
C. Ford	673		917
(S.P. Predictor Vehicle)			
Br. Morris (Original vehicle)	137		917
Now converted to Tractor 4 x 4 A/T			
17 pdr. Conversion details	670		917

TRAILERS

(Note: Semi-Trailers are included with load carrying vehicles)

STANDARDISED CHASSIS NOMENCLATURES FOR CANADIAN VEHICLES

Three different chassis nomenclatures have been used for Canadian-built vehicles:-

(A) Simplified Model Designation used in Maintenance Manuals and Parts Lists.

(B) The Manufacturers Chassis Serial Number also quoted in Parts Lists.

(C) The D.M.S. Chassis Model Code which used to be stamped on the chassis name plated mounted on the vehicle.

It was thus possible for the same chassis to be known by three different nomenclatures. It was, therefore, arranged with D.M.S. Canada to standardise in future on

(A) The Simplified Model Designation

for use in Drivers Handbooks, Maintenance Manuals, Parts Lists and the chassis plate attached to the actual vehicle.

This Designation was chosen as the simplest and the one most widely used in the Services. All Canadian vehicles are now coming through using the Standard Designation.

Here is a list of Canadian Chassis Types in W.D. Service under:-

(A) Their Standardised Codes with, for reference purposes,

(B) The Makers Chassis Serial No.
and
(C) The superseded D.M.S. Chassis Model Code.

The Standard Designations under (A) for Chevrolet and Ford will probably be familiar - the Dodge designations are new.

CANADIAN CHEVROLET MODELS

	(A)	(B)	(C)
8 Cwt. 4 x 2	C-8	8420	C.8421
Heavy Utility 4 x 4) Complete Vehicle)	C-8A	8445	C.H.U.441
Heavy Utility 4 x 4) Chassis and Cab)	"	8448	"
15 cwt. 4 x 2	C-15	8421	C.15421
15 cwt. 4 x 4	C-15A	8444	C.15441
30 cwt. 4 x 4	C-30	8441	C.30444
3 ton 4 x 2	CC60L/X2	1543X2	C.60420
3 ton 4 x 4 134" W.B.	C-60S	8442	C.60444
3 ton 4 x 4 158" W.B.	C-60L	8443	C.60448
F.A.T. 4 x 4	C-GT	8440	C.60441

G.M. CANADA MODELS

	(A)	(B)	(C)
15 cwt. 4 x 4) Armoured Truck)	C-15TA	8449	C.15441
3 ton 6 x 6	C-60X	8660	C.60660
G.M. Light Reconnaissance Car 4 x 4	RAC - Mk.I	8447	-

CANADIAN DODGE MODELS

	(A)	(B)	(C)
15 cwt. 4 x 2	D-15	T. 222	D-15428-C
3 ton 4 x 2 136" W.B.	D-60S	T.110L6	D-60426-C

CANADIAN DODGE MODELS (contd.)

	(A)	(B)	(C)
3 ton 4 x 2 160" W.B. (2 speed axle. Single Tyres)	D-60L	T.110L5	D-60420-C
3 ton 4 x 2 160" W.B. (Single speed axle. Twin tyres)	D-60L/D	T.110L9	-

CANADIAN FORD MODELS

	(A)	(B) Prior 1942	(B) 1942 and onwards	(C)
8 cwt. 4 x 2	F- 8	C011DF	C291D	F.8421
15 " "	F-15	C101WF	C291W	F.15421
15 " 4 x 4	F-15A	C011WQF	C291WQ	F.15441
30 " "	F-30	C01QF	C29Q	F.30444
Carrier S.P., 40 mm A.A.) S.P. Bofors 4 x 4)	F-60B	-	C39QB	F.60444-S
3 ton 4 x 4 134" W.B.	F-60S	C01QF	C29Q	F.60444
3 ton " 158" W.B.	F-60L	C018QF	C298QF	F.60448
3 ton " 115" W.B.	F-60T	-	C395Q	F.60445
3 ton 4 x 2 (Modified Conventional)	FC60L	*C098TFS	C298TFS	F.60428-C
3 ton " (Military Pattern)	F-602L	-	C298WFS	F.60428-M
3 ton 6 x 4	F-60H	C010QF	C290Q	F.60640
F.A.T. 4 x 4	F-GT.	C011QF	C291Q	F.60441

Above list <u>excludes</u> certain conventional and modified conventional chassis produced by Canada only to the order of Dominion Governments.

*Certain early types were also coded EC098TFS.

TYRE SUPPLEMENT

GENERAL NOTES ON W.D. TYRE EQUIPMENT

The following notes are intended as a general guide only to the main types of tyres fitted to W.D. Vehicles. Owing to the number of different makes and the variety of designs in different types there are exceptions to these notes, but they apply in the majority of cases.

1. Standard Commercial Tyre

(i.e. as normally supplied for Civilian Vehicles).

Purpose - Fitted to those W.D. Vehicles which are intended for use exclusively on roads. Occasionally fitted if sand tyres are not available as they are better than cross country for sand use.

Deflections - Operate at normal commercial deflections, i.e., from 11% to 14%.

2. Cross Country Type (C.C.)

Purpose - Fitted to W.D. Vehicles which may have to operate over roadless terrain.

Features - Tread usually made up of deep, widely spaced bars or chevrons which dig into the ground.

The weight carrying capacity of the cross country tyre is not always the same as that of the corresponding commercial tyre. The cross country in many cases has fewer plies to give more flexibility and thus permit greater deflection.

Deflections - Generally 18% deflection. Can operate at from 25% to 30% deflection but for short distances only.

IDENTIFICATION - Arrangements are being made for all makes of Cross Country to be marked "C.C." on the side wall.

3. Sand Type

Purpose - For desert use.
Features - Flexible casing to operate at low inflation pressures.
Flexible, shallow tread pattern, wide tread and flat contour, to give large contact area and low contact pressure on sand.
Deflections - 18% deflection, firm sand and cross country.
22% deflection, soft sand.
25%-30% deflection, for short distances only to get through very soft sand.

IDENTIFICATIONS - Arrangements are being made for all makes to be marked "Sand Type" on the side wall.

4. General Purpose Tyre (G.P.)

Purpose - This is a compromise between the cross-country and the sand tyre. In certain instances it has been used to replace the sand tyre owing to non-availability of the latter.
Features - This tyre combines some of the virtues of both parent types, but it does not give as good a performance as the parent patterns do on their own type of surface.
Deflections - 18%. For operations in sand, as for sand types.

IDENTIFICATION - Arrangements are being made for all makes to be marked "G.P." on the sidewall.

5. Runflat (or R.F.)

Purpose - For front line vehicles.
Features - Run Flat properties are such that, in the event of it being shot up, an R.F. tyre totally deflated will carry its given load a distance up to 50 miles so as to bring the vehicle to safety. With the inner tube completely deflated, the R.F. tyre only deflects 30% - 35%. In these conditions the vehicle is controllable at speeds up to 30 - 40 m.p.h.
Tread is cross country.
R.F. cover has special inner tube and a bead spacer; R.F. tyres have a smaller carrying capacity than other types of tyres in the same size.
Deflection - 18% same as cross country. Under exceptional circumstances, e.g. to get through very soft sand, the R.F. may be completely deflated, but must be reinflated as soon as possible thereafter. Deflation must not be effected by removing the valve insert. Can only be fitted to a W.D. type divided wheel.

IDENTIFICATION - Letters R.F. marked on tyre walls.

"DON'TS" FOR R.F. TYRES

DON'T overload.

DON'T fail to check inflation pressure daily.

DON'T run a yard further than necessary with tyres deflated.

DON'T forget that an R.F. tyre must be inspected by the workshop personnel when it has run 20 miles or more deflated.

DON'T attempt to fit an ordinary tube to an R.F. cover, you won't.

DON'T try to dismantle or fit an R.F. tyre unless you have the special tools for the job.

DON'T attempt to fit an R.F. tyre to any wheel other than the W.D. divided wheel.

DON'T fit an R.F. tyre without a bead lock on bead spacer.

LIST OF MAIN TYPES OF TYRES FITTED TO W.D. VEHICLES With Maximum Loads

I. LOW PRESSURE TYRES

SIZE	NEW ALTERNATIVE SIZE MARKING	TYPE	MAX. TYRE	LOAD PER IN CWTS.
15.00 - 20		C.C.		100
13.50 - 20	14.00 - 20	C.C.		74
13.50 - 20	Replaces	R.F.		65
13.50 - 20	13.50 - 20	Std.		74
12.00 - 20		Std.		56
12.00 - 20		C.C.		56
12.00 - 20		R.F.		48
10.50 - 16		C.C.		46
10.50 - 16		Sand		36*
10.50 - 16		R.F.		28
10.50 - 20	11.00 - 20	C.C.		48
10.50 - 20	Replaces	R.F.		34
10.50 - 20	10.50 - 20	Std.		45
10.50 - 13		Std.		40
9.75 - 20	10.00 - 20 Replaces 9.75 - 20	Std.		37½
9.25 - 16		R.F.		20
9.00 - 20		C.C.		31½
9.00 - 20		Std.		31½
9.00 - 16		C.C.		27½
9.00 - 16		R.F.		25
9.00 - 16		Std.		27½
9.00 - 13		C.C.		15
9.00 - 13		Std.		15
8.25 - 20		Std.		25

* 10 ply 36, 8 ply 28, 12 ply 46.

SIZE	NEW ALTERNATIVE SIZE MARKING	TYPE	MAX. LOAD PER TYRE IN CWTS.
8.25 - 10		Std.	21½
8.25 - 10		R.F.	10
7.50 - 20		C.C.	21
7.00 - 18		R.F.	16
7.00 - 16		Std.	12½
6.50 - 17		Std.	15
6.00 - 20		Std.	12½
6.00 - 16		C.C.	10
6.00 - 16		Std.	10
6.00 - 16		C.C.	10
5.50 - 20		Std.	11
5.50 - 16		Std.	8
5.25 - 16		Std.	7½
5.00 - 16		Std.	7
4.00 - 18		Std.	5 cwt.
3.25 - 19		Univ.	440 lbs.

Abbreviations: C.C. = Cross Country.
R.F. = Runflat. STD = Standard.
Univ. = Universal.

2. HIGH PRESSURE TYRES

40 x 8		Std.	40
36 x 8		Std.	40
34 x 7		Std.	27½
32 x 6		Std.	22½
32 x 6		T.T.	17½
30 x 5		Std.	15
29 x 8		Std.	35½
27 x 6		Std.	18

Abbreviations: T.T. = Truck Type.
Std. = Standard.

THE W.D. DIVIDED WHEEL

The W.D. type divided Wheel has played an essential part in the development of the special low pressure tyre equipment.

The wheel consists of two halves bolted together. The rim has tapered bead seats on to which the tyre beads are forced by the inflation pressure and the flanges are knurled, so that lower tyre pressures may be used than on the Standard commercial 3-piece wheel without creep.

Don't fit an R.F. tyre to any wheel other than the W.D. divided wheel.

- CROSS COUNTRY -

- DUNLOP -

MAKERS NAME "TRAK"

- CROSS COUNTRY -

- GOODYEAR -

MAKERS NAME "SUREGRIP" OR "HEAVY DUTY TRACTOR LUG TYPE"

CROSS COUNTRY CHARACTERISTICS

Heavily built tyres with Bold, Broad and Deep Treads.
All C.C. tyres carry the letters C.C. on sidewalls.

- CROSS COUNTRY -

- FIRESTONE -

MAKERS NAME "GROUND GRIP"

- CROSS COUNTRY -

- INDIA -

MAKERS NAME "SUPER TRACTION"

TYPICAL TREAD

PATTERNS

Here are the latest typical tread patterns of the main types of W.D. tyres as produced by four British makers.

← CROSS COUNTRY

(and on pages IND.049 & 050)

GENERAL PURPOSE →

SAND TYPE →

NOTE:- It is important to stress that, while these tread patterns are generally representative of their type, they will not serve as an infallible guide indetermining the type of tyre by tread pattern.

- GENERAL PURPOSE -

- DUNLOP -

MAKERS NAME - "G.P."

GENERAL PURPOSE TREAD

- GOODYEAR -

MAKERS NAME "ALL SERVICE SUREGRIP"

GENERAL PURPOSE CHARACTERISTICS

Tyres of lighter construction than Cross Country, with narrower, shallower treads and more rounded shoulders.

- GENERAL PURPOSE -

- FIRESTONE -

MAKERS NAME "G.P."GROUND GRIP"

- GENERAL PURPOSE -

- INDIA -

MAKERS NAME "G.P."

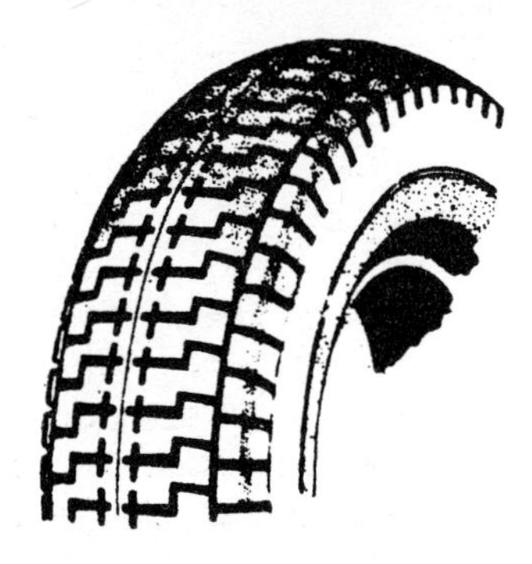

- SAND TYRE -

- DUNLOP -

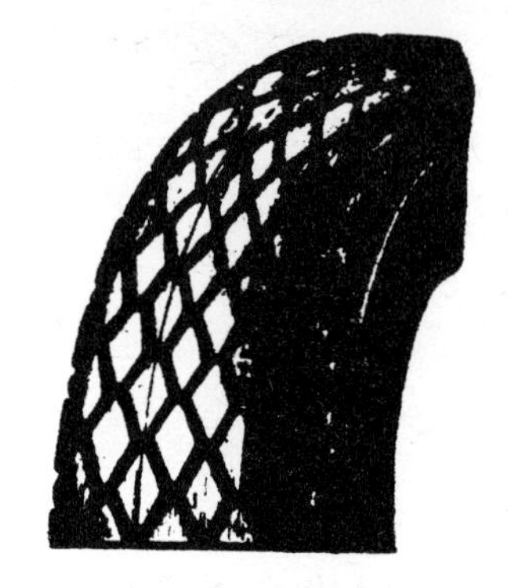

- SAND TYRE -

- GOODYEAR -

SAND TYRE CHARACTERISTICS

Tread often similar to the normal commercial design, but is broader and carcass is of special light construction. All Sand tyres have the word "Sand" marked on side walls of cover.

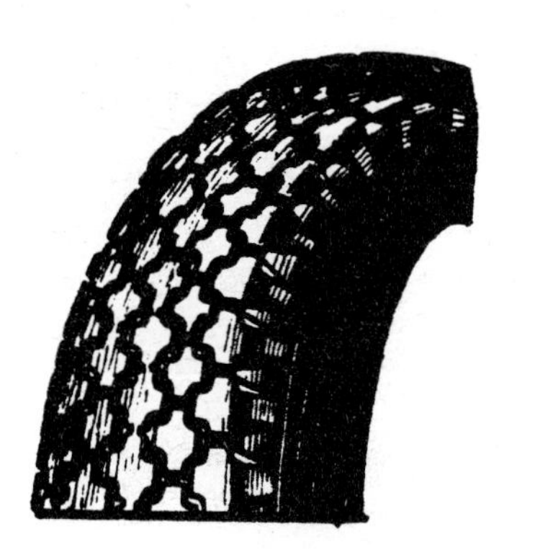

- SAND TYRE -

- FIRESTONE -

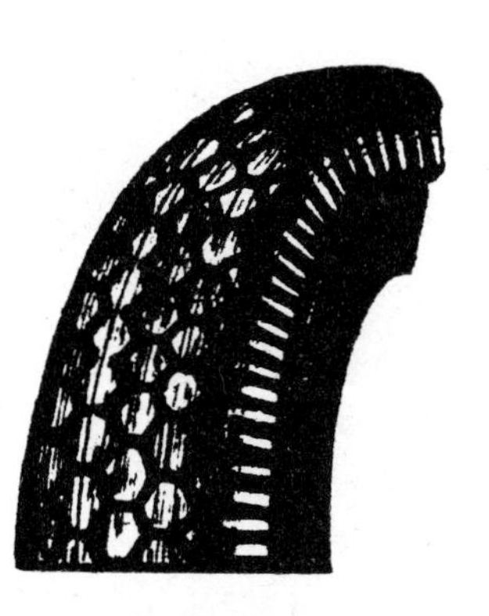

- SAND TYRE -

- INDIA -

DATA BOOK OF "B"

AND R.A.S.C. VEHICLES

PART I

BRITISH VEHICLES

FROM 8-CWT. UPWARDS

including

Tyre Supplement

INTRODUCTION

This Data Book contains details of the main load carrying vehicles and a number of specialised vehicles which have been produced by makers in this country since 1939. There are details of 50 different chassis and 44 different body types are illustrated and described. These specifications must be treated as typical and do not necessarily apply in all details to every contract relating to individual vehicles.

WEIGHTS

Where possible actual weighbridge weights are given throughout this Data Book. In certain instances it has been necessary to compute the weights, either from known chassis and body weights, or from comparable vehicles. Computed weights are distinguished from weighbridge weights by the symbol (E). See notes on page 4 for further details on weights.

INDEX BY TYPE

INDEX BY MAKE OF CHASSIS

IMPORTANT NOTES

ABBREVIATIONS

The following abbreviations are used:-

4 x 2 = four wheel rear wheel drive.
4 x 4 = four wheel four wheel drive.
6 x 4 = six wheel four rear wheel drive.

TANK CAPACITY DATA

Shows total fuel carried in vehicle tanks including reserve, but excluding spare fuel carried in cans.

GROUND CLEARANCE DATA

Data is for minimum clearance laden. Belly clearances are also laden.

WEIGHT DATA

UNLADEN WEIGHTS, complete vehicle but without load, manufacturers' tools and equipment, Fuel, W.D. Vehicle equipment, spare wheel and tyre.

LADEN WEIGHTS, with load, manufacturers' tools and equipment, W.D. vehicle equipment, full fuel, oil and water, spare wheel, driver and mate, and, in the case of personnel carriers, with all seats occupied.

ESTIMATED WEIGHTS, are distinguished from weighbridge weights by the symbol (E).

EQUIPMENT WEIGHTS, see standard W.D. allowances on page 96.

SHIPPING TONNAGE
40 cubic feet = 1 shipping ton.

GENERAL LOAD CARRYING VEHICLES

CHASSIS TYPE	BODY TYPE
8 cwt. 4 wheel	(P.U. (Wireless (Light Field (Ambulance
15 cwt. 4 wheel	(G.S. (Air Compressor (A.T. Tractor (A.T. Portee (Wireless (Water Tank (Office
30 cwt. 4 wheel	(G.S. (Armoured
30 cwt. 6 wheel	(G.S. (Winch Lorry (Breakdown (A.A. Tractor (Office
3 ton 4 wheel	(G.S. (Tipper (Searchlight
3 ton F.W.D.	G.S.
6 ton 4 wheel	G.S.
10 ton 6 wheel	G.S.

FORD W.O.C.I. 8-CWT. 4-WHEEL

4 x 2 Normal control Wheelbase 9' 4"

ENGINE:- Make, Ford. Petrol V.8. Bore 3.06". Stroke 3.75" Capacity 3.6 litres. Max. B.H.P. 85 at 3,800, governed to 2,500. Max. torque lbs/ins. 1,800 at 2,000.

GEARBOX:- 4 speeds and 1 reverse. Overall ratios, 26.3 to 1, 12.7 to 1, 6.9 to 1, 4.11 to 1, rev. 32.2 to 1.

REAR AXLE:- 3/4 floating, spiral bevel. Ratio, 4.11 to 1.

SUSPENSION:- Front and rear, single transverse cantilever, with double acting shock absorbers.

BRAKES:- Foot, Hydraulic on all wheels. Hand, mechanical on rear wheels.

TYRES:- 9.00 - 13.

TURNING CIRCLE:- 40 ft. Track, f. 5' 0 1/8", r. 5' 2 1/8".

GROUND CLEARANCE:- 9" r. axle.

TANK CAPACITY:- 12 gals. Radius 170 miles.

COOLING SYSTEM CAPACITY:- 4½ gals.

PERFORMANCE FIGURES (P.U. Body Laden)

B.H.P. per ton 41.0.
Tractive effort per ton (100%) 1615 lbs./ton.
Max. speed at governed r.p.m. 49 m.p.h.

HUMBER 8-CWT. 4-WHEEL CHASSIS

4 x 2 Normal Control Wheelbase 9' 6"

ENGINE:- Make, Humber. Petrol, 6-cylinder, side valves. Bore 85 m.m. Stroke 120 m.m. Capacity 4.1 litres. Max. B.H.P. 85 at 3,400 r.p.m. governed to 2,650. Max. Torque lbs./ins. 2,150 at 1,300 r.p.m.

GEARBOX:- 4 speed and 1 reverse. Overall ratio, 18.09 to 1, 12.07 to 1, 7.16 to 1, 4.88 to 1. Rev. 18.09 to 1.

REAR AXLE:- Semi-floating, spiral bevel drive. Ratio 4.88 to 1.

SUSPENSION:- Front: Independent, transverse semi-elliptic spring.
Rear: Semi-elliptic springs.
Hydraulic shock absorbers, front and rear.

BRAKES:- Foot, Hydraulic on all wheels. Hand, Mechanical on rear wheels.

TYRES:- 9.00-13.

TURNING CIRCLE:- 41 ft. Track, f, 5'0½", r, 5'1".

GROUND CLEARANCE:- 7½" engine sump.

TANK CAPACITY:- 16 gals. Radius 236 miles.

COOLING SYSTEM CAPACITY:- 3¾ gals.

PERFORMANCE FIGURES:- (Personnel Body).
B.H.P. per ton 39.0
Tractive Effort per ton (100%) 1248 lbs./ton.
Max. Speed Governed 45 m.p.h.

MORRIS P.U. 8-CWT. 4-WHEEL

4 x 2 Normal control Wheelbase 9' 0"

ENGINE:- Make, Morris. Petrol 6 cyl. Bore 82 m.m. Stroke 110 m.m. Capacity 3.48 litres. Max. B.H.P. 60 at 2,800, governed to 2,650. Max. Torque 1,665 lbs./ins. at 1,250 r.p.m.

GEARBOX:- 4 speeds and 1 reverse. Overall ratios, 27.4 to 1, 15.4 to 1, 8.3 to 1, 4.58 to 1, rev. 36.5 to 1.

REAR AXLE:- 3/4 floating, spiral bevel. Ratio 4.58 to 1.

SUSPENSION:- Front and rear, semi-elliptic, with double acting hydraulic shock absorbers.

BRAKES:- Foot, Hydraulic on all wheels. Hands, mechanical on rear wheels.

TYRES:- 9.00 - 13.

TURNING CIRCLE:- 33 ft. Track, f. 4' 10½", r. 5' 0".

GROUND CLEARANCE:- 7 1/8" r. axle. 13" Belly.

TANK CAPACITY:- 16 gals. Radius 220 miles.

COOLING SYSTEM CAPACITY:- 4 7/8 gals.

PERFORMANCE FIGURES (P.U. Body Laden)

B.H.P. per ton 27.2.
Tractive effort per ton (100%) 1,470 lbs./ton.
Max. speed at governed r.p.m. 46 m.p.h.

HUMBER F.W.D. 8-CWT. 4-WHEEL

4 x 4 Normal control Wheelbase 9' 3¾"

ENGINE:- Make, Humber. Petrol 6-cyl. Bore 85 m.m. Stroke 120 m.m. Capacity 4.08 litres. Max. B.H.P. 85 at 3,400 r.p.m. governed to 2,650 r.p.m. Max. torque 2,150 lbs./ins. at 1,300 r.p.m.

GEARBOX:- Main, 4-speeds and 1 reverse. Transfer case, direct and 1.48 to 1. Overall ratios (High): 18.1 to 1, 12.1 to 1, 7.1 to 1, 4.88 to 1, rev. 18.1 to 1. F.W. Drive on Low ratio only.

AXLES:- Semi-floating, spiral bevel drive. Ratio 4.88 to 1.

SUSPENSION:- Front, Independent, transverse semi-elliptic, Rear, semi-elliptic. Hydraulic shock absorbers front and rear.

BRAKES:- Foot, Hydraulic on all wheels. Hand, mechanical on rear wheels.

TYRES:- 9.25 - 16.

TURNING CIRCLE:- 46 ft. Track, f. and r. 5' 1".

GROUND CLEARANCES:- 9 1/8" f. spring, 13" Belly.

TANK CAPACITY:- 16 gals. Radius 200 miles

COOLING SYSTEM CAPACITY:- 4¾ gals.

PERFORMANCE FIGURES (Wireless Body Laden)

B.H.P. per ton 30.0.
Tractive effort per ton (100%) 1,337 lbs./ton.
Max. speed - governed to 50 m.p.h.

MORRIS P.U. 8/4 F.W.D.
8-CWT. 4-WHEEL

4 x 4 Normal control Wheelbase 8' 0"

ENGINE:- Make, Morris. Type O.H. Special. Petrol 6 cyl. Bore 82 m.m. Stroke 110 m.m. Capacity 3.48 litres. Max. B.H.P. 78.5 at 3,200 r.p.m. Max. Torque 1,800 lbs./ins. at 1,800 r.p.m.

GEARBOX:- 4 speeds and 1 reverse. Transfer case fitted, ratio direct. Overall ratios: 34.1 to 1, 19.2 to 1, 10.4 to 1, 5.7 to 1, rev. 45.5 to 1.

AXLES:- Front and rear, spiral bevel. Ratio 5.71 to 1.

SUSPENSION:- Semi-elliptic front and rear, with double acting hydraulic shock absorbers.

BRAKES:- Foot, Hydraulic on all wheels. Hand, mechanical on rear wheels.

TYRES:- 925 - 16.

TURNING CIRCLE:- 38 ft. Track, f. 5' 3", r. 5' 0".

GROUND CLEARANCES:- $8\frac{1}{2}$" f. and r. axles.

TANK CAPACITY:- 22 gals. Radius 240 miles.

COOLING SYSTEM CAPACITY:- 4 7/8 gals.

PERFORMANCE FIGURES (Wireless Laden)

B.H.P. per ton 28.4.
Tractive effort per ton (100%) 1,440 lbs./ton.
Max. governed speed (estimated) 50 m.p.h.

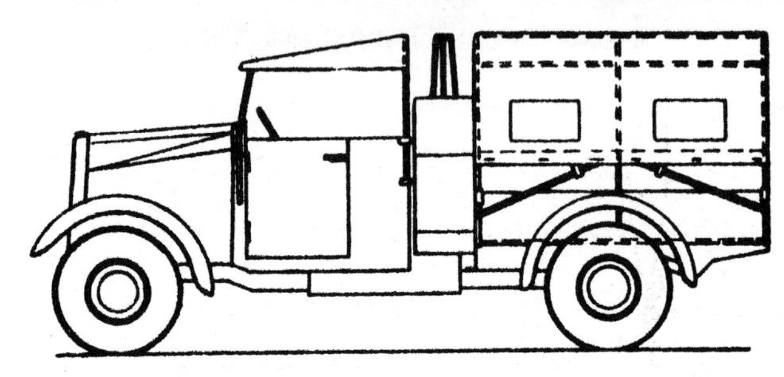

8-CWT. 4-WHEEL PERSONNEL BODY

On Morris 8-Cwt. P.U. 4 x 2 Chassis

(Similar body also fitted to Ford, Humber 4 x 2 and 4 x 4 and Morris 4 x 4 Chassis. For weights and dimensions of these, see data sheet at back).

DESCRIPTION:- Well type body, providing seating for 3 men, 2 facing offside, 1 the nearside. Lockers provided for kit and equipment and is interchangeable in this respect with 8-cwt. Wireless truck. Removable waterproof cover and tubular superstructure.

CAB:- Open type with fixed windscreen and foldup hood. Seating for driver and mate.

DIMENSIONS

	OVERALL	INSIDE BODY
Length	13'10"	4'$9\frac{3}{4}$"
Width	6' 4"	5'$8\frac{1}{2}$"
Height	6' 0"	3'$9\frac{1}{2}$"

WEIGHTS

	UNLADEN	LADEN
F.A.W.		$18\frac{1}{4}$ cwts.
R.A.W.		1 ton $7\frac{3}{4}$ cwts.
Gross	1 ton 16 cwts.	2 tons 6 cwts.

SHIPPING NOTES:- Tonnage, Standing 13.14; Cut Down 11.86. Height Cut Down 5' 5".

8-CWT. 4-WHEEL WIRELESS BODY

ON HUMBER 8-CWT. 4 x 4 CHASSIS

(Similar body also fitted to Ford, Humber 4 x 2, Morris-Commercial 4 x 2 and 4 x 4 Chassis. For weights and dimensions of these, see data sheet at back).

DESCRIPTION:- For the carriage of a No. 11 Wireless set. Similar to Personnel body described on previous page, except seating for 2 men only. Table for wireless transmitting and fittings for batteries and equipment. Locker along nearside of body for aerial bags, signalling gear and wireless mast. Auxiliary Dynamo driven from power take-off is fitted for charging Wireless batteries.

CAB:- Open type with fixed windscreen, fold-up hood, seating for driver and mate.

DIMENSIONS

	OVERALL	INSIDE BODY
Length	13'5"	4'9¾"
Width	6'4"	5'8½"
Height	6'4"	3'9½"

WEIGHTS (E)

	*UNLADEN	LADEN
F. A. W.		1 ton 4¼ cwt.
R. A. W.		1 ton 11 cwt.
Gross	1 ton 18¾ cwt.	2 tons 15¼ cwt.

* Unladen weight excludes all Wireless equipment.

SHIPPING NOTES:- Tonnage, Standing, 13.5 Cut Down 13.0. Height Cut Down 6'1½".

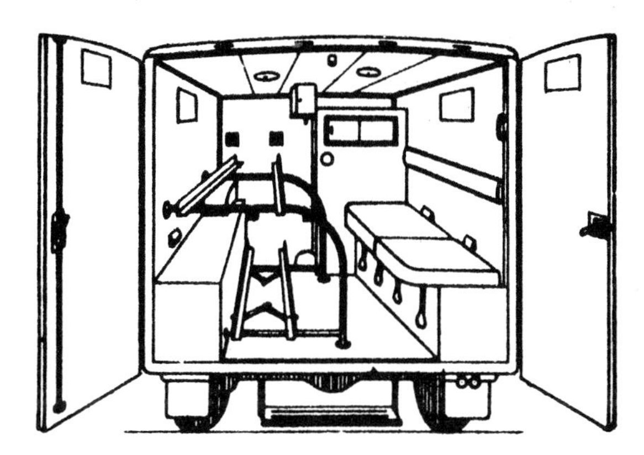

8-CWT. 4-WHEEL LIGHT FIELD AMBULANCE ON HUMBER 8-CWT. F.W.D. CHASSIS

Designed to carry driver and attendant with 2 stretcher and 3 sitting cases, or with 5 sitting cases. Seats for attendant and 3 sitting cases on offside. Stretcher runners on nearside can be collapsed to make room for 2 additional sitting cases. Equipment includes water tank, 2 interior lights, Clayton heater, 2 roof and 4 side ventilators. Rear doors fold back flush to body sides.

TYRE EQUIPMENT:- 9.00 - 16 runflat. (Note, in cases where runflat tyres are not fitted, a spare wheel can be carried in N/S locker).

OVERALL DIMENSIONS

Length 13' 9" Width 6' 0". Height 7' 5".

WEIGHTS (E)

	UNLADEN	LADEN (10-cwt. load)
F. A. W.	1 ton 2½ cwt.	1 ton 3-cwt.
R. A. W.	1 ton 2¼ cwt.	1 ton 11¾ cwt.
GROSS	2 tons 4¾ cwt.	2 tons 14¾ cwt.

SHIPPING NOTES:- Tonnage standing 15.3.

BEDFORD MW 15 CWT. 4-WHEEL CHASSIS

4 x 2 Normal Control Wheelbase 8' 3"

ENGINE:- Make, Bedford. Petrol 6-cyl. o.h.v. Bore 3.3/8". Stroke 4". Capacity 3.5 litres. Max. B.H.P. 72 @ 3,000. 67 B.H.P. @ governed speed 2,425 r.p.m. Max. Torque lbs./ins. 1932 @ 1200 r.p.m.

GEARBOX:- 4 speeds and 1 reverse. Overall Ratios: 44.7 to 1, 21.5 to 1, 10.6 to 1, 6.2 to 1, reverse 44.3 to 1.

REAR AXLE:- Full floating spiral bevel. Ratio 6.2 to 1.

SUSPENSION:- Front and rear, semi-elliptic. Hydraulic shock absorbers all round.

BRAKES:- Foot, Hydraulic on all wheels. Hand, Mechanical on rear wheels.

TYRES:- 9.00 - 16.

TURNING CIRCLE:- 44 ft. Track, f. 5' 3", r. 4' 11½".

GROUND CLEARANCES:- Rear Axle 9", Belly 16".

TANK CAPACITY:- 20 gals. Radius 236 miles.

COOLING SYSTEM CAPACITY:- 4 1/8 gals.

PERFORMANCE FIGURES (G.S. Body):-

B.H.P. per Ton 22.0.
Tractive Effort per Ton (100%) 1628 lbs./Ton.
Max. Speed at governed r.p.m. 40 m.p.h.

FORD W.O.T.2. 15-CWT. 4-WHEEL

4 x 2 Normal Control Wheelbase 8' 10"

ENGINE:- Make, Ford. Petrol V.8. Bore 3.06" Stroke 3.75". Capacity 3.6 litres. Max. B.H.P. 85 @ 3,800 governed to 3,000 r.p.m. Max. Torque 1,800 @ 2,000 r.p.m.

GEARBOX:- 4 speeds and 1 reverse. Overall ratios: 43.7 to 1, 21.1 to 1, 11.5 to 1, 6.83 to 1, rev. 53.4 to 1.

REAR AXLE:- Full floating, spiral bevel. Ratio 6.83 to 1.

SUSPENSION:- Front, transverse. Rear, semi-elliptic. Hydraulic shock absorbers front and rear.

BRAKES:- Foot, Mechanical on all wheels. Hand, Mechanical on all wheels.

TYRES:- 9.00 - 16.

TURNING CIRCLE:- 45 ft. left; 55 ft. right.

TRACK:- Front 4' 11½", rear 5' 2½".

GROUND CLEARANCE:- 11" r. axle.

TANK CAPACITY:- 23 gals. Radius 250 miles.

COOLING SYSTEM CAPACITY:- 4¾ gals.

PERFORMANCE FIGURES (G.S. Body Laden)

B.H.P. per ton 28.3
Tractive effort per ton (100%) 1550 lbs./Ton.
Max. speed at governed r.p.m. 40 m.p.h.

GUY "ANT" 15-CWT. 4-WHEEL

4 x 2 Normal Control Wheelbase 8' 5"

ENGINE:- Make, Meadows 4.E.L.A. Petrol 4-cyl. Bore 95 m.m. Stroke 130 m.m. Capacity 3.68 litres. Max. B.H.P. 55 @ 2,600 r.p.m. governed to 2,700 r.p.m. Max. Torque 1,890 lbs./ins. @ 1,400 r.p.m.

GEARBOX:- 4 speeds and 1 reverse. Overall ratios, 43.9 to 1, 21.2 to 1, 11.5 to 1, 6.76 to 1, rev. 62.1 to 1.

REAR AXLE:- Full floating, spiral bevel drive. Ratio 6.87 to 1.

SUSPENSION:- Semi-elliptic front and rear. Hydraulic shock absorbers front and rear.

BRAKES:- Foot, Mechanical on all wheels. Hand, Mechanical on all wheels.

TYRES:- 9.00 - 16.

TURNING CIRCLE:- 39 ft. Track, f. 5' 0 7/8", r. 4' 11".

GROUND CLEARANCES:- r. axle 9¼", Belly 14".

TANK CAPACITY:- 20 gals. Radius 255 miles.

COOLING SYSTEM CAPACITY:- 4 gals.

PERFORMANCE FIGURES (G.S. Body Laden)

B.H.P. per ton 18.3.
Tractive effort per ton (100%) 1660 lbs./Ton.
Max. speed at governed r.p.m. 39 m.p.h.

MORRIS C.S.8. 15-CWT. 4-WHEEL

4 x 2 Normal control Wheelbase 8' 2"

ENGINE:- Make, Morris. Petrol 6 cyl. Bore 82 m.m. Stroke 110 m.m. Capacity 3.48 litres. Max. B.H.P. 60 @ 2,800 governed to 2,650 r.p.m. Max. Torque 1,665 lbs./ins. @ 1,250 r.p.m.

GEARBOX:- 4 speeds and 1 reverse. Overall Ratios: 39.3 to 1, 22.3 to 1, 12.0 to 1, 6.57 to 1, rev. 52.4 to 1.

REAR AXLE:- Full floating, spiral bevel drive. Ratio 6.57 to 1.

SUSPENSION:- Semi-elliptic, front and rear, with double acting hydraulic shock absorbers.

BRAKES:- Foot, Hydraulic on all wheels. Hand, mechanical on rear wheels.

TYRES:- 9.00 - 16.

TURNING CIRCLE:- 38 ft. Track, F. 5' 3", r. 4' 11¾".

GROUND CLEARANCES:- r. axle 9½", Belly 12".

TANK CAPACITY:- 22 gals. Radius 250 miles.

COOLING SYSTEM CAPACITY: 4 7/8 gals.

PERFORMANCE FIGURES (G.S. Body Laden)

B.H.P. per ton 20.0.
Tractive effort per ton (100%) 1320 lbs./Ton.
Max. speed at governed r.p.m. 40 m.p.h.

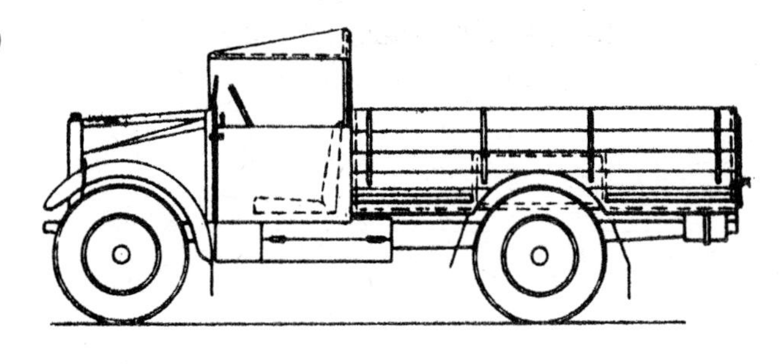

15 CWT. 4 WHEEL G.S. BODY

ON MORRIS C.S.8. CHASSIS

(Similar body also fitted to Bedford M.W.D., Ford and Guy 15 cwt. chassis. For weights and dimensions of these see Data Sheet at back).

BODY:- Standard G.S. with W.D. type drawbar gear at rear.

Open driving compartment with folding detachable canvas top. Seating for one man and driver.

DIMENSIONS

	Overall	Inside Body
Length	13' 10½"	6' 5½"
Width	6' 6"	5' 10½"
Height	6' 6"	1' 10"

WEIGHTS

	Unladen	Laden
F.A.W.		1 ton 2½ cwt.
R.A.W.		1 ton 17½ cwt.
Gross	1 ton 18¾ cwt.	3 tons 0 cwt.

SHIPPING NOTES:- Tonnage Standing 14.65; Cut Down 12.4; Height Cut Down 5' 6".

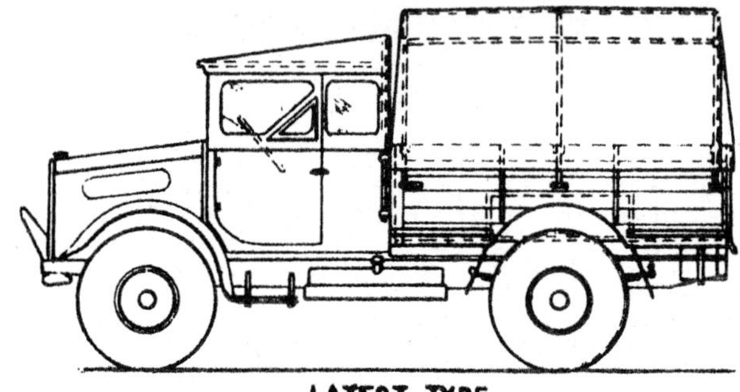

LATEST TYPE
15 cwt. 4 wheel G.S. Body

With Enclosed Cab and Tilt.
on Bedford M.W. Chassis

(Similar bodies and cabs also to be fitted to Ford W.O.T.2.H, Guy Ant and Morris C.S.8. chassis).

BODY:- Fixed sides and tailboard. Detachable canvas tilt carried on 3 hoops.

CAB:- Enclosed. Half doors fitted with detachable side curtains. Full width divided windscreen with 2 wipers. Removable canvas hood with gauze covered rear windows.

DIMENSIONS

	OVERALL	INSIDE BODY
Length	14' 4½"	6' 5½"
Width	6' 6½"	5' 10½"
Height	7' 6"	4' 7"

WEIGHTS (E)

	UNLADEN	LADEN
F.A.W.		1 ton 5 cwt.
R.A.W.		2 tons 4½ cwt.
Gross	2 tons 5 cwt.	3 tons 9½ cwt.

SHIPPING NOTES:- Tonnage Standing 17.6. Stripped Shipping. Height (Canvas cab hood and body superstructure removed) 6' 5".

15 CWT. 4 WHEEL AIR COMPRESSOR

On Morris C.S.8. Chassis

BODY - Generally the same as the G.S. Body, but with underframe modified to permit fitment of Compressor Unit on to Chassis. Compressor is driven by power take-off from gearbox and has a piston displacement of 150 cu. ft. @ 1200 R.P.M. Boxes provided to carry the following pneumatic tools: Saw, woodboring machine, concrete breaker, picks, rock drills together with detachable bits and accessories.

Folding, detachable canvas top to driving compartment. Seating for 1 man and driver.

DIMENSIONS

as G.S. Body

WEIGHTS

	UNLADEN	LADEN
F.A.W.		1 ton 4¾ cwt.
R.A.W.		1 ton 18¾ cwt.
Gross 2 tons 18 cwt.		3 tons 3½ cwt.

SHIPPING NOTES: Tonnage standing 14.65, Cut Down 13.8, Height Cut Down 6' 2".

15 CWT. ANTI-TANK TRACTOR

ON BEDFORD MWT CHASSIS

(Similar body also fitted to Morris C.S.8. chassis. For weights and dimensions of this, see Data Sheet at back).

BODY:- Similar to G.S. body described on previous page, but with detachable cover and superstructure and seats for personnel. W.D. drawbar gear at rear.

DIMENSIONS

As G.S. Body, except:- Overall Height 7' 6½"; Inside Height 4' 6"; Cut Down Height 6' 10".

WEIGHTS

	(E) UNLADEN	LADEN
F.A.W.		1 ton 3 cwt.
R.A.W.		2 tons 4 cwt.
GROSS 2 tons 3 cwts.		3 tons 7 cwt.

15 CWT. ANTI-TANK PORTEE

ON BEDFORD MWG CHASSIS

BODY:- Similar to G.S., but with gun loading ramps carried in outrigger brackets either side of body. Gun wheels rest in curved channels in fixed ramp on floor of body.

DIMENSIONS:- As G.S. except overall width 6' 11½".

WEIGHTS

	UNLADEN	LADEN
F.A.W.	1 ton 1 cwt.	1 ton 9½ cwt.
R.A.W.	1 ton 3 cwt.	1 ton 17 cwt.
GROSS	2 tons 4 cwt.	3 tons 16½ cwt.

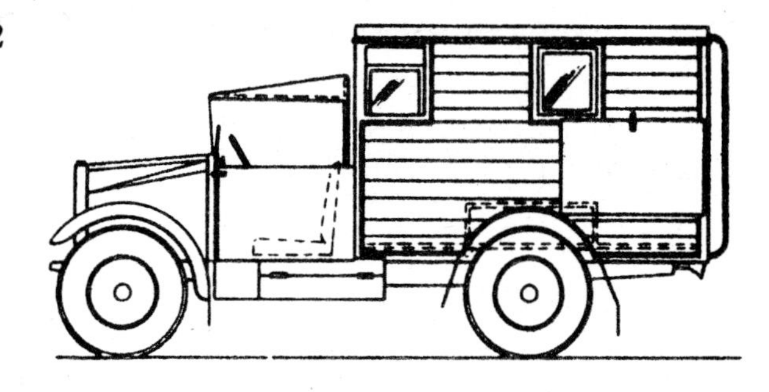

15 CWT. 4 WHEEL WIRELESS (HOUSE TYPE)

ON MORRIS C.S.8. CHASSIS

BODY:- House type for carriage of a No.9 Wireless Set. Removable table at front of body to carry wireless set. Seating for 3 men in body, driver and mate in front. Compartment for generating set located offside rear of body with separate door so that generator may be taken and operated outside vehicle. Folding map table outside on nearside. Auxiliary dynamo fitted driven from gearbox power take-off.

DIMENSIONS

	OVERALL	INSIDE BODY
Length	14' 0"	6' 10½"
Width	6' 8½"	5' 6"
Height	7' 4"	4' 8"

WEIGHTS

	UNLADEN	LADEN
F.A.W.		1 Ton 5 Cwt.
R.A.W.		1 Ton 17½ Cwt.
GROSS	2 Tons 6¼ Cwt.	3 Ton 2½ Cwt.

SHIPPING NOTES: Tonnage Standing 17.5.

"Latest types, pump is driven from P.T.O. on gear box"

15 CWT. 4 WHEEL 200 GAL. WATER TANK

ON BEDFORD M.W.C. CHASSIS

(Similar tanker on Morris C.S.8. 15 cwt. chassis. For weights and dimensions see data sheet at back).

BODY:- 200 gallon (nett) water tank. Pump mounted at front of chassis and driven by starter dog is used to pick up water from any available source. Two hand operated pumps located either side of chassis at front end of tank. Water is passed through filters and chlorinators for purification.
DRIVERS COMPARTMENT:- Fitted with two seats Folding detachable canvas canopy.

DIMENSIONS

Overall length, pump in position 15' 3½".
" " pump removed 14' 4 3/8".
Overall width 6' 6½", height 6' 4".

WEIGHTS

	UNLADEN	LADEN
F.A.W.	1 Ton 1 Cwt.	1 Ton 5¾ Cwt.
R.A.W.	1 Ton 8¼ Cwt.	2 Ton 11¾ Cwt.
GROSS	2 Ton 9¼ Cwt.	3 Ton 17½ Cwt.

SHIPPING NOTES:- Tonnage Standing 14.75.

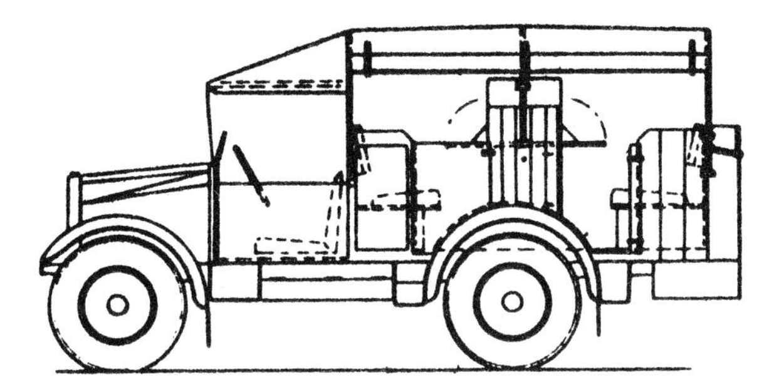

15 CWT. 4 WHEEL OFFICE

ON MORRIS C.S.8 CHASSIS

BODY:- Office Type body with side "Tent" extensions. When erected extensions form enclosed tents either side of body each measuring 6' 10" long x 5' 0" wide x 5' 6" high. Central locker divides body into two compartments and is fitted with folding tables etc. 2 seats in front compartment and 3 at rear. Two loose folding tables and chairs detachable side curtains and tent supports carried in rear lockers. Drivers compartment fitted with two seats and folding canvas canopy.

DIMENSIONS

	OVERALL	INSIDE BODY
Length	14' 8"	6' 10"
Width	6' 10"	5' 10"
Height	7' 4"	4' 8"

WEIGHTS

	UNLADEN	LADEN
F.A.W.		1 Ton $0\frac{1}{4}$ Cwt.
R.A.W.		2 Ton $5\frac{3}{4}$ Cwt.
GROSS	2 Ton 6 Cwt.	3 Ton 6 Cwt.

SHIPPING NOTES: Tonnage Standing 18.3.

AUSTIN 30 CWT. 4-WHEEL

4 x 2 Semi-forward Control Wheelbase 11' 2"

ENGINE:- Make, Austin. Petrol 6-cyl. Bore 3.34". Stroke 4". Capacity 3.46 litres. Max. B.H.P. 60 @ governed speed of 3,000 r.p.m. Max. Torque 1,830 lbs./ins. @ 1,200 r.p.m.

GEARBOX:- 4 speeds and 1 reverse. Overall ratios: 46.7 to 1, 20.4 to 1, 10.0 to 1, 5.85 to 1, rev. 42.9 to 1.

REAR AXLE:- Full floating, spiral bevel drive. Ratio 5.85 to 1.

SUSPENSION:- Semi-elliptic, front and rear.

BRAKES:- Foot, Hydraulic on all wheels. Hand, mechanical on rear wheels.

TYRES:- 10.50 - 16.

TURNING CIRCLE:- 53 ft. Track, f.5' 1". r.5' $1\frac{1}{4}$".

GROUND CLEARANCE:- $10\frac{3}{4}$" rear axle.

TANK CAPACITY:- 24 gals. Radius 240 miles.

COOLING SYSTEM CAPACITY:- $3\frac{1}{2}$ gals.

PERFORMANCE FIGURES (G.S. Body Laden)

B.H.P. per ton 12.8.
Tractive effort per ton (100%) 990 lbs./Ton.
Max. speed at governed r.p.m. 48 m.p.h.

BEDFORD OX 30 CWT. 4 WHEEL

4 x 2 Semi-forward control Wheelbase 9' 3"

ENGINE:- Make, Bedford. Petrol 6-cyl, O.H.V. Bore 3 3/8". Stroke 4". Capacity 3.5 Litres. Max. B.H.P. 72 @ 3000 r.p.m. governed to 2580 r.p.m. Max. torque lbs./ins. 1932 @ 1200 r.p.m.

GEARBOX:- 4 speeds and 1 reverse. Overall ratios: 49.3 to 1, 23.7 to 1, 11.6 to 1, 6.8 to 1, reverse 48.8 to 1.

REAR AXLE:- Full floating, spiral bevel. Ratio 6.8 to 1.

SUSPENSION:- Front, semi-elliptic. Rear, semi-elliptic. Hydraulic shock absorbers all round.

BRAKES:- Foot, Hydraulic, vacuum servo assisted, an all wheels. Hand, mechanical on rear wheels.

TYRES:- 10.50 - 16.

TURNING CIRCLE:- 49' 6". TRACK f. 5' 3" r. 5' 2½".

GROUND CLEARANCE:- 9" r. axle. 19¼" Belly.

TANK CAPACITY:- 24 gals. Radius 240 miles.

COOLING SYSTEM CAPACITY:- 4 1/8 gals.

PERFORMANCE FIGURES (G.S. Body)

B.H.P. per Ton 16.45.
Traçtive Effort per Ton (100%) 1216 lbs./Ton.
Max. speed at governed r.p.m. 40 m.p.h.

COMMER Q.2. 30 CWT. 4 WHEEL

4 x 2 Normal Control Wheelbase 10' 0".

ENGINE:- Make, Commer. Petrol 6 cyl. Bore 75m.m. Stroke 120 m.m. Capacity 3.18 litres. Max. B.H.P. 65 @ 3,200 r.p.m. Max. Torque 1,610 lbs./ins. @ 1,200/1,400 r.p.m.

GEARBOX:- 4 speeds and 1 reverse. Overall Ratios: 44.0 to 1, 23.1 to 1, 12.2 to 1, 6.86 to 1, rev. 56.5 to 1.

REAR AXLE:- Full floating, spiral bevel drive. Ratio 6.86 to 1.

SUSPENSION:- Semi-elliptic, front and rear. Hydraulic shock absorbers front and rear.

BRAKES:- Foot, Hydraulic on all wheels. Hand, mechanical on rear wheels.

TYPES:- 32" x 6", twin rear.

TURNING CIRCLE:- 42 ft. Track, f. 5' 5¼", r. 5' 3¾".

GROUND CLEARANCES:- r. axle 8 5/8", Belly 14".

TANK CAPACITY:- 16 gals. Radius 160 miles.

COOLING SYSTEM CAPACITY:- 4¼ gals.

PERFORMANCE FIGURES (G.S. Body Laden)

B.H.P. per ton 16.0.
Tractive effort per ton (100%) 1,080 lbs./Ton.
Max. speed 40 m.p.h.

FORD W.O.T. 3 30 CWT. 4 WHEEL

4 x 2 Normal Control Wheelbase 11' 11½"

ENGINE:- Make, Ford. Petrol V.8. Bore 3.06". Stroke 3.75. Capacity 3.62 litres. Max. B.H.P. 85 @ 3,800 governed to 3,000 r.p.m. Max. Torque 1,800 lbs./ins. @ 2,000 r.p.m.

GEARBOX:- 4 speeds and 1 reverse. Overall Ratios. 42.6 to 1, 20.6 to 1, 11.2 to 1, 6.67 to 1, rev. 52.1 to 1.

REAR AXLE:- Full floating, spiral bevel drive Ratio 6.67 to 1.

SUSPENSION:- Semi-elliptic, front and rear.

BRAKES:- Foot, mechanical on all wheels. Hand, mechanical on rear wheels.

TYRES:- 10.50 - 16.

TURNING CIRCLE:- 53 ft. Track, f. 5' 9¼", r. 5' 5¼".

GROUND CLEARANCES:- 10¼" r. axle.

COOLING SYSTEM CAPACITY:- 4¾ gals.

TANK CAPACITY:- 15½ gals. Radius 170 miles.

PERFORMANCE FIGURES (G.S. Body Laden)

B.H.P. per ton 18.4.
Tractive effort per ton (100%) 980 lbs./Ton.
Max. speed at governed r.p.m. 45 m.p.h.

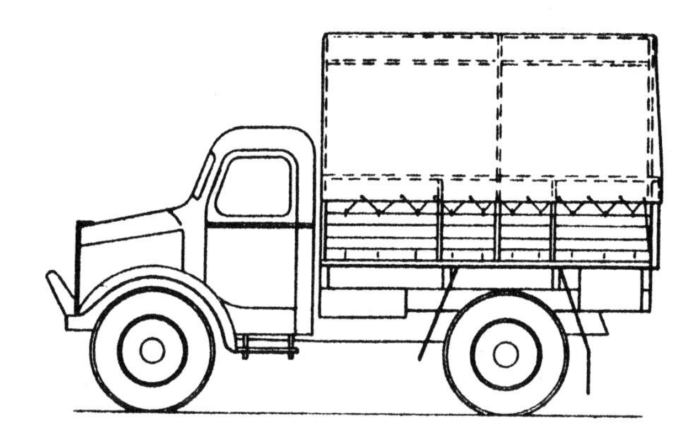

30-CWT. 4 WHEEL G. S. BODY

ON BEDFORD OX CHASSIS

(Similar body also fitted to Austin, Commer and Ford 30 Cwt. Chassis. For weights and dimensions of these, see data sheet at back).

BODY:- Standard G.S. body with flat floor detachable cover and superstructure.

CAB:- Enclosed steel panel cab seating 1 man and driver.

DIMENSIONS

	OVERALL	INSIDE BODY
Length	16' 2¾"	8' 8"
Width	7' 1½"	6' 6"
Height	9' 8¾"	5' 6"

WEIGHTS

	UNLADEN (E)	LADEN
F.A.W.		1 Ton 6¾ cwt.
R.A.W.		3 Tons 5 cwt.
Gross	2 tons 11¾ cwt.	4 tons 11¾ cwt.

SHIPPING NOTES:- Tonnage Standing 28.2, Cut Down 23.98. Height Cut Down 8' 4".

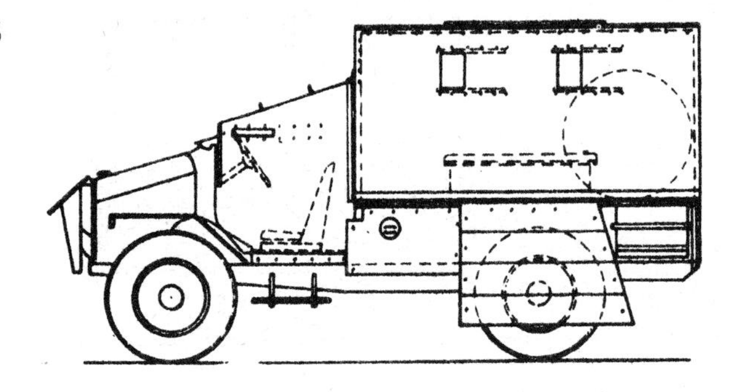

30 CWT. 4 WHEEL ARMOURED BODY

ON BEDFORD OXA. CHASSIS

BODY:- Armoured. Two apertures with sliding steel shutters, fitted front, either side and in rear entrance door. Entrance to drivers compartment from main body. Seat measuring 3' 10" x 1' 8" in centre of body. Roof aperture measuring 4' 1" long by 3' 5" wide. Tool box, spare petrol can carriers and spare wheel carried inside main body at rear.

TYRES:- 10.50-16, Runflat front, cross country rear and spare.

DIMENSIONS

	OVERALL	INSIDE BODY
Length	16' 1 3/8"	8' 6"
Width	6' 9 1/8"	5' 6"
Height	7' 8¾"	4' 0"

LADEN WEIGHT

F.A.W. 1 ton 16¾ cwts.
R.A.W. 4 tons 12¾ cwts.
GROSS 6 tons 9½ cwts.

SHIPPING NOTES:- Tonnage Standing 21.25.

MORRIS CDF 30 CWT. 6 WHEEL

6 x 4 Forward Control Wheelbase 10' 7½"
Bogie Centres 3' 4"

ENGINE:- Make, Morris. Petrol 4-cyl. Bore 100 mm. Stroke 112 mm. capacity 3.5 litres. Max. B.H.P. 55 @ 2750. Max. torque lbs./ins. 1630 @ 1500.

GEARBOX:- 5 speeds & 1 reverse. Overall ratios:- 69.6 to 1, 39.1 to 1, 21.0 to 1, 12.0 to 1, 6.75 to 1. Reverse 69.6 to 1.

REAR AXLES:- Full Floating, overhead worm drive. Ratio 6.75 to 1.

SUSPENSION:- Front, semi-elliptic. Rear, semi-elliptic inverted twins.

BRAKES:- Foot, Hydraulic on front and intermediate axle drums. Hand, Mechanical on four rear wheels.

TYRES:- 7.50 - 20.

TURNING CIRCLE:- 50' TRACK, f. 4' 9¾", r. 4' 11¾".

GROUND CLEARANCE:- 10" rear axle.

FUEL CAPACITY:- 15 gals. Radius 160 miles.

COOLING SYSTEM CAPACITY:- 5 gals.

PERFORMANCE FIGURES:- (G.S. Body)

B.H.P. per Ton 12.3
Tractive Effort per Ton (100%) 1500 lbs./Ton.
Max. speed at governed r.p.m. 37 m.p.h.

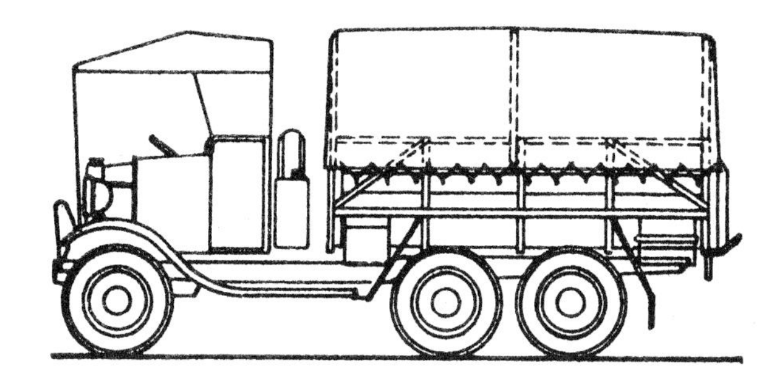

30 CWT. 6 WHEEL G.S. BODY

ON MORRIS C.D.F. CHASSIS

BODY:- G.S. Body with well floor. Detachable cover and hoops.

CAB:- Open type, seating 1 man and driver, with folding canopy and storm aprons. W.D. type Drawbar gear at rear.

DIMENSIONS

	OVERALL	INSIDE BODY
Length	17' 4"	10' 0"
Width	6' 2"	5' 6"
Height	8' 8"	5' 4"

WEIGHTS

	UNLADEN	LADEN
F.A.W.	14½ cwt.	1 ton 2 cwt.
R.A.W.	1 ton 19½ cwt.	3 tons 10 cwt.
GROSS	2 tons 14 cwt.	4 tons 12 cwt.

SHIPPING NOTES:- Tonnage Standing 23.27; Cut Down 20.58, Height Cut Down 7' 8".

Also MORRIS CDFW WINCH LORRY

Fitted 4 ton Winch with 120 ft. of .63 diam. rope. Body similar to G.S. Body above. Weights see Data Sheet.

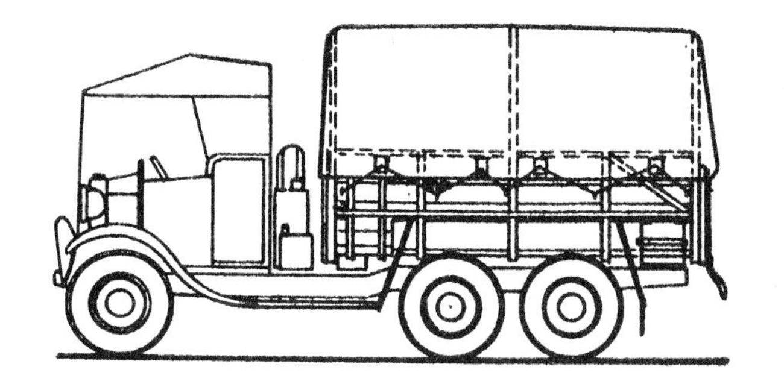

30 CWT. 6 WHEEL OFFICE LORRY

ON MORRIS C.D.F. CHASSIS

BODY:- Externally this body is similar to the G.S. vehicle. The canvas cover (mounted on steel superstructure) has side extensions which can be extended to form canopies both sides of the body, the four extreme corners of the canopy being supported by steel tubes. Seats are provided inside the body with folding tables. Two collapsible tables and chairs are carried for use under the canopies.

DIMENSIONS

	OVERALL	INSIDE BODY
Length	17' 9"	10' 2¾"
Width	6' 3"	5' 6"
Height	9' 4"	6' 0"

LADEN WEIGHTS (E)

F.A.W.	1 ton 4 cwt.
R.A.W.	2 tons 9½ cwt.
GROSS	3 tons 13½ cwt.

SHIPPING NOTES:- Tonnage standing 25.8.

MORRIS CD/SW 30 CWT. 6 WHEEL

6 x 4 Normal Control Wheelbase 9' 7½"
Bogie Centres 3' 4"

ENGINE:- Make, Morris. Petrol 6 cyl.
Bore 82 mm. Stroke 110 mm. Capacity
3.48 Litres. Max. B.H.P. 60 @ 2800 r.p.m.
Max. torque lbs./ins. 1665 @ 1250 r.p.m.

GEARBOX:- 5 speeds and 1 reverse.
Overall ratios:- 69.6 to 1, 39.1 to 1,
21.0 to 1, 12.0 to 1, 6.75 to 1, and
reverse 69.6 to 1.

REAR AXLES:- Full Floating, overhead worm
drive. Ratio 6.75 to 1.

BRAKES:- Foot, Hydraulic on front and
intermediate axle drums.
Hand, Mechanical on four rear wheels.

TYRES:- 9.00 - 16.

TURNING CIRCLE:- 48' TRACK, f, 5' 3" r, 4' 11¾".

GROUND CLEARANCES:- r. axle 10 5/8".
belly 14".

TANK CAPACITY:- 22 gals. Rad. of action
180 miles.

COOLING SYSTEM CAPACITY:- 5 gals.

PERFORMANCE FIGURES:- (Bofors Tractor)

B.H.P. per Ton 11.4.
Tractive Effort in lbs. per Ton (100%) 1340.

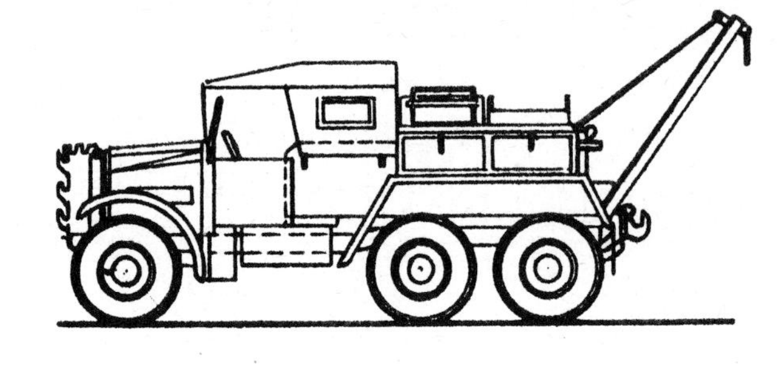

30 CWT. 6 WHEEL BREAKDOWN

ON MORRIS CD/SW CHASSIS

BODY:- Seating for driver and mate. 1 ton hoist operated by pulley blocks is attached to a jib fixed to the back of chassis frame. Counterbalance weights for use with hoist can be fitted to racks at front of vehicle; when not in use they are carried in floor of body. Triangulated towing bar provided at rear of chassis. Four lockers in body for carrying wide range of Breakdown equipment.

WINCH:- 4 Ton Winch with 120 ft. .63" diam. rope. W.D. Drawbar gear at rear.

DIMENSIONS

OVERALL

Length 17' 4½", Width 6' 8", Height 7' 9".

WEIGHTS

	UNLADEN	LADEN
F.A.W.		
R.A.W.		
GROSS		

SHIPPING NOTES:- Tonnage Standing 22.4.

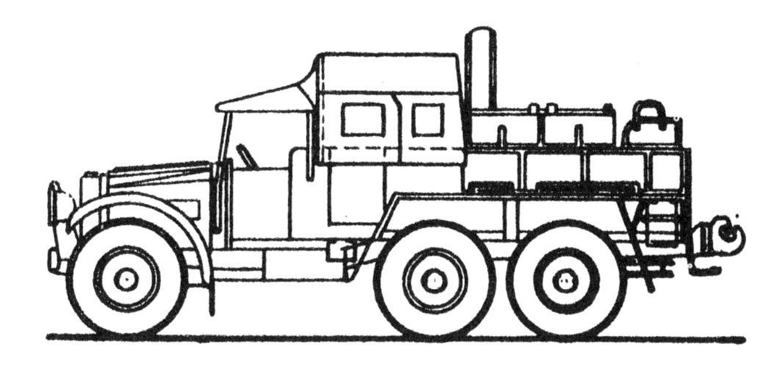

30 CWT. 6 WHEEL BOFORS TRACTOR

ON MORRIS CD/SW CHASSIS

BODY:- Seating for driver and a crew of six in front portion of body, covered by detachable tubular hood superstructure. Lockers above wheel arches carry three cases of ammunition each side of body, two further cases carried in lockers at rear. Spare gun barrel case carried in lower portion of well. Spare gun wheel behind crews compartment.

WINCH:- 4 ton Winch with 120 ft. .63" diam. rope. W.D. Drawbar gear at rear.

DIMENSIONS:- Overall length 17' 2½", width 7' 4", height 7' 6".

WEIGHTS

	UNLADEN (E)	LADEN
F.A.W.	1 ton 3¾ Cwt.	1 ton 2¼ Cwt.
R.A.W.	2 ton 9 Cwt.	4 ton 3¼ Cwt.
GROSS	3 tons 12¾ Cwt.	5 ton 5½ Cwt.

SHIPPING NOTES:- Tonnage Standing 23.6.

4 x 2 Semi-forward control Wheelbase 13' 1½"

ENGINE:- Make, Austin. Petrol 6 cyl. Bore 85 m.m. Stroke 101.6 m.m. Capacity 3.46 litres. Max. B.H.P. 60 @ governed speed of 3,000 r.p.m. Max. Torque 1,830 lbs./ins. @ 1,200 r.p.m.

GEARBOX:- 4 speeds and 1 reverse. Overall Ratios 57.4 to 1, 25.0 to 1, 12.3 to 1, 7.2 to 1, rev. 52.7 to 1.

REAR AXLE:- Full floating, spiral bevel drive. Ratio 7.2 to 1.

SUSPENSION:- Semi-elliptic, front and rear.

BRAKES:- Foot, Hydraulic on all wheels. Hand, mechanical on rear wheels.

TYRES:- 10.50 - 16.

TURNING CIRCLE:- 62 ft. Track, f.5' 5½", r.5' 4".

GROUND CLEARANCES:- r.axle 9¾", Belly 15¾".

TANK CAPACITY:- 32 gals. Radius 280 miles.

COOLING SYSTEM CAPACITY:- 3 7/8 gals.

PERFORMANCE FIGURES (G.S. Body laden)

B.H.P. per ton 9.7.
Tractive effort per ton (100%) 990 lbs./Ton.
Max. speed at governed r.p.m. 40 m.p.h.

BEDFORD OY 3 TON 4 WHEEL

4 x 2 Semi-forward control Wheelbase 13' 1"

ENGINE:- Make, Bedford. Petrol 6 cyl. Bore 3 3/8". Stroke 4". Capacity 3.5 Litres. Max. B.H.P. 72 @ 3000 r.p.m. Governed to 2,800 r.p.m. Max. torque lbs./ins. 1932 @ 1,200 r.p.m.

GEARBOX:- 4 speed and 1 reverse. Overall ratios: 53.4 to 1, 25.6 to 1, 12.6 to 1, 7.4 to 1, reverse 52.9 to 1.

REAR AXLE:- Full floating, spiral bevel. Ratio 7.4 to 1.

SUSPENSION:- Front and rear, semi-elliptic. Hydraulic shock absorbers front and rear.

BRAKES:- Foot, Hydraulic vacuum servo assisted on all wheels.
Hand, Mechanical on rear wheels.

TYRES:- 10.50 - 16.

TURNING CIRCLE:- 66' TRACK f. and r. 5' 4".

GROUND CLEARANCES:- R. Axle 9". Belly 16".

FUEL CAPACITY:- 32 gals. Radius 280 miles.

COOLING SYSTEM CAPACITY:- 4 1/8 gals.

PERFORMANCE FIGURES (G.S. Body Laden)

B.H.P. per Ton 11.2.
Tractive Effort per Ton (100%) 940 lbs./Ton.
Max. speed at governed r.p.m. 40 m.p.h.

COMMER Q.4 3 TON, 4 WHEEL

4 x 2 Normal Control Wheelbase 13' 9"

ENGINE:- Make, Commer. Petrol 6 cyl. Bore 85 m.m. Stroke 120 m.m. Capacity 4.08 litres. Max. B.H.P. 81 @ 3,200 r.p.m. governed to 3,000. Max. Torque 2,140 lbs./ins. @ 1,200 r.p.m.

GEARBOX:- 4 speeds and 1 reverse. Overall Ratios 45.8 to 1, 24.0 to 1, 12.76 to 1, 7.14 to 1, rev. 58.8 to 1.

REAR AXLE:- Full floating, spiral bevel drive. Ratio 7.14 to 1.

SUSPENSION:- Semi-elliptic, front and rear, with hydraulic shock absorbers.

BRAKES:- Foot, Hydraulic on all wheels.
Hand, mechanical on rear wheels.

TYRES:- 10.50 - 16.

TURNING CIRCLE:- 56 ft. Track, f. 5' 4", r. 5' 8".

GROUND CLEARANCES:- r. axle 9", Belly 17¼".

TANK CAPACITY:- 24 gals. Radius 210 miles.

COOLING SYSTEM CAPACITY:- 4¼ gals.

PERFORMANCE FIGURES: (G.S. Body Laden)

B.H.P. per ton 12.6.
Tractive effort per ton (100%) 890 lbs./Ton.
Max. speed @ governed r.p.m. 40 m.p.h.

DENNIS 3 TON 4 WHEEL (TIPPER CHASSIS)

4 x 2 Semi-forward control Wheelbase 9' 8"

ENGINE:- Make Dennis. Petrol 4 cyl. Bore 100 m.m. Stroke 120 m.m. Capacity 3.76 litres. Max. B.H.P. 75 @ 3,000 r.p.m. governed to 2,800 r.p.m. Max Torque 1,850 lbs./ins. @ 1,500 r.p.m.

GEARBOX:- 4 speeds and 1 reverse. Overall Ratios. 44.0 to 1, 22.1 to 1, 12.2 to 1, 7.0 to 1, rev. 56.9 to 1.

REAR AXLE:- Full floating, spiral bevel drive. Ratio 7.0 to 1.

SUSPENSION:- Semi-elliptic, front and rear.

BRAKES:- Foot, Hydraulic on all wheels. Hand, mechanical on rear wheels.

TYRES:- 32 x 6, twin rear.

TURNING CIRCLE:- 37 ft. Track, f. 5' 6¼", r. 5' 3¾".

GROUND CLEARANCE:- r. axle 8".

TANK CAPACITY:- 20 gals. Radius 160 miles.

COOLING SYSTEM CAPACITY:- 5½ gals.

PERFORMANCE FIGURES (Tipper Laden)

B.H.P. per ton 12.4.
Tractive effort per ton (100%) 814 lbs./Ton.
Max. road speed at governed r.p.m. 40 m.p.h.

LEYLAND "LYNX" 3 TON, 4 WHEEL

4 x 2 Semi forward control Wheelbase 12' 0"

ENGINE:- Make, Leyland. Petrol 6 cyl. Bore 3½". Stroke 5". Capacity 4.73 litres. Max. B.H.P. 76.6 @ governed speed of 2,450 r.p.m. Max. Torque 2,600 lbs./ins. @ 1,200 r.p.m.

GEARBOX:- 5 speeds and 1 reverse. Overall Ratios:- 46.2 to 1, 27.3 to 1, 15.2 to 1, 10.8 to 1, 6.5 to 1, rev. 48.4 to 1.

REAR AXLE:- Full floating, worm and wheel drive. Ratio 6.5 to 1.

SUSPENSION:- Semi-elliptic, front and rear.

BRAKES:- Foot, Hydraulic on all wheels. Hand, mechanical on rear wheels.

TYRES:- 8.25 - 20, twin rear.

TURNING CIRCLE:- 57 ft. Track, f. 5' 5 3/8", r. 5' 6 3/8".

GROUND CLEARANCE:- 10¼" r. axle.

TANK CAPACITY:- 24 gals. Radius 200 miles.

COOLING SYSTEM CAPACITY:- 4½ gals.

PERFORMANCE FIGURES (G. S. Body Laden)

B.H.P. per ton 11.1
Tractive effort per ton (100%) 985 lbs./Ton.
Max. speed at governed r.p.m. 40 m.p.h.

THORNYCROFT ZS/TC4 3-TON, 4-WHEEL (Searchlight)

4 x 2 Semi-forward control Wheelbase 13' 4"

ENGINE:- Make, Thornycroft T.C.4. Petrol 4 cyl. Bore 98.4 m.m. Stroke 120 m.m. Capacity 3.86 litres. Max. B.H.P. 60 at 2,400 r.p.m. governed to 1,980 r.p.m. Max. Torque 1930 lbs./ins. at 1,350 r.p.m.

GEARBOX:- Main, 4 speeds and 1 reverse. Auxiliary, direct and 1.51 to 1. Overall Ratios (High) 49.6 to 1, 24.3 to 1, 13.9 to 1, 7.75 to 1, rev. 71.8 to 1.

REAR AXLE:- Full floating, overhead worm drive Ratio 7.75 to 1.

SUSPENSION:- Semi-elliptic, front and rear.

BRAKES:- Foot, Mechanical servo-assisted on all wheels. Hand Mechanical on all wheels.

TYRES:- 9.00 - 20, twin rear.

TURNING CIRCLE:- 55 ft. Track, f. 5' 7 7/8", r. 5' 6".

GROUND CLEARANCE:- r. axle, 11½", Belly 16½".

TANK CAPACITY:- 22 gals. Radius 185 miles.

COOLING SYSTEM CAPACITY:- 6 gals.

PERFORMANCE FIGURES (Searchlight Body Laden)

B.H.P. per ton 7.8.
Tractive effort per ton (100%) 1,040 lbs./ton.
Max. speed at governed r.p.m. 29 m.p.h.

TILLING STEVENS T.S. 20/2 3-TON 4-WHEEL

4 x 2 Forward Control Wheelbase 12' 0"

ENGINE:- Make Tilling Stevens 5V. Petrol 4 cyl. Bore 102 m.m. Stroke 140 m.m. Capacity 4.57 litres. Max. B.H.P. 70 at 2,400 r.p.m., governed to 2,020 r.p.m. Max. Torque 2,520 lbs./ins. at 1,000 r.p.m.

GEARBOX:- Main, 4 speeds and 1 reverse. Auxiliary, direct and 1.5 to 1. Overall Ratios (High) 49.8 to 1, 26.9 to 1, 14.8 to 1, 7.5 to 1, rev. 64.5 to 1.

REAR AXLE:- Full floating, overhead worm drive. Ratio 7.5 to 1.

SUSPENSION:- Semi-elliptic, front and rear.

BRAKES:- Foot, Mechanical servo-assisted on all Wheels. Hand, Mechanical on rear wheels.

TYRES:- 9.00 - 20 cross country, twin rear.

TURNING CIRCLE:- 57 ft. Track, f. 5' 7", r. 5' 5½".

GROUND CLEARANCES:- r. axle 12", Belly 20".

TANK CAPACITY:- 20 gals. Radius 140 miles.

PERFORMANCE FIGURES (Searchlight Body Laden)

B.H.P. per ton 9.1.
Tractive effort per ton (100%) 1,310 lbs./ton.
Max. speed at governed r.p.m. 28 m.p.h.

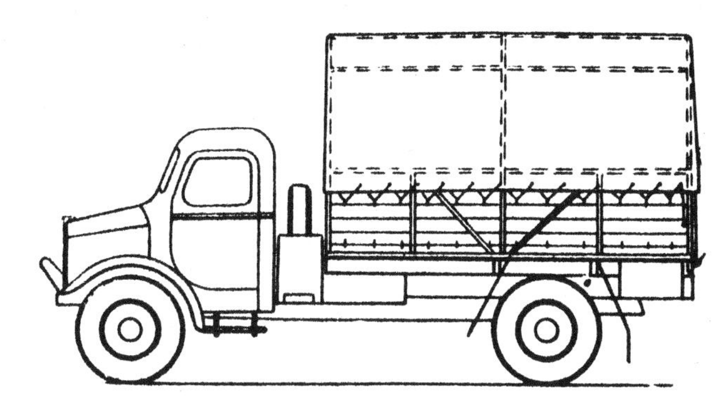

3 TON, 4 WHEEL G.S. BODY ON BEDFORD OY CHASSIS

(Similar body also fitted to Austin, Commer, Dennis, Leyland and Thornycroft "Sturdy" chassis. For weights and dimensions of these, see data sheet at back).

BODY - Standard G.S. body with flat floor, detachable cover and hoops.

CAB - Enclosed steel panel cab seating two.

DIMENSIONS

	OVERALL	INSIDE BODY
Length	20'4¾"	11'6"
Width	7'1½"	6'6"
Height	10'2¼"	6'0"

WEIGHTS

	UNLADEN			LADEN		
F.A.W.	1 ton,	5¾	cwt.	1 ton,	19	cwt.
R.A.W.	1 "	9¾	"	4 "	9	"
Gross	2 "	15¾	"	6 "	8	"

SHIPPING NOTES:- Tonnage standing 36.6, Cut down 31.8. Height cut down 8'10".

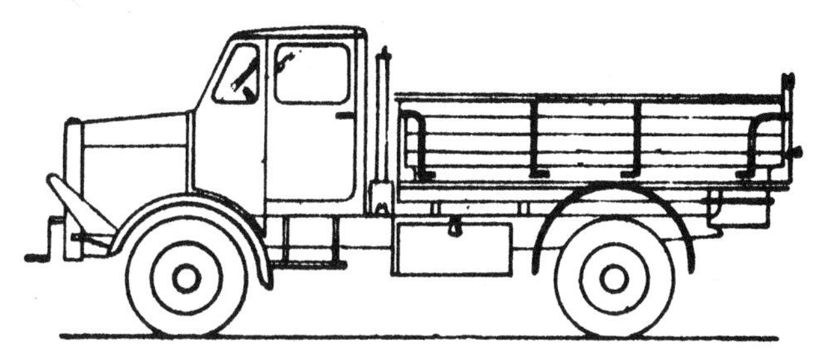

3-TON 4-WHEELED HYDRAULIC END TIPPER ON DENNIS 3-TON CHASSIS

BODY:- Dropsides, hinged tail board.

CAB:- All enclosed, seats for driver and mate.

TIPPING GEAR:- "Edbro" Hydraulic tipping gear. Pump operated by power "take-off". Telescopic ram.

DIMENSIONS

	OVERALL	INSIDE BODY
Length	17' 9"	9' 0"
Width	7' 3"	6' 9½"
Height	7' 4"	1' 8½"

WEIGHTS

	UNLADEN (E)	LADEN (E)
F.A.W.		1-ton 12½ cwt.
R.A.W.		4-tons 8½ cwt.
Gross	2-tons 15-cwt.	6-tons 1 cwt.

SHIPPING NOTES:- Tonnage standing 23.3.

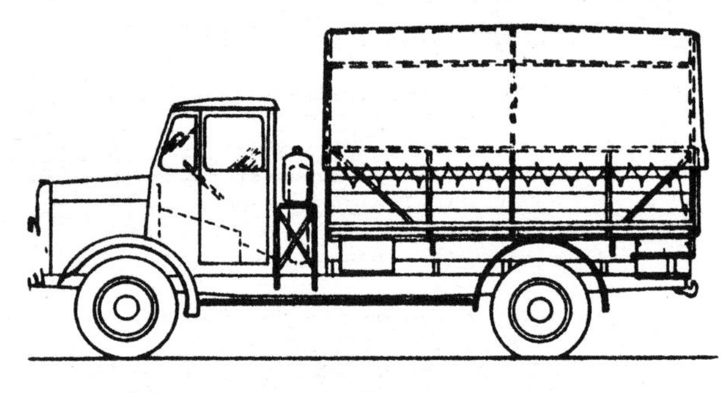

3-TON, 4-WHEELED SEARCHLIGHT LORRY

ON THORNYCROFT ZS/T.C.4. CHASSIS

(similar body also fitted to Tilling-Stevens T.S. 20/2 chassis. For weights and dimensions of this see Data Sheet at back).

BODY:- Designed to carry a 90 cm. projector. Loading ramps for projector carried under floor. Detachable canvas cover and superstructure.

CAB:- All enclosed, seating 1 man and driver.

SPECIAL FITMENTS:- 24 k.w. generator mounted at front end of engine. W.D. drawbar gear at rear.

DIMENSIONS

	OVERALL	INSIDE BODY
Length	21'7½"	12'0"
Width	7'4"	6'7"
Height	10'8"	6'6"

WEIGHTS (E)

	UNLADEN		LADEN	
F.A.W.	1 ton,	17½ cwt.	2 tons,	9½ cwt.
R.A.W.	2 "	3½ "	5 "	3 "
Gross	4 "	1 "	7 "	12½ "

SHIPPING NOTES:- Tonnage Standing 42.2

ALBION F.T.11. 3-TON F.W.D. 4-WHEEL

4 x 4 Forward Control Wheelbase 12' 0"

ENGINE:- Make, Albion E.N.280. Petrol 6 cyl. Bore 3 5/8" Stroke 4½". Capacity 4.56 litres. Max. B.H.P. 96 at 2,900 r.p.m. Max. Torque 2,520 lbs./ins. at 1,200 r.p.m.

GEARBOX:- Main, 4 speeds and 1 reverse. Transfer case ratios .993 to 1 and 2.04 to 1. Third differential, capable of being locked, located in transfer case. Overall Ratios (High): 54.5 to 1, 26.7 to 1, 14.1 to 1, 8.27 to 1, rev. 69.3 to 1.

AXLES:- Worm drive. Ratio 9.33 to 1.

SUSPENSION:- Semi-elliptic, front and rear, with Hydraulic shock absorbers.

BRAKES:- Foot, hydraulic vacuum servo-assisted on all wheels. Hand, mechanical on rear wheels.

TYRES:- 12.00 - 20.

TURNING CIRCLE:- 55 ft. Track, f.6'2", r. 6' 3 3/8".

GROUND CLEARANCE:- 12¾" under axles.

TANK CAPACITY:- 31 gals. Radius 240 miles.

COOLING SYSTEM CAPACITY:- 3¾ gals.

PERFORMANCE FIGURES (G.S. Body Laden)

B.H.P. per ton 12.3.
Tractive effort per ton (100%) 1,975 lbs./ton.
Max. speed at Max. B.H.P. 38 m.p.h.

AUSTIN 3-TON F.W.D. 4-WHEEL

4 x 4 Forward Control Wheelbase 12'0"

ENGINE:- Make, Austin. Petrol 6 cyl. Bore 3 7/16" Stroke 4 3/8". Capacity 3.99 litres. Max. B.H.P. 85 at governed speed of 2,900 r.p.m. Max. Torque 2,100 lbs./ins. at 1,600 r.p.m.

GEARBOX:- Main, 4 speeds and 1 reverse. Transfer case, ratios 1.148 to 1 and 2.075 to 1. Overall ratios (High) 54.5 to 1, 28.8 to 1, 14.1 to 1, 8.27 to 1, rev. 54.4 to 1,

F.W.D. Low ratio. Rear wheeldrive High ratio.

AXLES:- Spiral bevel drive. Ratio 7.2 to 1.

SUSPENSION:- Semi-elliptic, front and rear, with Hydraulic shock absorbers.

BRAKES:- Foot, Hydraulic servo-assisted on all wheels. Hand, mechanical on rear wheels.

TYRES:- 10.50 - 20.

TURNING CIRCLE:- 57 ft. Track, f. and r. 5'11".

GROUND CLEARANCE:- r. axle 10 5/8, f. axle 11 1/8.

TANK CAPACITY:- 32 gals. Radius 250 miles.

COOLING SYSTEM CAPACITY:- 4 gals.

PERFORMANCE FIGURES (G.S. Body Laden)

B.H.P. per ton 11.7.
Tractive effort per ton (100%) 1,490 lbs./ton.
Max. speed at governed r.p.m. 38 m.p.h.

BEDFORD QL 3-TON F.W.D.4-WHEEL

4 x 4 Forward Control Wheelbase 11' 11"

ENGINE:- Make, Bedford. Petrol 6 cyl. O.H.V. Bore 3 3/8" Stroke 4". Capacity 3.5 litres Max. B.H.P. 72 at 3,000. 68 B.H.P. at governed speed of 2,500 r.p.m. Max. torque lbs./ins. 1,932 at 1,200 r.p.m.

GEARBOX:- Main, 4 speeds and reverse. Transfer case, 1.18 and 2.29 to 1. Overall Ratios: (High) 52.7 to 1, 25.3 to 1, 12.5 to 1, 7.3 to 1, Reverse 52.2 to 1.

F.W.D. Low Ratio. Rear Wheel drive High Ratio.

AXLES:- Full Floating, spiral bevel ratio 6.16 to 1.

SUSPENSION:- Semi-elliptic, front and rear, with double acting shock absorbers.

BRAKES:- Foot, Hydraulic servo-assisted on all wheels. Hand, Mechanical on rear wheels.

TYRES:- 10.50 - 20.

TURNING CIRCLE:- 55' Track f.5' 8" r.5' 6½".

GROUND CLEARANCES:- Axles, front and rear, 12".

FUEL CAPACITY:- 28 gals. Radius 230 miles.

COOLING SYSTEM CAPACITY:- 3 5/8 gals.

PERFORMANCE FIGURES:- (G.S. Lorry)

B.H.P. per ton 10.6.
Tractive effort per ton (100%) 1,490 lbs./ton.
Max. speed at governed r.p.m. 38 m.p.h.

FORD W.O.T.8. 30 CWT. F.W.D.

4 x 4 FORWARD CONTROL. WHEELBASE 9' 10"

ENGINE:- Make, Ford. Petrol V.8. Bore 3.08". Stroke 3.75" Capacity 3.62 litres. Max. B.H.P. 85 @ 3,800. Max. torque 1800 lbs./ ins. @ 2000 r.p.m.

GEARBOX:- Main, 4 speeds and reverse. Transfer case ratios, Direct and 2.42 to 1. Overall Ratios (High), 42.6 to 1, 20.6 to 1, 11.25 to 1, 6.66 to 1, rev. 52.1 to 1. F.W.D. Low ratio. Rear wheel drive High ratio.

AXLES:- Full floating, spiral bevel drive. Ratio 6.66 to 1.

SUSPENSION:- Semi-elliptic front and rear.

BRAKES:- Foot, Mechanical on all wheels. Hand, Mechanical on rear wheels.

TYRES:- 10.50 - 20.

TURNING CIRCLE:- 49 ft. Track, f. and r. 5' 10".

GROUND CLEARANCE:- 11½" r. axle.

TANK CAPACITY:- 35 gals. Radius 300 miles.

COOLING SYSTEM CAPACITY:- 5 gals.

PERFORMANCE FIGURES (G.S. Laden)

B.H.P. per ton.
Tractive effort per ton (100%) 1700 lbs./ Ton.
Max. speed @ governed r.p.m. 45 m.p.h.

BODY DATA:- G.S. Body with low wheel arches. Enclosed cab (non-detachable cab roof).

DIMENSIONS:- Overall, 16'8½" x 7'6" x 9'1¼". Inside Body, 9'6" x 6'10" x 6'0".

WEIGHTS:- Unladen 3 tons 16 cwts. Laden 5 tons - 13 cwt.

(48A)

FORD W.O.T.6. 3 TON F.W.D.

4 x 4 Forward Control. Wheelbase 11' 11½"

ENGINE:- Make, Ford. Petrol V.8. Bore 3.08". Stroke 3.75". Capacity 3.62 litres. Max. B.H.P. 85 @ 3,800 Max. torque 1800 lbs./ins. @ 2000 r.p.m.

GEARBOX:- Main, 4 speeds and reverse. Transfer case ratios, direct and 2.42 to 1. Overall ratios (High): 48.5 to 1, 23.4 to 1, 12.8 to 1, 7.6 to 1, rev. 59.4 to 1. F.W.D. Low ratio. Rear wheel drive High ratio.

AXLES:- Full floating, spiral bevel drive. Ratio 7.6 to 1.

SUSPENSION:- Semi-elliptic front and rear.

BRAKES:- Foot, Mechanical on all wheels. Hand, Mechanical on rear wheels.

TYRES:- 10.50 - 20.

TURNING CIRCLE:- 58½ ft. Track, f. & r. 5' 10".

GROUND CLEARANCE:- 11½" r. axle.

TANK CAPACITY:- 35 gals. Radius 280 miles.

COOLING SYSTEM CAPACITY:- 5 gals.

PERFORMANCE FIGURES (G.S. Laden).

B.H.P. per Ton 11.3.
Tractive effort per ton (100%) 1480 lbs./ton.
Max. speed at governed r.p.m. 45 m.p.h.

BODY DATA

G.S. Body with low wheel arches. Enclosed cab (detachable cab roof to give stripped shipping height of 6'9").

DIMENSIONS:- Overall 19'10½" x 7'6" x 10'6". Inside body, 12'6" x 6'10" x 6'0".

WTS:- Unladen 3 tons 18 cwt. Laden 7 tons 10¼ cwt.

(48B)

KARRIER K.6. 3-TON F.W.D. 4-WHEEL

4 x 4 Forward Control Wheelbase - 13' 0"

ENGINE:- Make, Karrier Petrol. 6 cyl. Bore 85 m.m. Stroke 120 m.m. Capacity 4.08 litres. Max. B.H.P. 80 at governed speed of 3,200 r.p.m. Max. Torque 2,140 lbs./ins. at 1,200 r.p.m.

GEARBOX:- Main, 4 speeds and 1 reverse transfer case, 1.26 to 1, 1.96 to 1, Overall Ratios (High): 55.7 to 1, 29.2 to 1, 15.5 to 1, 8.69 to 1, rev. 71.7 to 1.

Four wheel drive in either High or Low ratio of transfer case. Rear wheel only drive in High ratio only."

AXLES:- Full floating, spiral bevel drive. Ratio: 6.86 to 1.

SUSPENSION:- Semi-elliptic, front and rear.

BRAKES:- Foot, Hydraulic on all wheels. Hand, mechanical on rear wheels.

TYRES:- 10.50 - 20.

TURNING CIRCLE:- 57 ft. Track, f. 5' 10", r. 5' 8".

GROUND CLEARANCE:- Axles 12", Belly 19".

TANK CAPACITY:- 30 gals. Radius 240 miles.

COOLING SYSTEM CAPACITY:- 5¼ gals.

PERFORMANCE FIGURES (G.S. Body Laden)

B.H.P. per ton 11.2
Tractive effort per ton (100%) 1,360 lbs./ton.
Max. speed at governed r.p.m. 40 m.p.h.

THORNYCROFT TF/AC4/1 3-TON F.W.D. 4-WHEEL

4 x 4 Forward Control Wheelbase 12' 0"

ENGINE:- Make Thornycroft AC4/1. Petrol 4 cyl. bore 111 m.m. Stroke 133.4 m.m. Capacity 5.17 litres. Max. B.H.P. 85 at governed to 2,500 r.p.m. Max. torque 2,652 lbs./ins. at 1,200 r.p.m.

GEARBOX:- Main, 4 speeds and 1 reverse. Transfer case, 1 to 1 and 2.26 to 1. Overall Ratios (High) 39.6 to 1, 19.6 to 1, 11.8 to 1, 7.48 to 1.

F.W.D. on Low Ratio. Rear wheel drive High Ratio.

AXLES:- Full floating, spiral bevel drive with epicyclic hub reduction gears. Ratio 7.48 to 1.

SUSPENSION:- Semi-elliptic, front and rear.

BRAKES:- Fott, Hydraulic vacuum-servo assisted on all wheels. Hand, Mechanical on all wheels.

TYRES:- 12.00 - 20.

TURNING CIRCLE:- 55 ft. track, f. 6' 1¼", r. 6' 1½".

GROUND CLEARANCE:- Axles 13", Belly 18½".

TANK CAPACITY:- 30 gals. Radius 200 miles.

PERFORMANCE FIGURES (G.S. Body Laden)

B.H.P. per ton 10.
Tractive effort per ton (100%) 1,390 lbs./ton.
Max. speed 39 m.p.h.

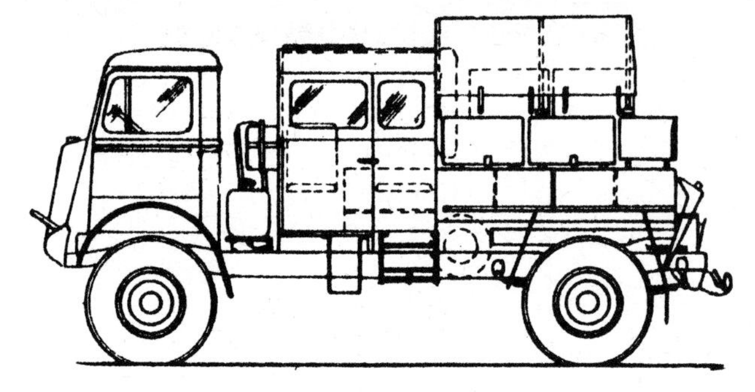

3 Ton F.W.D. BOFORS TRACTOR

(ON BEDFORD Q.L.B. CHASSIS)

BODY:- Seating for driver and a crew of 8; driver anf 1 man in cab, five in crew's compartment, and 2 more in the rear section of body. Doors either side with detachable side curtains to crew's compartment, sliding panel in roof. Rear section of body carries 4 ammunition cases each side, kit containers, spare gun barrel case, spare gun wheel behind crew compartment, and equipment lockers. Detachable canvas cover and hoops.

CAB:- All enclosed steel panelled cab, with detachable top. Crews compartment also has detachable top to give a max. shipping ht. of 7' 6".

SPECIAL FITMENTS:- 5 ton winch with 175 ft. of cable. W.D. drawbar gear at rear.

OVERALL DIMENSIONS:- Length 18' 8", Width 7' 6", Height 9' 0".

WEIGHTS

	UNLADEN	LADEN (E)
F.A.W.	2 tons 4¼ cwt.	
R.A.W.	2 tons 0 cwt.	
Gross	4 tons 4¼ cwt.	6 tons 12 cwt.

SHIPPING NOTES:- Tonnage standing 31.8. Stripped shipping height 7' 6" (Super-structure, top of cab and crews compartment detached).

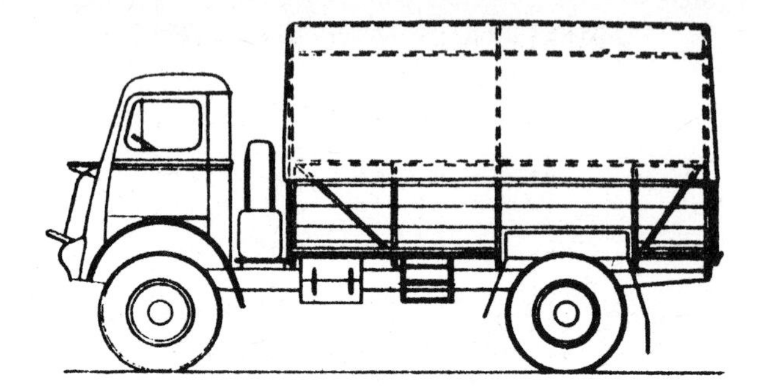

3-TON F.W.D. 4-WHEEL G.S. BODY

On Bedford Q.L. Chassis

(Similar body also fitted to Albion, Austin, Karrier and Thornycroft F.W.D. Chassis. For weights and dimensions of these, see data sheet at back).

BODY:- G.S. Body with flat floor except for low wheel arches. Detachable cover and superstructure.

CAB:- Enclosed steel panel cab seating 1 man and driver. Spare wheel carried behind cab.

DIMENSIONS

	OVERALL	INSIDE BODY
Length	19' 8"	12' 4¾"
Width	7' 4¾"	7' 0"
Height	9' 8¾"	6' 0"

WEIGHTS

	UNLADEN (E)	LADEN (E)
F.A.W.	1 ton 16¾ cwt.	2 tons 16½ cwt.
R.A.W.	1 ton 8 cwt.	4 tons 1¾ cwt.
Gross	3 tons 4½ cwt.	6 tons 18¼ cwt.

SHIPPING NOTES:- Tonnage Standing 35.1, Cut Down 29.8 Height Cut Down 8'3½".

DENNIS "MAX" 6-TON 4-WHEEL

4 x 2 Forward Control Wheelbase 14' 0"

ENGINE:- Make Dennis 04. C.I. 4 cyl. Bore 117.5 m.m. Stroke 150 m.m. Capacity 6.5 litres. Max. B.H.P. 77 at 1,800 r.p.m. Max. Torque lbs./ins. 2,988 at 1,000 r.p.m.

GEAR BOX:- 4 speeds and 1 reverse. Overall ratios: 43.9 to 1, 23.0 to 1, 13.9 to 1, 8.25 to 1, rev. 51.9 to 1.

REAR AXLE:- Full floating, overhead worm drive. Ratio 8.25 to 1.

SUSPENSION:- Semi-elliptic, front and rear.

BRAKES:- Foot, Mechanical servo-assisted on all wheels. Hand, mechanical on rear wheels.

TYRES:- 36" x 8" twin rear.

TURNING CIRCLE:- 50 ft. Track, f.6' 3¼", r.5' 10".

GROUND CLEARANCES:- r. axle 10 1/8", Belly 17½".

TANK CAPACITY:- 35 gals. Radius 370 miles.

PERFORMANCE FIGURES (G.S. Body Laden)

B.H.P. per ton 6.8
Tractive effort per ton (100%) 628 lbs./ton.
Max. speed at governed r.p.m. 24 m.p.h.

COOLING SYSTEM CAPACITY: 8½ gals.

E.R.F. "2.C.I.4". 6-TON 4-WHEEL

4 x 2 Forward Control Wheelbase 12' 5½"

ENGINE:- Make, Gardner 4.L.W. C.I. 4 cyl. Bore 4¼", Stroke 6". Capacity 5.6 litres. Max. B.H.P. 68 at 1,700 r.p.m. Max. Torque 2,765 lbs./ins. at 1,250 r.p.m.

GEARBOX:- 5 speeds and 1 reverse. Overall Ratios: 41.1 to 1, 21.8 to 1, 12.2 to 1, 6.75 to 1 (direct), 5.06 to 1, (overdrive), rev. 40.3 to 1.

REAR AXLE:- Full floating, overhead worm drive. Ratio 6.75 to 1.

SUSPENSION:- Semi-elliptic, front and rear.

BRAKES:- Foot, Hydraulic servo-assisted on all wheels. Hand, Mechanical on rear wheels.

TYRES:- 36" x 8", twin rear.

TURNING CIRCLE:- 54 ft. Track, f.6', 4½", r.5'7".

GROUND CLEARANCE:- r. axle 9¾", Belly 18".

TANK CAPACITY:- 32 gals. Radius 400 miles.

COOLING SYSTEM CAPACITY:- 5 gals.

PERFORMANCE FIGURES (G.S. Body Laden)

B.H.P. per ton 6.1.
Tractive effort per ton (100%) 542 lbs./ton.
Max. speed at governed r.p.m. 32 m.p.h.

FODEN 6-TON 4-WHEELED CHASSIS

4 x 2 Forward Control Wheelbase 12' 7"

ENGINE:- Make, Gardner 4 L.W. C.I. 4 cyl. Bore 4¼". Stroke 6". Capacity 5.5 litres. Max. B.H.P. 68 at governed speed of 1,700 r.p.m. Max. torque lbs./ins. 2,765 at 1,250 r.p.m.

GEARBOX:- Main, 4 speed and 1 reverse. Auxiliary, for 1st and reverse, ratio 2.16 to 1. Overall Ratios: 38.3 to 1, 18.9 to 1, 10.2 to 1, 6.25 to 1, reverse 38.3 to 1.

REAR AXLE:- Full floating, worm drive. Ratio 6.25 to 1.

SUSPENSION:- Front, semi-elliptic. Rear, semi-elliptic.

BRAKES:- Foot, Hydraulic on all wheels. Hand, mechanical on rear wheels.

TYRES:- 36 x 8 twin rear.

TURNING CIRCLE:- 55'. TRACK f. 6' 3 3/8", r. 5' 5¼".

GROUND CLEARANCE:- Minimum 7¼" under rear spring bolts.

FUEL CAPACITY:- 50 gals. Radius 600 miles.

COOLING SYSTEM CAPACITY:- 7 gals.

PERFORMANCE FIGURES:-

B.H.P. per ton 6.2.
Tractive effort per ton 1,138 lbs./ton.
Max. speed at governed r.p.m. 28 m.p.h.

MAUDSLAY "MILITANT" 6-TON, 4-WHEEL

4 x 2 Forward Control Wheelbase 14" 0"

ENGINE:- Make, Gardner 4 L.W. C.I. 4 cyl. Bore 4¼", Stroke 6", Capacity 5.6 litres Max. B.H.P. 68 at 1,700 r.p.m. Max. Torque 2,765 lbs./ins. at 1,250 r.p.m.

GEARBOX:- 4 speeds and 1 reverse. Overall Ratios: 42.4 to 1, 22.1 to 1, 13.2 to 1, 7.75 to 1, rev. 41.2 to 1.

REAR AXLE:- Full floating, overhead worm drive Ratio 7.75 to 1.

SUSPENSION:- Semi-elliptic, front and rear.

BRAKES:- Foot, Dewandre vacuum-servo to all four wheels. Hand, Mechanical on rear wheels.

TYRES:- 36" x 8", twin rear.

TURNING CIRCLE:- 54 ft. Track, f. 6' 4", r. 5' 10".

GROUND CLEARANCE:- r. axle 9¼".

TANK CAPACITY:- 36 gals. Radius 430 miles.

COOLING SYSTEM CAPACITY:- 5 gals.

PERFORMANCE FIGURES (G.S. Body Laden)

B.H.P. per ton 6.2.
Tractive effort per ton (100%) 586 lbs./ton.
Max. speed at governed r.p.m. 24 m.p.h.

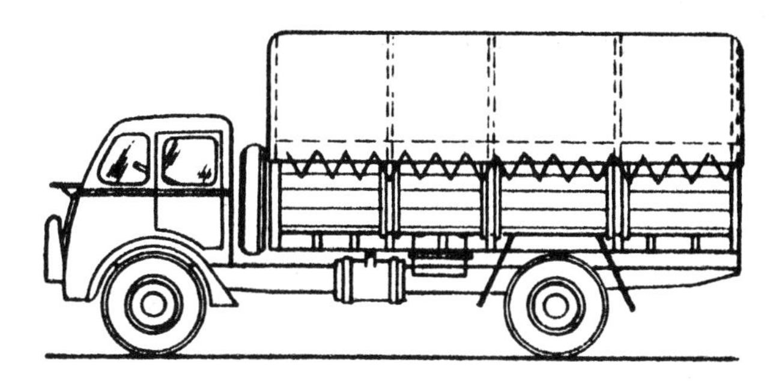

6-TON 4-WHEEL G.S. BODY

ON FODEN 6-TON CHASSIS

(Similar bodies also available on Dennis, E.R.F. and Maudslay 6-ton 4-wheel chassis. For weights and dimensions of these, see Data Sheet at back).

BODY:- Standard G.S. body with flat-floor, detachable hoops and tilt.

W.D. type drawbar gear at rear.

CAB:- All enclosed, steel panel cab, seating two.

DIMENSIONS

	OVERALL	INSIDE BODY
Length	21' 3 3/8"	14' 4 3/8"
Width	7' 6"	6' 11¼"
Height	9'11"	6' 0"

WEIGHTS (E)

UNLADEN	LADEN
F.A.W.	3-tons 15¼ cwt.
R.A.W.	7-tons 3 cwt.
GROSS	10-tons 18¼ cwt.

SHIPPING NOTES:- Tonnage Standing 39.5. Tonnage cut down 32.9, Height cut down 8' 7".

ALBION CX6N. 10 TON 6 WHEEL

6 x 4 Normal Control Wheelbase 16' 0"
Bogie Centres 4' 6".

ENGINE:- Make, Albion E.N. 244. C.I. 6 cyl. Bore 4 5/8". Stroke 5½". Capacity 9.08 litres Max. B.H.P. 100 @ governed speed of 1,750 r.p.m. Max. torque 4,400 lbs./ins. @ 1,000 r.p.m.

GEARBOX:- Main, 4 speeds and 1 reverse. Auxiliary direct and 1.84 to 1. Overall Ratios (High) 40.2 to 1, 21.7 to 1, 12.2 to 1, 7.5 to 1, rev. 48.9 to 1.

REAR AXLES:- Full floating, overhead worm drive. Ratio 7.5 to 1.

SUSPENSION:- Front, semi-elliptic. Rear, twin semi-elliptic with rocking bar anchorage.

BRAKES:- Foot, Hydraulic servo assisted on all wheels. Hand, Mechanical on rear wheels.

TYRES:- 40" x 8" single front, twin rear.

TURNING CIRCLE:- 65 ft. Track, f.6' 4", r. 5' 0½".

GROUND CLEARANCES:- 13" r. axle, 16½" Belly.

TANK CAPACITY:- 33 gals. Radius 250 miles.

COOLING SYSTEM CAPACITY:- 7 gals.

PERFORMANCE FIGURES (G.S. Body Laden)

B.H.P. per ton 5.6
Tractive effort per ton (100%) 888 lbs./Ton.
Max. speed at governed r.p.m. 28 m.p.h.

FODEN DG/6/10 10 TON 6 WHEEL

6 x 4 Forward Control Wheelbase 15' 6½". Bogie Centres 4' 0".

ENGINE:- Make, Gardner 6 L.W. C.1. 6-cyl. Bore 4¼" Stroke 6". Capacity 8.36 litres. Max. B.H.P. 102 @ 1700 r.p.m. Max torque lbs./ins. 4164 @ 1100 r.p.m.

GEARBOX:- Main 4 speed & 1 reverse Auxiliary, superlow, ratio 2.16 to 1, on 1st & reverse only. Overall Ratios (High), 38.6 to 1, 18.9 to 1, 10.2 to 1, 6.25 to 1, reverse 42.4 to 1.

REAR AXLES:- Full floating, overhead worm drive. Ratio 6.25 to 1.

SUSPENSION:- Front, semi-elliptic. Rear semi-elliptic free ends.

BRAKES:- Foot, Hydraulic on all wheels. Hand, mechanical on four rear wheels.

TYRES:- 36 x 8, twins rear.

TURNING CIRCLE:- 60 ft. Track f. 6' 3 3/8", r. 5' 10¼".

GROUND CLEARANCES:- r. axle 8", belly 12½".

FUEL CAPACITY:- 50 galls. Radius 400 miles.

COOLING SYSTEM CAPACITY:- 8 gals.

PERFORMANCE FIGURES

B.H.P. per ton 5.8
Tractive effort per ton. (100%) 1072 lbs./ton.
Max. Speed @ max. B.H.P. 30 m.p.h.

ALBION CX23N 10 TONS 6 WHEEL

6 x 4 Forward Control. Wheelbase 16'0". Bogie Centres 4' 6".

ENGINE:- Make, Albion E.N.244 C.1. 6-cyl. Bore 4 5/8". Stroke 5½". Capacity 9.08 litres. Max. B.H.P. 100 @ governed speed of 1750 r.p.m. Max. torque 4,400 lbs./ins. @ 1000 r.p.m.

GEARBOX:- Main, 4 speeds and 1 reverse. Auxiliary, Ratio Direct and 1.84 to 1. Overall Ratios (High) 44.6 to 1, 24.1 to 1, 13.6 to 1, 8.33 to 1, rev. 54.4 to 1.

REAR AXLES:- Fully articulated r. axle bogie. Both axles driven. Overhead worm drive. Ratio 8.33 to 1.

SUSPENSION:- Front, semi-elliptic. Rear, two semi-elliptic springs, one mounted above other, pivotted at centre.

BRAKES:- Foot, Hydraulic vacuum servo assisted on 4 rear wheels, Hand Mechanical on 4 rear wheels.

TYRES:- 13.50-20, singles front and rear.

TURNING CIRCLE:- 65 ft.

GROUND CLEARANCES:- 10" spring clip bolt, 17¾" Belly.

TANK CAPACITY:- 33 gals. Radius 240 miles (E).

BODY DATA:- G.S. Body with well floor, detachable tilt and hoops. All enclosed cab. Inside Body Dimensions: 17'6" x 7'2" x 6'7".

WEIGHTS:- Unladen. Laden.

LEYLAND "HIPPO" 10 TON 6 WHEEL

6 x 4 Forward Control Wheelbase 16' 0"
Bogie Centres 4' 7".

ENGINE:- Make, Leyland, C.I. 6 cyl. Bore 4½" stroke 5½". Capacity 8.99 litres. Max. B.H.P. 97.2 @ governed speed of 1,880 r.p.m. Max. Torque 3,720 lbs./ins. @ 1,140 r.p.m.

GEARBOX:- 5 speeds and 1 reverse. Overall ratios: 56.4 to 1, 36.4 to 1, 22.8 to 1, 12.7 to 1, 7.33 to 1, rev. 67.3 to 1.

REAR AXLES:- Full floating, overhead worm drive. Ratio 7.33 to 1.

SUSPENSION:- Front, semi-elliptic. Rear, twin semi-elliptic with rocking bar anchorage.

BRAKES:- Foot, Hydraulic servo-assisted on all wheels. Hand, mechanical on 4 rear wheels.

TYRES:- 36" x 8" twin rear.

TURNING CIRCLE:- 63 ft. Track, f. 6' 3½". r. 5' 8½".

GROUND CLEARANCE:- 10", steering track rod.

TANK CAPACITY:- 70 gals. Radius 520 miles.

COOLING SYSTEM CAPACITY:- 6 3/4 gals.

PERFORMANCE FIGURES (G.S. Body Laden

B.H.P. per ton 5.3
Tractive effort per ton (100%) 610 lbs./Ton.
Max. speed @ governed r.p.m. 27 m.p.h.

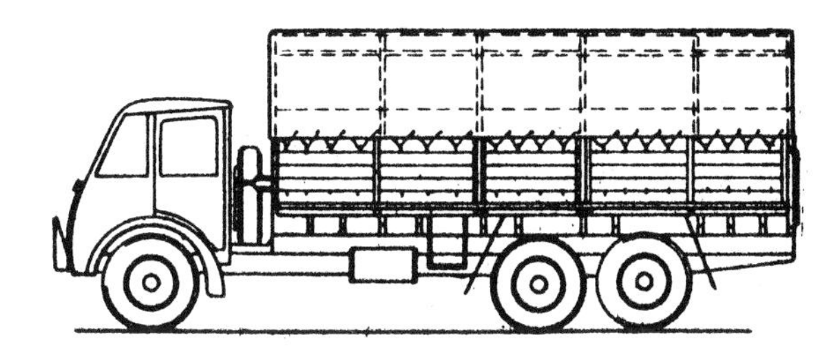

10 TON 6 WHEEL G.S. LORRY

ON FODEN DG/6/10 CHASSIS

(Similar body also fitted to Albion CX6N & Leyland Hippo Chassis. For weights and dimensions of these see Data Sheet at back.)

BODY:- G.S. Body with Flat Floor, detachable hoops and tilt. W.D. Drawbar gear at rear.

CAB:- Enclosed Steel panel cabs seating 2. (Note: Leyland Hippo fitted with open type cab with folding canopy).

DIMENSIONS

	OVERALL	INSIDE BODY
Length	26' 7"	19' 3"
Width	7' 6"	6' 11¼"
Height	10' 2½"	6' 0"

WEIGHTS

	UNLADEN (E)	LADEN
F.A.W.		4 Tons 2 Cwt.
P.A.W.		13 Tons 9 Cwt.
GROSS	7 Tons 4 Cwt.	17 Tons 11 Cwt.

SHIPPING NOTES:- Tonnage Standing 50.5, Cut Down 43.2 Height Cut Down 8' 8".

3 Ton 6 Wheeled Chassis with Specialised Bodies.

BODY TYPES

G.S.
Breakdown
Machinery (House)
Workshop
Derrick
Folding Boat Equipment
Pontoon
Trestle or Sliding Bay
Small Box Girder Bridge
Searchlight
Wireless (House)
Coles Crane

NOTE - A specification for a typical 3 ton 6 wheel chassis - the Leyland "Retriever" - is given on the next page. The main points of difference between this and other makes of 3 ton 6 wheel chassis are shown on the following pages.

LEYLAND RETRIEVER
3 TON, 6 WHEEL

6 x 4 Forward Control Wheelbase 13' 0"
Bogie Centres 4' 0".

ENGINE:- Make, Leyland Petrol 4-cyl. Bore 4 9/16". Stroke 5½". Capacity 5.9 litres. B.H.P. 73 @ 2120 r.p.m. governed to 2150. Max. torque lbs./ins. 3120 @ 1150 r.p.m.

GEARBOX:- Main, 4 speed and 1 reverse.
Auxiliary: Direct and 2.22 to 1.

Overall Ratios:		
	38.8 to 1	78.9 to 1
	22.5 " "	49.5 " "
	12.9 " "	28.9 " "
	7.5 " "	16.6 " "
Reverse	41.8 " "	92.0 " "

REAR AXLES:- Full floating, overhead worm drive. Ratio 7.5 to 1. Drive to all four rear wheels.

SUSPENSION:- Front: semi-elliptic, Rear: Inverted semi-elliptic, pivotted at centre.

BRAKES:- Foot: Hydraulic, servo assisted, all wheels. Hand: Mechanical, four rear wheels.

TYRES: 9.00-20 all round.

TURNING CIRCLE:- 55ft. Track f. 5' 11½". r. 5' 7¾".

GROUND CLEARANCES:- R. Axle 11½". Belly 17½".

FUEL CAPACITY:- 31 gals. Radius 195 miles.

COOLING SYSTEM CAPACITY:- 5½ gals.

PERFORMANCE FIGURES (G.S. Body Laden):-

B.H.P. per Ton 8.6.
Tractive Effort per Ton (100%) 1552 lbs./Ton.
Max. speed - at Governed r.p.m. 33 m.p.h.

Specifications generally similar to Leyland Retriever, except as indicated below:-

ALBION B.Y.I. (Wheelbase 13 ft. Bogie Centres 4 ft.)

ENGINE:- Make Albion. Petrol 4 cyl. Bore 3¾" Stroke 5 3/8". Capacity 3.89 litres. Max. B.H.P. 63.5 @ governed speed of 2,460 r.p.m. Max. Torque 2,050 lbs./ins. @ 1,500 r.p.m.

GEARBOX:- Aux. Ratios. Direct and 2.29 to 1, Overall Ratios (High) 46.9 to 1, 24.6 to 1, 14.7 to 1, 8.3 to 1, rev. 65.1 to 1.

BRAKES:- Foot, mechanical, servo assisted on all wheels.

TRACK:- f. 5' 10 11/16" r. 5' 10¼".

GROUND CLEARANCE:- 10½" at spring bolts, r. bogie.

TANK CAPACITY:- 24 gals. Radius 170 miles.

COOLING SYSTEM CAPACITY:- 4¼ gals.

ALBION B.Y.3. Same as B.Y.I. except.

ENGINE:- Make, Albion. Petrol 6 cyl. Bore 3½" Stroke 4½". Capacity 4.25 litres. Max. B.H.P. 80 @ governed speed of 3,000 r.p.m. Max. Torque 2,150 lbs./ins. at 1,300 r.p.m.

GEARBOX:- Overall Ratios (High) 61.5 to 1, 30.2 to 1, 15.9 to 1, 9.33 to 1, rev. 78.2 to 1.

TANK CAPACITY:- 24 gals. Radius 195 miles.

COOLING SYSTEM CAPACITY:- 5¼ gals.

ALBION B.Y.5. Same as B.Y.3 except,

ENGINE:- Make, Albion E.N. 280. Petrol 6 cyl. Bore 3 5/8". Stroke 4½". Capacity 4.56 litres. Max. B.H.P. 96 @ 2,900 r.p.m. Max. Torque 2,520 lbs./ins. @ 1,200 r.p.m.

Specifications generally similar to Leyland Retriever, except as indicated below:-

A.E.C. MARSHALL (Wheelbase 12' 8½", Bogie Centres 4 ft.)

ENGINE:- Make, A.E.C. Petrol 4 cyl. Bore 112 mm. Stroke 130 mm. Capacity 5.1 litres Max. B.H.P. 70 @ 2000 governed to 2,200. Max. torque 2640 @ 1200.

GEARBOX:- Auxiliary Ratios, Direct and 2.59 to 1. Overall Ratios (High) 31.8 to 1, 19.5 to 1, 11.5 to 1, 7.25 to 1, reverse 38.6 to 1.

BRAKES:- Foot, Mechanical servo-assisted on all wheels.

TRACK:- f. 5' 9 3/8", r. 6' 0 3/8".

GROUND CLEARANCES:- R. axle 10". Belly 13".

TANK CAPACITY:- 30 gals. Radius 180 miles.

COOLING SYSTEM CAPACITY:- 7 gals.

AUSTIN 3 TON 6 WHEEL. (Wheelbase 12' 9". Bogie Centres 4').

ENGINE:- Make, Austin. Petrol 6-cyl. Bore 3.43". Stroke 4.37". Capacity 3.99 litres. Max. B.H.P. 72 @ 2800. Max. Torque 2094 @ 1300.

GEARBOX:- Auxiliary Ratios Direct and 1.87 to 1. Overall Ratios (High), 51.0 to 1, 26.9 to 1, 13.2 to 1, 7.75 to 1, reverse 50.9 to 1.

GROUND CLEARANCE: 9½", r. bogie pivot.

TRACK:- f. and r. 5' 8 1/8".

TANK CAPACITY:- 30 gals. Radius 210 miles.

COOLING SYSTEM CAPACITY:- 4 gals.

Specifications generally similar to Leyland Retriever, except as indicated below:-

CROSSLEY IGL8:- (Wheelbase 12' 10". Bogie Centres 4').

ENGINE:- Make, Crossley. Petrol 4-cyl. Bore 4.5/16". Stroke 5½". Capacity 5.2 litres. Max. B.H.P. 75 at 2,200 governed to 2,000 r.p.m. Max. Torque 2,940 at 1,000.

GEARBOX:- Aux. Ratios Direct and 3.18 to 1. Overall Ratios (High) 37.3 to 1, 20.9 to 1, 11.9 to 1, 7.0 to 1, reverse 32.9 to 1.

BRAKES:- Foot, Mechanical servo assisted on middle axle. Hand, Mechanical on rear axle.

TRACK:- f.5' 5¾", r. 5' 2¼".

GROUND CLEARANCES:- Rear axle 9½", belly 16½".

TANK CAPACITY:- 21 gals. Radius 120 miles.

COOLING SYSTEM CAPACITY:- 9½ gals.

GUY F.B.A.X.:- (Wheelbase 12' 6". Bogie Centres 4').

ENGINE:- Make, Guy. Petrol 4 cyl. Bore 4¼". Stroke 5½". Capacity 5.1 litres. Max. B.H.P. 76 at 2,400 governed to 2,350. Max. Torque 2,520 at 1,200.

GEARBOX:- Aux. Ratios Direct and 2.64 to 1. Overall Ratios (High), 43.2 to 1, 24.8 to 1, 14.4 to 1, 8.33 to 1, reverse 43.2 to 1.

BRAKES:- Foot, mechanical, servo assisted on middle axle. Hand, mechanical on rear axle.

TRACK:- f.5' 9", r.5' 8.5/16".

GROUND CLEARANCES:- r. axle 10½", belly 16½".

TANK CAPACITY:- 25 gals. Radius 140 miles.

COOLING SYSTEM CAPACITY:- 4¼ gals.

Specifications generally similar to Leyland Retriever, except as indicated below:-

KARRIER CK6:- (Wheelbase 12' 6". Bogie Centres 4').

ENGINE:- Make, Karrier. Petrol 6-cyl. Bore 3.35". Stroke 4.72". Capacity 4.08 litres. Max. B.H.P. 80 at 3,100 governed to 3,000. Max. Torque 2,140 at 1,250.

GEARBOX:- 5 forward speeds. Auxiliary not fitted. Overall Ratios: 101.0 to 1, 69.8 to 1, 38.6 to 1, 17.9 to 1, 10.33 to 1, reverse 87.3 to 1.

BRAKES:- Foot, Mechanical servo assisted all wheels. Hand, Mechanical, all wheels.

TRACK:- f.5' 7¼", r.5' 8½".

GROUND CLEARANCES:- R. Axle 10". Belly 17".

TANK CAPACITY:- 30 gals. Radius 175 miles.

COOLING SYSTEM CAPACITY:- 6 gals.

THORNYCROFT WOF/AC4:- (Wheelbase 13'. Bogie Centres 4').

ENGINE:- Make, Thornycroft. Petrol 4-cyl. Bore 4.3/8". Stroke 5¼". Capacity 5.1 litres. Max. B.H.P. 77 at 2,400 governed to 2,200. Max. Torque 2,592 at 1,000.

GEARBOX:- Aux. Ratios Direct and 2.06 to 1. Overall Ratios (High) 40.9 to 1, 20.3 to 1, 12.2 to 1, 7.75 to 1, reverse 63.9 to 1.

BRAKES:- Foot, Mechanical servo-assisted on all wheels.

TRACK:- f. 5' 7.5/8", r. 5' 4½".

GROUND CLEARANCES:- R. Axle 10.3/4". Belly 19½".

TANK CAPACITY:- 22 gals. Radius 180 miles.

COOLING SYSTEM CAPACITY:- 9.7/8 gals.

THORNYCROFT WOF/DC4:- Same chassis as above, but fitted with C.I. engine.

ENGINE:- Make, Thornycroft. C.I. 4 cyl. Bore 4.1/8". Stroke 6". Max. B.H.P. 60. Max. Torque 2,628 at 1,000. Radius of Action 300 miles.

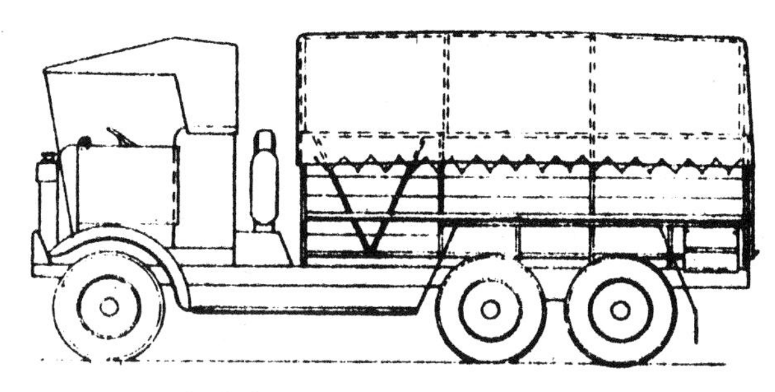

3-TON, 6-WHEEL G.S. LORRY

ON LEYLAND RETRIEVER CHASSIS

(Similar bodies also fitted to Albion, A.E.C., Austin, Crossley, Guy, Karrier and Thornycroft 3-ton 6-wheel chassis. For weights and overall dimensions of these, see data sheet at back).

BODY:- G.S. body with flat floor at front, well construction over rear bogie. Detachable hoops and canvas cover.

W.D. Drawbar gear at rear.

CAB:- Open, with folding canopy, seats 2. (Note, Austin 3-ton 6-wheel has enclosed steel panel cab).

DIMENSIONS

	OVERALL	INSIDE BODY
Length	22' 5½"	14' 0"
Width	7' 5¼"	6' 6"
Height	10' 4½"	6' 6"

WEIGHTS

	UNLADEN	LADEN
F.A.W.	2-tons 0 cwt.	2-tons 10½ cwt.
R.A.W.	2-tons 16¼ cwt.	5-tons 19½ cwt.
GROSS	4-tons 16¼ cwt.	8-tons 10 cwt.

SHIPPING NOTES:- Tonnage Standing 42.4, Cut Down 32.8. Height Cut Down 8' 0".

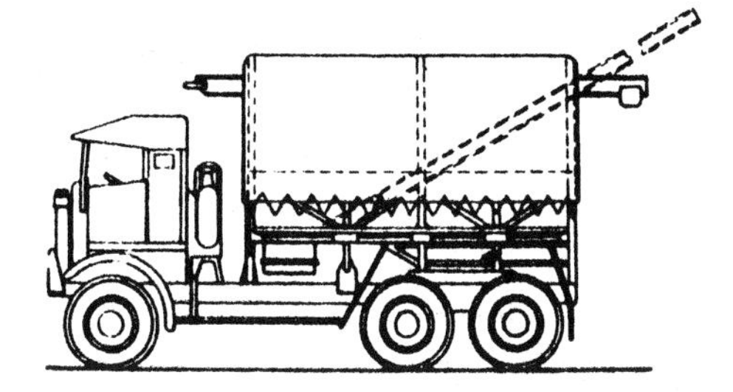

3-TON 6-WHEEL BREAKDOWN

ON LEYLAND RETRIEVER CHASSIS

(Similar bodies also available on Crossley and Guy chassis. For weights and dimensions of these, see Data Sheet at back).

BODY:- Floor partly flat, partly well type. Superstructure supports longitudinal runway with hand-operated travelling block. Runway capacity is 2½ tons at 3' from end of body, 1½ tons at 5' and 1 ton at 6'. Height of lift can be increased by dropping front end of runway to floor. Max. load then 15 cwt. W.D. Drawbar gear at rear.

SPECIAL FITMENTS:- Breakdown is fitted with 5 ton winch.)

CAB:- Semi-enclosed, seats 2, folding canopy, fixed windscreen. Note, this is a later type of Cab to that shown on facing page.

DIMENSIONS

	OVERALL	INSIDE BODY
Length	20' 3"	12' 0"
Width	7' 6"	6' 8½"
Height	11' 4"	* 5' 4½"

* flat floor to inside runway joist.

WEIGHTS (E)

	UNLADEN	LADEN
F.A.W.		2 tons 10 cwt.
R.A.W.		6 tons 2 cwt.
GROSS	5 tons 14 cwt.	8 tons 12 cwt.

SHIPPING NOTES:- Tonnage Standing 43.04.

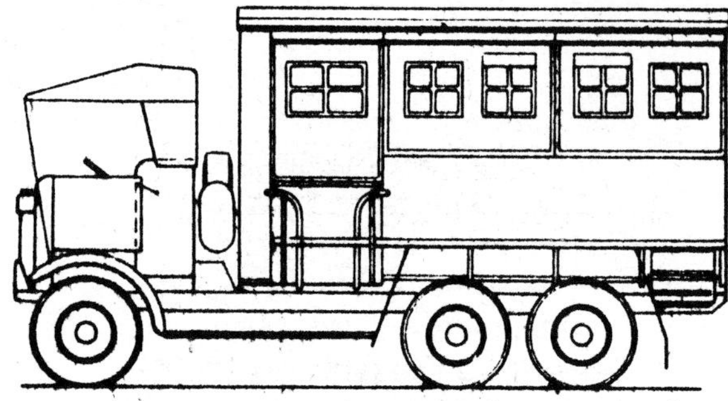

3-TON 6-WHEEL MACHINERY (HOUSE TYPE) TYPE NO.7. MK.I

ON LEYLAND RETRIEVER CHASSIS

(Similar bodies also available on Albion and Thornycroft chassis. For weights and dimensions of these, see Data Sheet at back).

BODY:- Well type floor with flat portion in front. Windows and top part of body sides hinged to swing outwards for ventilation. Bottom part of front nearside door hinged to swing downwards to form a platform for battery charging. Provision made for machinery, lathes etc.

SPECIAL FITMENTS:- 7.5 k.w. generator driven from P.T.O. on gearbox. W.D. Drawbar gear at rear.

CAB:- Open, seats 2, folding canopy.

DIMENSIONS

	OVERALL	INSIDE BODY
Length	20' 0"	14' 0"
Width	7' 5"	7' 0"
Height	10' 3"	6' 9"

WEIGHTS (E)

	UNLADEN	LADEN
F.A.W.		2 tons 15 cwt.
R.A.W.		6 tons 0 cwt.
GROSS		8 tons 15 cwt.

SHIPPING NOTES:- Tonnage Standing 38.4.

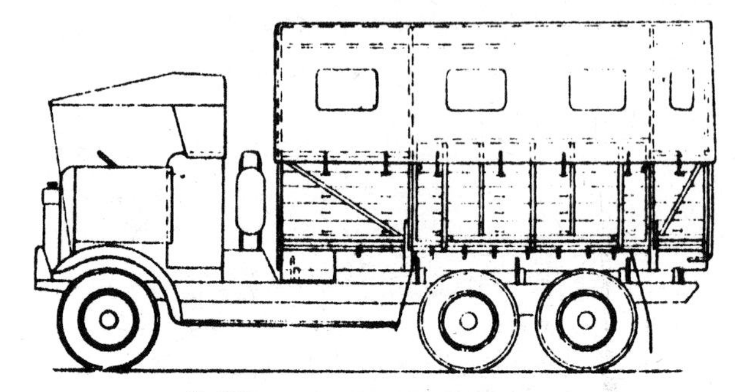

3-TON 6-WHEEL WORKSHOP NO.4. MK.III

ON LEYLAND RETRIEVER CHASSIS

(Similar bodies also fitted to Albion, Crossley, Guy, Karrier and Thornycroft chassis. For weights and dimensions of these, see Data Sheet at back).

BODY:- Flat floor. Tubular steel superstructure with canvas cover. Centre part of body sides hinged to fold to 2 positions:- (1) Horizontal, to give extra floor space. (2) Double fold, to form benches for working on from ground level.

SPECIAL FITMENTS:- 7.5 kw. generator and W.D. drawbar gear.

CAB:- Open type, seats 2, folding canopy.

DIMENSIONS

	OVERALL	INSIDE BODY
Length	22' 5½"	14' 0"
Width	7' 5¼"	6' 6"
Height	11' 6"	5' 9"

WEIGHTS (E)

	UNLADEN	LADEN
F.A.W.		
R.A.W.		
GROSS		8 tons 9 cwt.

SHIPPING NOTES:- Tonnage Standing 47.8 (Note: Shackles are provided for slinging the body).

3-TON 6-WHEEL DERRICK

ON GUY F.B.A.X. CHASSIS

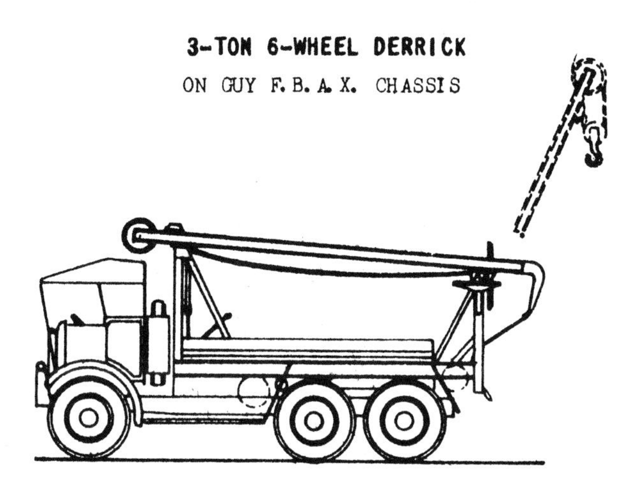

(Similar bodywork also fitted to Crossley and Leyland chassis. For weights and dimensions of these, see Data Sheet at back).

BODY:- Flat platform body 11' 9" x 7'. 3 ton Derrick consisting of single jib and tie. Jib counterbalanced by spring for ease of handling. Swing up legs provided at rear to relieve chassis of vertical loads when lifting.

WINCH:- Hoisting by 5 ton winch with 160 ft. 3/4" diameter rope. W.D. drawbar gear at rear.

CAB:- Open type with folding canopy, seats 2.

OVERALL DIMENSIONS

Length 25' 0"   Width 7' 3"   Height 10' 8½"

WEIGHTS (E)

	UNLADEN	LADEN
F.A.W.		
R.A.W.		
GROSS	6 tons 2¼ cwt.	

SHIPPING NOTES:- Tonnage Standing 46.8.

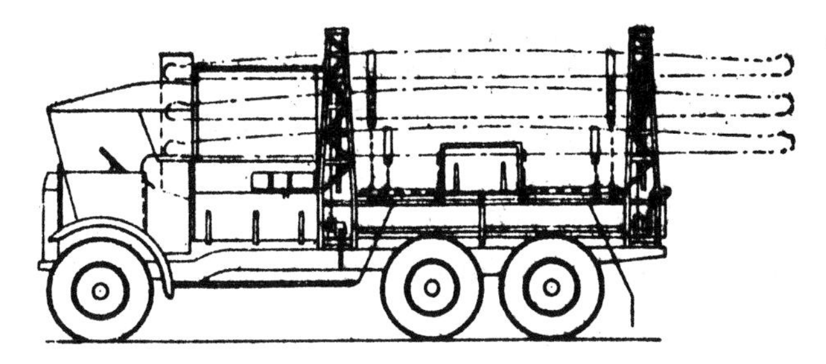

3-TON 6-WHEEL BRIDGING EQUIPMENT

FOLDING BOAT BODY NO.6 MK.I
ON ALBION BY1 CHASSIS

(Only fitted to Albion 3 ton 6 wheel chassis)

Designed to carry two different loadings, either Landing Bay Unit, or Floating Bay Unit as shown in drawing above. There are four towers with hand winches for loading purposes.

W.D. Drawbar gear at rear.

CAB:- Open type, seats 2, folding canopy, but modified by a recess at the back to take the boat noses.

DIMENSIONS

	OVERALL	OVERALL INCLUDING LOAD
Length	21' 9"	28' 4"
Width	7' 6"	7' 6"
Height	10' 3"	10' 3"

WEIGHTS

UNLADEN (E)	4 tons 10 cwt.
LADEN (Boat Unit)	8 tons 11½ cwt.
LADEN (Landing Stage)	8 tons 9¾ cwt.

SHIPPING NOTES:- Tonnage Standing 40.8.

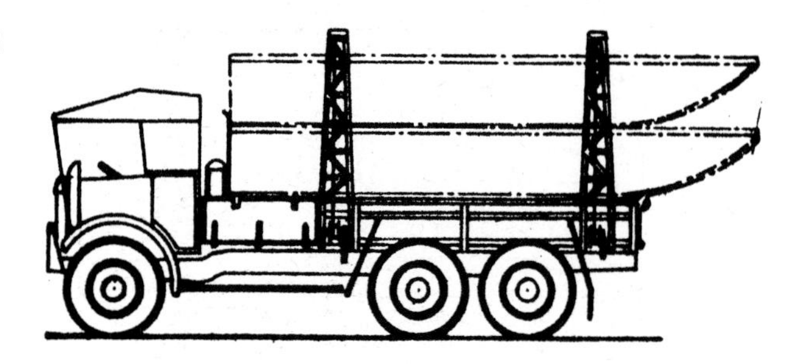

3 TON 6 WHEEL BRIDGING EQUIPMENT

PONTOON BODY NO. 5 Mk. I (RAFT UNIT)
ON ALBION BYI CHASSIS

(Similar bodywork also fitted to Leyland chassis. For weights and dimensions of these, see Data Sheet at back).

Designed to carry the two halves of a pontoon and certain other equipment. Four towers fitted with hand operated winches for loading purposes.

W. D. Drawbar gear at rear.

CAB:- Open type with folding canopy, seats 2.

DIMENSIONS

	OVERALL	OVERALL INCLUDING LOAD
Length	22' 3"	27' 3"
Width	7' 6"	7' 6"
Height	10' 10"	10' 10"

WEIGHTS

	UNLADEN	LADEN
GROSS	4 tons 10 cwt.	7 tons 8 cwt.

SHIPPING NOTES:- Tonnage Standing 45.2.

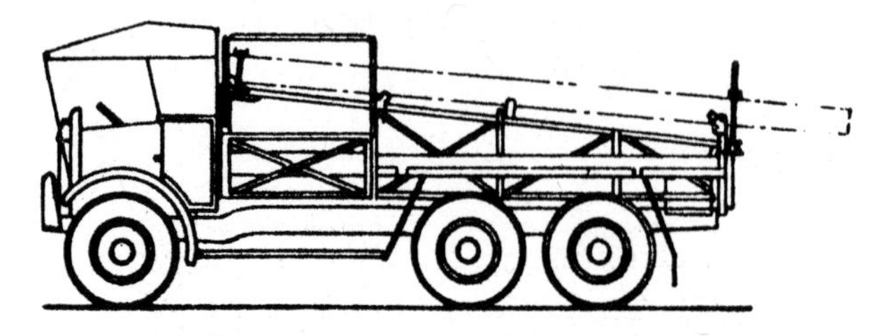

3 TON 6 WHEEL BRIDGING EQUIPMENT

TRESTLE OR SLIDING BAY UNIT
ON ALBION BYI CHASSIS

(Similar body also fitted to Leyland Chassis. For weights and dimensions of this see Data Sheet at back).

This Unit has been designed to carry either the Trestle or Sliding Bay equipment. At the forward end of the body is a compartment for carrying some of the smaller items of equipment.

W. D. Drawbar gear at rear.

CAB:- Open type, seats 2, folding canopy.

DIMENSIONS

	OVERALL	OVERALL INCLUDING LOAD
Length	22' 4"	27' 4"
Width	7' 6"	7' 6"
Height	8' 3"	8' 3"

WEIGHTS

UNLADEN	4 tons 10¼ cwt.
LADEN (Trestle Unit)	8 tons 0 cwt.
LADEN (Sliding Bay Unit)	8 tons 3¼ cwt.

(Also Long Landing Bay Unit, carried on 3 ton 6 wheel G. S. Lorry, Laden Weight 8 tons 17 cwt.)

SHIPPING NOTES:- Tonnage Standing 34.7.

3 TON 6 WHEEL BRIDGING EQUIPMENT

SMALL BOX GIRDER BRIDGE ON ALBION BY1 CHASSIS

(Similar bodywork also fitted to A.E.C. and Karrier chassis. For weights and dimensions of these see Data Sheet at back).

The superstructure has been designed to carry the hornbeams, centre sections and launching nose of the Box Girder Bridge.

W.D. Drawbar gear at rear.

CAB:- Open type with folding canopy, seats 2.

DIMENSIONS

	OVERALL	OVERALL INCLUDING LOAD
Length	21' 10¾"	26' 3"
Width	7' 7½"	7' 7½"
Height	8' 3"	10' 3"

WEIGHTS

	UNLADEN	LADEN (E)
GROSS	4 tons 1¼ cwt.	7 tons 11 cwt.

SHIPPING NOTES:- Tonnage Standing 34.0.

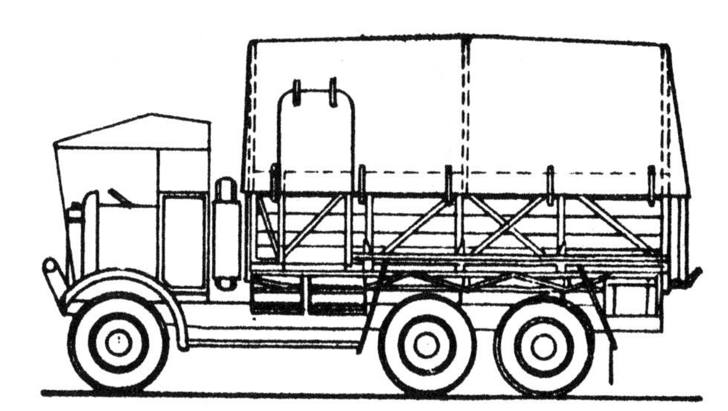

3 TON 6 WHEEL SEARCHLIGHT

ON GUY F.B.A.X. CHASSIS

(Similar bodywork also fitted to Leyland Chassis. For weights and dimensions of this, see Data Sheet at back).

BODY:- Designed to carry 90 c.m. projector. Flat floor, detachable canvas cover. Seating for 8 men with entrance doors either side. Ramps for loading projector stored under floor of body.

SPECIAL FITMENTS:- 24 k.w. generator driven from P.T.O. on gearbox. W.D. Drawbar gear at rear.

CAB:- Open type with folding canopy, seats 2.

DIMENSIONS

	OVERALL	INSIDE BODY
Length	21' 3"	14' 4 3/8"
Width	7' 6"	6' 7¼"
Height	11' 8"	6' 9"

WEIGHTS

	UNLADEN	LADEN
F.A.W.		2 tons 8½ cwt.
R.A.W.		7 tons 13¼ cwt.
GROSS		10 tons 1¾ cwt.

SHIPPING NOTES:- Tonnage Standing 46.4, Cut Down 33.3 Height Cut Down 8' 4".

3 TON 6 WHEEL WIRELESS (HOUSE TYPE)

ON GUY F.B.A.X. CHASSIS

(Similar bodywork also fitted to Leyland chassis. For weights and dimensions of this, see Data Sheet at Back).

BODY:- House type for No.3 W./T. Set. Body divided into two compartments. Front compartment with flat floor houses generator set and batteries. Rear compartment well type, houses W./T. set with seats for operator and 6 personnel, and equipment lockers. 70 ft. wireless masts carried outside in racks.

W.D. Drawbar gear at rear.

DIMENSIONS

OVERALL		INSIDE BODY	
		Front Comp.	Rear Comp.
Length	21' 1"	4' 1½"	9' 8½"
Width	7' 5"	6' 2¼"	6' 2¼"
Height	10' 6"	5' 7½"	5' 7½"

WEIGHTS

	UNLADEN	LADEN
F.A.W.		2 tons 14¼ cwt.
R.A.W.		5 tons 10¾ cwt.
GROSS	5 tons 15¼ cwt.	8 tons 5 cwt.

SHIPPING NOTES:- Tonnage Standing 41.1

ON CROSSLEY I.G.L8 CHASSIS
Also fitted to Leyland Retriever

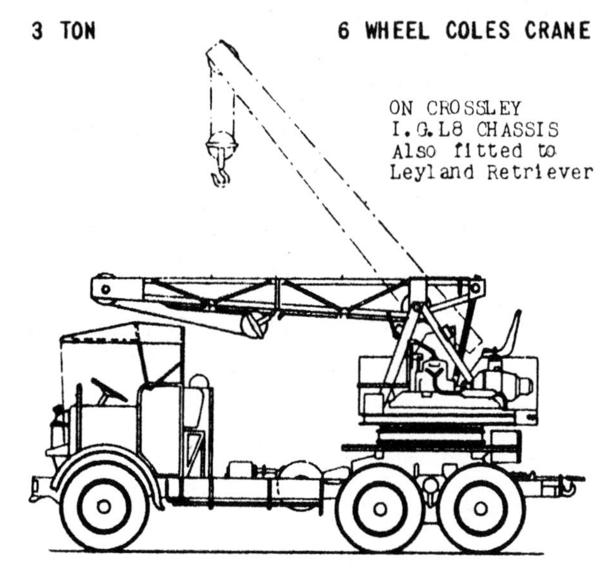

Coles E.M.A. Mk.VI Petrol-Electric Crane. Ford 10 h.p. petrol engine drives variable voltage generator which supplies power for a 6 h.p. Hoist and Derrick Motor and 1 h.p. Slewing Motor.

DUTY:- 2 tons at 7' 9" radius
1½ " " 10' 0" "
1 ton " 13' 0" "

SPECIAL FITMENTS: Winch and W.D. Drawbar gear at rear.

DIMENSIONS (Jib Horizontal)

Length 21' 9". Width 7' 2" Ht. 11' 0".
Max. Height of Lift 16' 6".

WEIGHTS

Weight of Crane in Working Order 3 ton 15 cwt.
" " " on Crossley Chassis 9 tons 4 cwt.

SHIPPING NOTES:- Tonnage Standing 43.3.

SPECIALISED VEHICLES

MAKE	TYPES
Austin	4 Wheel Heavy Ambulance
Morris	" Field Artillery Tractor
Guy	" Field Artillery Tractor
Morris	" Anti Tank Gun Portee
A. E. C.	" Heavy Tractor
Scammell	6 " Heavy Tractor
"	" Heavy Breakdown
"	30 ton Tank Transporter
A. E. C.	Armoured Command
"	Armoured Demolition
"	Armoured Personnel

4 x 2 Semi-forward control Wheelbase 11' 2"

ENGINE:- Make, Austin, Petrol 6-cyl. Bore 3.35" Stroke 4" Capacity 3.46 Litres. B.H.P. 60 @ governed speed 3000 r.p.m. Max. torque lbs./ins. 1830 @ 1200 r.p.m.

GEARBOX:- 4 speeds 1 reverse.
Overall ratios: 46.7 to 1, 20.4 to 1, 10.0 to 1, 5.85 to 1, rev. 42.9 to 1.

REAR AXLE:- Full Floating Spiral Bevel.
Ratio 5.85 to 1.

SUSPENSION:- Semi-elliptic, front and rear, with hydraulic shock absorbers.

BRAKES:- Foot: Hydraulic on all wheels.
Hand: Mechanical on rear wheels.

TYRES:- 10.50 - 16.

TURNING CIRCLE:- 55 ft. TRACK, f. 5' 1", r. 5' 1¼".

GROUND CLEARANCES:- R. Axle 8¼", Belly 14".

FUEL CAPACITY:- 24 gals. Radius 240 miles.

COOLING SYSTEM CAPACITY:- 3 7/8 gals.

PERFORMANCE FIGURES:- (Ambulance Laden)
B.H.P. per Ton 15.
Tractive Effort per Ton (100%) 1242 lbs./Ton.
Max. speed at governed r.p.m. 55 m.p.h.

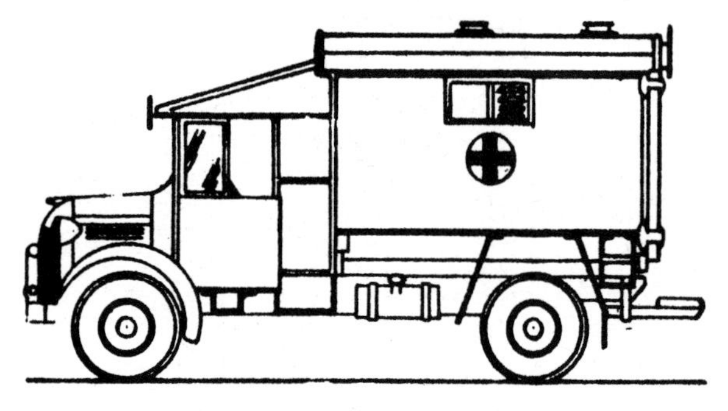

HEAVY AMBULANCE ON AUSTIN K2

(Similar body also fitted to Bedford ML Ambulance chassis. For weights and dimensions of this, see Data Sheet at back).

BODY:- Accommodation for Driver and Mate in cab, Attendant and 4 stretcher or 10 sitting cases in main body. Communicating door between cab and body, double swing doors and folding step at rear. Air conditioning by ventilators in body sides, rear doors and front partition, 2 roof extractor ventilators, Clayton Dewandre heating installed.

DIMENSIONS

	OVERALL	INSIDE MAIN BODY
Length	18' 0"	8' 7"
Width	7' 3"	6' 5"
Height	9' 2"	5' 5¾"

WEIGHTS

	UNLADEN	LADEN
F.A.W.	1 Ton 7¼ Cwt.	1 Ton 10 Cwt.
F.A.W.	1 Ton 14¼ Cwt.	2 Ton 10 Cwt.
GROSS	3 Tons 1½ Cwt.	4 Tons 0 Cwt.

SHIPPING NOTES:- Tonnage Standing 30.0.

MORRIS C.8 4 WHEEL F.A.T. TRACTOR CHASSIS

4 x 4 Normal Control Wheelbase 8' 3"

ENGINE:- Make, Morris. Petrol 4 cyl. Bore 100 mm. Stroke 112 mm. Capacity 3.5 litres. Max. B.H.P. 70 @ 3000. Governed to 2500. Max. torque lbs./ins. 1800 @ 1750 r.p.m.

GEARBOX:- 5 speeds and 1 reverse. Overall Ratios: 67.7 to 1, 38.2 to 1, 20.4 to 1, 11.7 to 1, 6.57 to 1, reverse 67.7 to 1. (Drive to all four wheels) On C.8 Mk.III model front wheel drive can be disengaged on all gears save 1st and reverse.

AXLES:- Front and rear, full floating, spiral bevel. Ratio 6.57 to 1.

SUSPENSION:- Front and rear, semi-elliptic with hydraulic shock absorbers.

BRAKES:- Foot, Hydraulic on all wheels. Hand, Mechanical on rear wheels.

TYRES:- 10.50 - 20. (Later models 10.50 - 16).

TURNING CIRCLE:- 56 ft. Track, f. 5' 3¾" r. 5' 9".

GROUND CLEARANCES:- Front axle 11½". Rear Axle 12 3/8". Belly 19½".

TANK CAPACITY:- 30 gals. Radius 160 miles towing. 25 pdr. gun and trailer.

COOLING SYSTEM CAPACITY:- 4½ gals.

PERFORMANCE FIGURES (Laden):-
B.H.P. per Ton 14.6.
Tractive Effort per Ton (100%) 1455 lbs./ Ton.
Max. Speed at governed r.p.m. 42 m.p.h.

GUY "QUAD ANT" F.W.D. F.A.T.

4 x 4 Normal Control. Wheelbase 8' 5".

ENGINE: Make, Meadows, 4 E.L.A. Petrol 4-cyl. Bore 95 mm. Stroke 130 mm. Capacity 3.68 litres. Max B.H.P. 58 @ 2400, governed to 2700. Max torque 1980 @ 1400.

GEARBOX:- 4 speeds and 1 reverse. Transfer case ratio direct. Overall ratios:- 54.7 to 1, 24.7 to 1, 13.8 to 1, 7.5 to 1, rev: 78.0 to 1. Drive to all wheels or rear wheels only).

AXLES:- Full floating, spiral bevel drive. Ratio 7.5 to 1.

SUSPENSION:- Semi-elliptic front and rear, with hydraulic shock absorbers.

BRAKES:- Front and rear, Mechanical on all Wheels.

TYRES:- 10.50 - 20. (Later models 10.50 - 16).

TURNING CIRCLE:- 55 ft. TRACK f. 5'9¼", r. 5'9¼".

GROUND CLEARANCES:- r. axle 10", belly 15½".

TANK CAPACITY:- 28 gals. Radius 160 miles towing, 25 pdr. gun & trailer.

COOLING SYSTEM CAPACITY:- 4¾ gals.

GUY FIELD ARTILLERY TRACTOR BODY

Seating for driver, Commander and a crew of four. Lockers for 96 rounds of amn. (shells and cartridges) and 8 rds. of A.P. shot. (Gun Tractor Loading).

For weights and dimensions of GUY F.A.T. see Data Sheet at back.

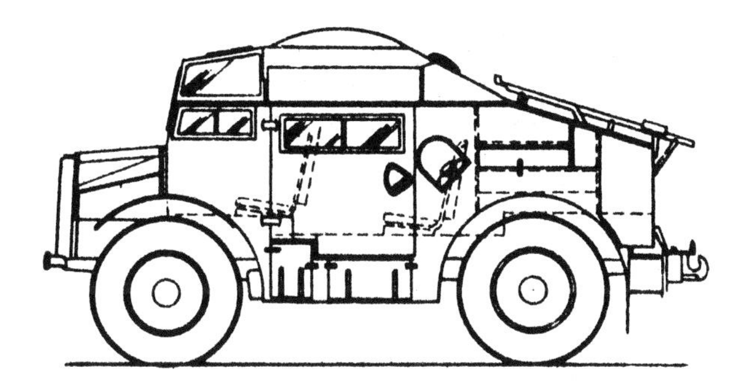

MORRIS C.8 4 WHEEL F.A.T.

TRACTOR BODY

BODY:- All metal gun tractor body with roll-up canvas roof. Accommodation for driver, Commander and four men. Fitments at rear to carry either a spare gun wheel or gun firing platform.

AMMUNITION ACCOMMODATION:- 2 Trays measuring 31 1/16" x 10¼" x 5¼". 2 Cartridge Compartments measuring 19" x 9.7" x 12.1". 4 Projectile Compartments measuring 19.15" x 8.5" x 7.85". 8 rounds of A.P. shot.

SPECIAL FITMENTS:- 4-ton winch to pull from front or rear of tractor. W.D. draw-bar gear at rear.

DIMENSIONS

	OVERALL	INSIDE BODY
*Length	14' 8¾"	6' 10¼"
Width	7' 3"	6' 11"
Height	7' 5"	4' 0"
*Overall length with gun platform	15' 5¼"	

WEIGHTS

	UNLADEN	LADEN
F.A.W.	1 ton 13¾ cwt.	1 ton 18 cwt.
R.A.W.	1 ton 13¼ cwt.	2 tons 18 cwt.
GROSS	3 tons 7 cwt.	4 tons 16 cwt.

SHIPPING NOTES:- Tonnage Standing 19.71.

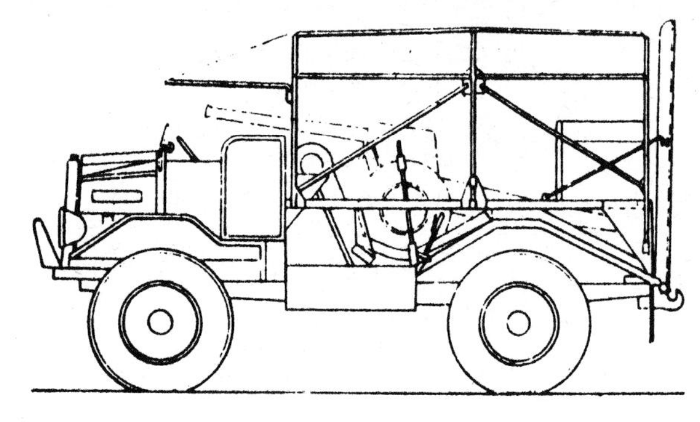

MORRIS C.8 4-WHEEL
ANTI TANK PORTEE

BODY:- Designed for the carriage of a 2 pdr. Anti-tank gun. Gun is hoisted into vehicle by hand operated which and then rests in curved channels secured by spring loaded clamps. On taking in the winch rope with gun in position the entire superstructure is raised 15" at front end so giving wheel guides a rearward cant. Body is then maintained in this position by a pair of folding supports. Seating for driver and crew of four. Detachable canvas cover.

W.D. Drawbar gear at rear.

OVERALL DIMENSIONS

Length 14' 8". Length, ramp extended, 19' 6". Width 7' 0". Height 8' 2".

WEIGHTS

	UNLADEN	LADEN
F.A.W.	1 Ton 14¾ Cwt.	2 Tons 5 Cwt.
R.A.W.	1 Ton 11 Cwt.	2 Tons 14¾ Cwt.
GROSS	33 Tons 5¾ Cwt.	4 Tons 19¾ Cwt.

SHIPPING NOTES:- Tonnage standing 20.9.

A.E.C. MATADOR 4 WHEEL
MEDIUM TRACTOR

4 x 4 Forward Control Wheelbase 12' 7½"

ENGINE:- Make, A.E.C. C.I. 6-cyl. Bore 105 mm. Stroke 145 mm. Capacity 7.58 litres. Max. B.H.P. 95 @ 1780 r.p.m. Governed to 1800 r.p.m. Max. torque lbs./ins. 3720 @ 1200 r.p.m.

GEARBOX:- Main, 4 speeds and reverse. Auxiliary direct and 2.91 to 1. Overall Ratios (High) 27.4 to 1, 16.8 to 1, 9.9 to 1, 6.25 to 1, reverse 33.3 to 1.

AXLES:- Full floating, spiral bevel and double helical gears. Ratio 6.25 to 1.

SUSPENSION:- Semi-elliptic, front and rear.

BRAKES:- Foot, Hydraulic with compressed air servo on all wheels. Hand, Mechanical on rear wheels. Warner electric braking attachment also fitted.

TYRES:- 13.50-20.

TURNING CIRCLE:- 60 ft. Track, f. 6' 3 1/8", r. 5' 10¾".

GROUND CLEARANCES:- F. axle 13", R. axle 15¼", belly 18½".

FUEL CAPACITY:- 40 gals. Radius 360 miles.*

COOLING SYSTEM CAPACITY:- 6 gals.

PERFORMANCE FIGURES:-
B.H.P. per Ton 8.7.
Tractive Effort per Ton (100%) 1250 lbs. / Ton.
Max. Speed at governed r.p.m. 36 m.p.h.

Later contracts: Axle front, Spiral bevel, ratio 7.9 to 1. Axle rear, worm, ratio 7.75 to 1.
Difference in axle ratio adjusted in transfer box.
Transfer ratios 1.00 and 2.3 to 1.
BRAKES:- Compressed Air (no hydraulic lines) 2 line Air trailer brake connections are fitted.

* Indicates vehicle towed 6½ ton trailer.

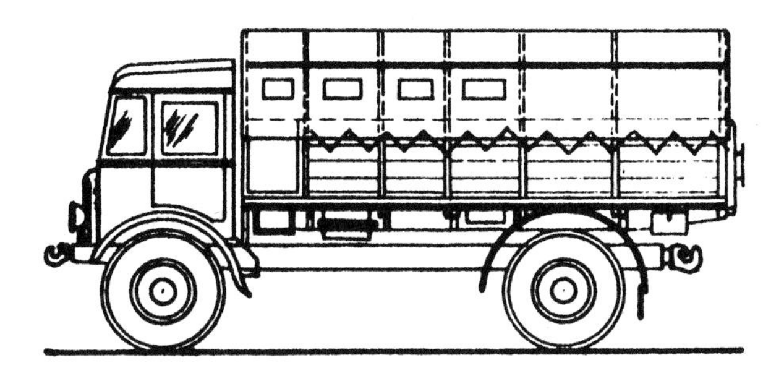

A.E.C. 4 WHEEL MEDIUM TRACTOR

BODY:- Flat floor, detachable hoops and canvas cover. Entrance door on nearside. Seats for 10 men provided at front of body with lockers and equipment racks. Adjustable shell carrier runners on floor. Four seats can be folded away to take alternative loading of A.A. equipment.

CAB:- All enclosed, seats driver and mate.

SPECIAL FITMENTS:- 7 ton winch with 250 ft. ¾" diam. rope. W.D. drawbar gear front and rear.

DIMENSIONS

	OVERALL	INSIDE BODY
Length	20' 9"	14' 6"
Width	7' 10½"	7' 4"
Height	10' 2"	5' 6"

WEIGHTS

	UNLADEN	LADEN (E)
F.A.W.	3 tons 14 cwt.	4 tons 4¼ cwt.
R.A.W.	3 tons 7½ cwt.	6 tons 12¾ cwt.
GROSS	7 tons 1½ cwt.	10 tons 17 cwt.

SHIPPING NOTES:- Tonnage Standing 41.8.

SCAMMELL 6 WHEEL HEAVY GUN TRACTOR

6 x 4 Normal Control Wheelbase 12' 2"
Bogie Centres, 4' 3¼".

ENGINE:- Make, Gardner 6 L.W. C.I. 6-cyl. Bore 4¼". Stroke 6". Capacity 8.4 litres. Max. B.H.P. 102 @ governed speed 1750 r.p.m. Max. Torque 4164 lbs.ins. @ 1100 r.p.m.

GEARBOX:- 6 speeds and 1 reverse. Overall Ratios: 111.0 to 1, 69.0 to 1, 42.9 to 1, 25.1 to 1, 15.6 to 1, (overspeed) 9.67 to 1, reverse 141.8 to 1.

REAR AXLE:- Worm drive Ratio 7.4 to 1. Final Drive by gear train bogie, Ratio 2.1 to 1. Overall ratio to road wheels 15.6 to 1.

SUSPENSION:- Front, Transverse inverted semi-elliptic. Rear, semi-elliptic spring on final drive gear case capable of swivelling about axle centre.

BRAKES:- Foot, on all wheels assisted by compressed air servo.
Hand: (1) Neate brake on driving wheels.
(2) Transmission brake.

TYRES:- 13.50-20.

TURNING CIRCLE:- 70 ft. Track, f. 6' 10", r. 7' 1".

GROUND CLEARANCES:- F. axle 18½", Belly 21½", R. Axle 12¾".

FUEL CAPACITY:- 50 gals. Radius 355 miles.*

COOLING SYSTEM CAPACITY:- 8¼ gals.

PERFORMANCE FIGURES:-

B.H.P. per Ton 8.3.
Tractive Effort Per Ton (100%) 1668 lbs./Ton.
Max. Speed at governed r.p.m. 24 m.p.h.

"Later types have 2 line air Trailer Brake connections".

(* Indicates vehicle towed 6½ ton Trailer).

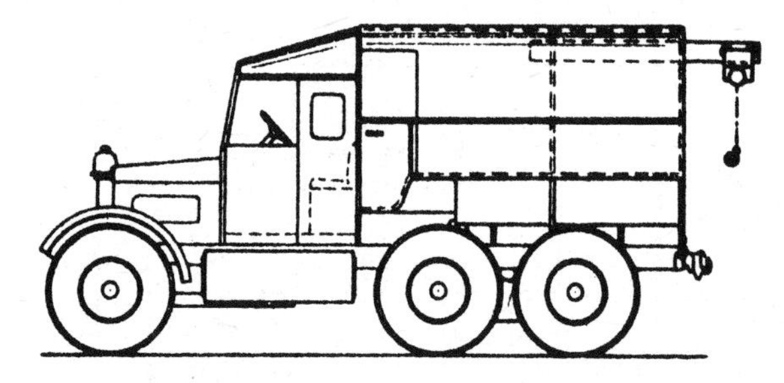

SCAMMELL 6 WHEEL HEAVY GUN TRACTOR

BODY:- Steel panelled body with wooden well type floor, providing accommodation for 9 men, kit, ammunition and W.D. vehicle equipment. Overhead runway and 10 cwt. hoist for loading and unloading.

CAB:- Enclosed steel panelled, open to body at rear, seating for driver and 2 men.

SPECIAL FITMENTS:- 8 ton vertical spindle winch with 430 ft. of 7/8" diam. rope W.D. Drawbar gear at rear.

DIMENSIONS

	OVERALL	INSIDE BODY
Length	20' 3"	9' 8"
Width	8' 6"	8' 3"
Height	9' 9"	5' 0"

WEIGHTS

	UNLADEN	LADEN (R.A.O.C. loading)
F.A.W.	2 tons 9½ cwt.	3 tons 10¼ cwt.
R.A.W.	5 tons 18 cwt.	8 tons 12 cwt.
GROSS	8 tons 7½ cwt.	12 tons 2¾ cwt.

SHIPPING NOTES:- Tonnage standing 42.1.

SCAMMELL 6 WHEEL HEAVY BREAKDOWN

6 x 4 Normal Control Wheelbase 12' 2"
Bogie Centres 4' 3½"

ENGINE:- Make, Gardner 6 L.W. C.I. 6-cyl. Bore 4¼". Stroke 6". Capacity 8.4 litres. Max. B.H.P. 102 @ governed speed, 1750 r.p.m. Max. Torque 4164 lbs./ins. @ 1100 r.p.m.

GEARBOX:- 6 speeds and 1 reverse. Overall Ratios: 111.0 to 1, 69.0 to 1, 42.9 to 1, 25.1 to 1, 15.6 to 1, over-speed 9.67 to 1, rev. 141.8 to 1.

REAR AXLE:- Worm drive, Ratio 7.4 to 1. Final drive by gear train bogie, Ratio 2.1 to 1. Overall ratio to road wheels 15.6 to 1.

SUSPENSION:- Front, Transverse inverted semi-elliptic. Rear, semi-elliptic spring on final drive gear case capable of swivelling about axle centre.

BRAKES:- Foot on all wheels assisted by compressed air servo.
Hand: (1) Neate brake on driving wheels.
(2) Transmission brake.

TYRES:- 13.50-20.

TURNING CIRCLE:- 70 ft. Track, f. 6' 10", r. 7' 1".

GROUND CLEARANCES:- F. axle 18½", Belly 21½", R. Axle 12¾".

FUEL CAPACITY:- 50 gals. Radius 535 miles.

COOLING SYSTEM CAPACITY:- 8¼ gals.

PERFORMANCE FIGURES:- B.H.P. per Ton 9.9. Tractive Effort per Ton (100%) 2042 lbs./ton. Max. Speed at governed r.p.m. 24 m.p.h. Latest type have 2 line air Trailer Brake connections.

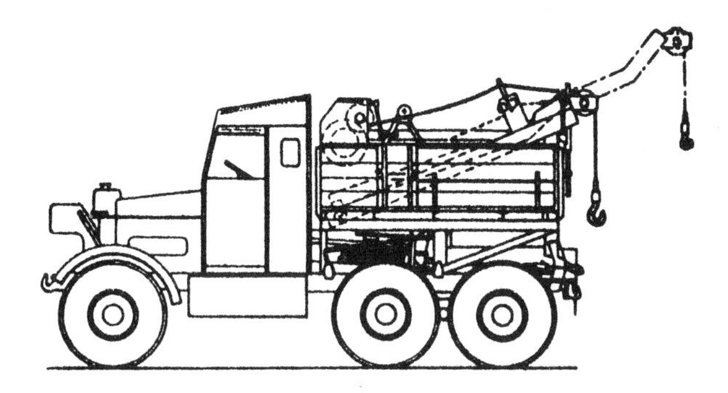

SCAMMELL 6 WHEEL HEAVY BREAKDOWN

(Model SV/2S)

BODY:- Sliding jib extended by means of small hand winch, with three positions. (1) Long Lift, 2 ton load. (2) Short Lift 3 ton load. (3) Folded home for travelling. Worm driven hand operated hoist. Detachable ballast weights for front end of vehicle. Enclosed cab, seating driver and two men.

SPECIAL FITMENTS:- 8 ton vertical winch with 450 ft. of 7/8" diam. rope. W.D. Drawbar gear front and rear.

OVERALL DIMENSIONS

Length 20' .3". Width 8' 6". Height 9' 5". Overall length, jibs fully extended, 25' 4".

WEIGHTS

	UNLADEN	LADEN (E)
F.A.W.	2 tons 17 cwt.	3 tons 14½ cwt.
R.A.W.	6 tons 16¼ cwt.	7 tons 2½ cwt.
GROSS	9 tons 13¼ cwt.	10 tons 17 cwt.

SHIPPING NOTES:- Tonnage Standing 40.8.

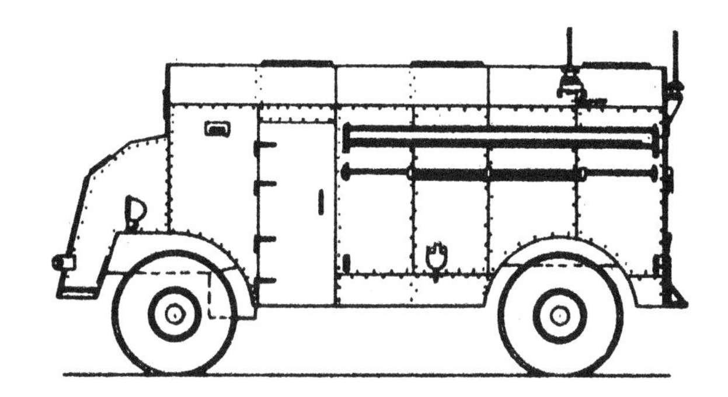

A.E.C. ARMOURED COMMAND VEHICLE

On A.E.C. Matador F.W.D. Chassis

CHASSIS:- Generally the same as A.E.C. Matador chassis on page 86, except:- Tyres, 13.50-20 Runflat, Modified instrument panel, Winch and drawbar gear not fitted.

BODY:- Armoured on a 12 mm. basis. Internal seats, desks and lockers for 4 officers and 2 wireless operators in addition to driver's seat. Fitments include two No. 14 and one No. 9 wireless sets, cipher machine, necessary batteries and electrical equipment. Spring loaded blinds are fitted outside vehicle on either side, they can be opened out to form awnings on folding supports.

Extra belt-driven dynamo supplies current for wireless sets.

OVERALL DIMENSIONS

Length 20' 0" Width 8' 6" Height 9' 5"

WEIGHTS

	UNLADEN	LADEN (E)
F.A.W.	4 Tons 19½ cwt.	5 tons 3 cwt.
F.A.W.	4 Tons 16¼ cwt.	5 tons 10 cwt.
GROSS	9 Tons 16 cwt.	10 tons 13 cwt.

SHIPPING NOTES:- Tonnage Standing 40.0.

Generally, this vehicle is similar to the A.E.C. Armoured Command, except that the wireless sets and cipher ma hine are not fitted. The wiring for Wireless and the extra dynamo are fitted so that, if necessary, the vehicle may quickly be converted to an Armoured Command.

A.E.C. ARMOURED DEMOLITION

Similar armoured body to the Armoured Command Vehicle except that there is a special hatch in the roof to allow operation of pile-driver. Equipped with a compressor (driven by power take-off from the main drive of the vehicle and pile-driver. The function of the pile-driver is to drive holes in the road for inserting charges for demolition. A wide range of pneumatic tools are carried, including, Hammer drills and borehole filling equipment.

WEIGHTS

	UNLADEN	LADEN
F.A.W.	5 tons 4¾ cwt.	5 tons 5¼ cwt.
R.A.W.	5 tons 7¾ cwt.	5 tons 18¼ cwt.
GROSS	10 tons 11¾ cwt.	11 tons 3¾ cwt.

The weights of W.D. Vehicle Equipment and manufacturers special tools very with the make of chassis, type of body and tyre equipment. As a general guide the following table is appended, which represents the standard allowances as set out under W.D. Spec. 36E for test purposes.

Chassis Type	15 Cwt.		30 Cwt.		3 Ton 4 Whl.		3 Ton 6 Whl.	
	T.	Cwt.	T.	Cwt.	T.	Cwt.	T.	Cwt.
Body & Cab	0	6	0	10	0	15½	0	17
W.D. Vehicle Equipment	0	0½	0	1¼	0	1	0	1
Makers Tools	0	0¼	0	0½	0	0½	0	0½
Non Skid Chains	0	1½	0	1½	0	2	0	3
Personnel with Kit etc.	0	3½	0	3½	0	3½	0	5¼
Spare Petrol Oil & Water	0	0¼	0	1¼	0	1½	0	1¼
Useful Load	1	0	1	10	3	0	3	0
GROSS	1	12	2	8	4	4	4	8

ADDITIONAL
BRITISH
TYPES

TRUCK 15-CWT 4 x 2 FITTED FOR WIRELESS

On Bedford M.W.R. 15-cwt. 4 x 2 Chassis

CHASSIS:- Bedford M.W.R. is similar to standard Bedford M.W. chassis on page 14, except for the following modifications:-

1. Special screening of ignition and electrical equipment.
2. Lead from vehicle lighting to 2 pin socket in body.
3. P.T.O. on gearbox driving auxiliary dynamo.
4. Control board for auxiliary dynamo.
5. Hand throttle control.

BODY:- Fitted out to carry the following wireless sets:-

Nos. 11, or 19 or 21.

Three seats in body. Plan view shows (dotted) alternative position of two rear seats when carrying speiial R.A. equipment.

CAB:- Enclosed. Removable side screens and top. Hatch in top.

SPECIAL FITMENTS:- Chore Horse Battery Charger mounted in Chore Box behind offside front wing. Provision for spare tyre on special carrier at rear.

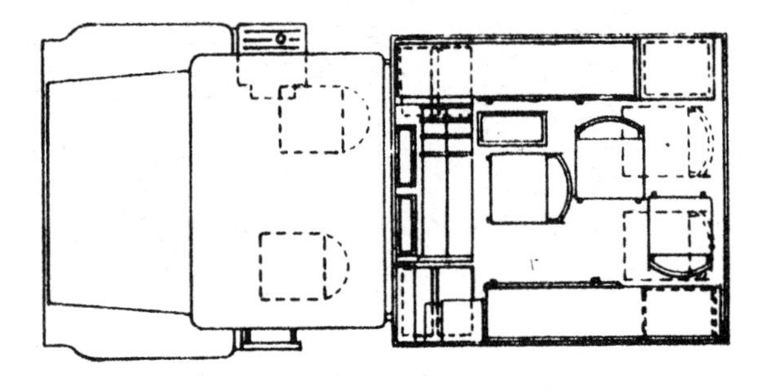

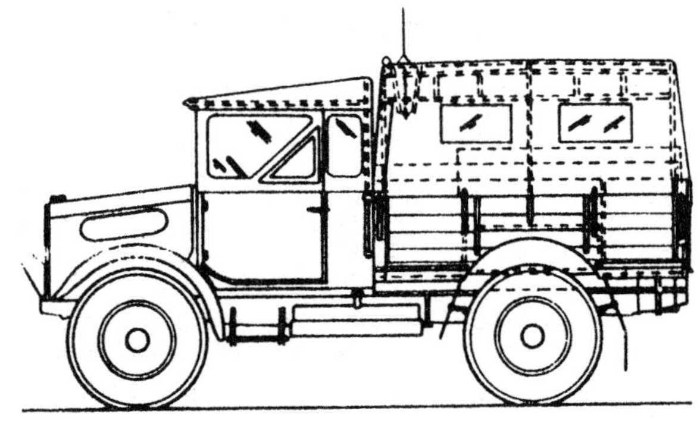

TRUCK 15-CWT. 4 x 2 FITTED FOR WIRELESS

DIMENSIONS

	OVERALL	INSIDE BODY
Length	14' 4½"	6' 5½"
Width	6' 6½"	5' 10½"
Height	7' 6"	4' 7"

WEIGHTS

THORNYCROFT STURDY WZ/TC4.
3-TON 4 x 2

Forward Control   Wheelbase (G.S.) 13'4"
"   (Tipper) 9'0"

ENGINE:- Make Thornycroft T.C.4. Petrol 4 cyl. Bore 3 7/8" stroke 5" Capacity 3.86 litres. Max. torque 1930 lbs./ton @ 1,400 r.p.m. Max. B.H.P. 60.

GEARBOX:- 4 speed and reverse. Ratios, 6.4 to 1, 3.14 to 1, 1.79 to 1, 1.00 to 1, and 9.27 to 1.

REAR AXLES:- Full floating, overhead worm drive,

Ratio (G.S. L.W.B.) 7.25 to 1.
" (Tipper S.W.B.) 7.75 to 1.

SUSPENSION:- Semi-elliptic, front and rear.

BRAKES:- Foot, Hydraulic on all wheels. Hand, mechanical on rear wheels.

TYRES:- 10.50-20.

TURNING CIRCLES:- L.W.B. 56'. S.W.B. 44'

GROUND CLEARANCES:- R. axle 13".

CAPACITIES:- Fuel 20 galls. Cooling 4½ galls.

PERFORMANCE FIGURES (G.S.)

B.H.P. Per Ton 8.9
Tractive Effort per Ton (100%) 677 lbs. Ton
Max. speed, 35 m.p.h.
M.P.G. 8. Radius 160 miles.

LORRY 3-TON 4 x 2 G.S.

On Thornycroft Sturdy WZ/TC4 (Long) Chassis

BODY:- Standard G.S. with tilt 13' 4" x 6' 6" x 6' 0" high.

CAB:- All enclosed non-detachable top.

OVERALL DIMENSIONS:- 21' 8" x 7' 4" x 9' 8" Stripped Shipping Height 7' 9".

WEIGHTS:- Unladen, F.A. 1 ton 15¼ cwt. R.A. 1 ton 11¼ cwt., Laden, F.A. 2 tons 5 cwt. R.A. 4 tons 10 cwt.

LORRY 3-TON 4 x 2 TIPPING (3-way)

On Thornycroft Sturdy WZ/TC.4 (Short) Chassis

BODY:- 3 way Tipping gear, hydraulic operated from P.T.O. on gear box. Direction of tip selected by removing appropriate pins from body brackets. Hinged sides and Tailboard. Dimensions 8' 10" x 6' 7" x 1' 6" (3 cu.yds.).

CAB:- All enclosed with detachable Top.

OVERALL DIMENSIONS:- 15' 4" x 7' 1" x 7' 9".

WEIGHTS:- Laden, F.A. 2 tons 8 cwt., R.A. 4 tons 6 cwt.

STRIPPED SHIPPING HEIGHT:- 7' 4" cab top detached.

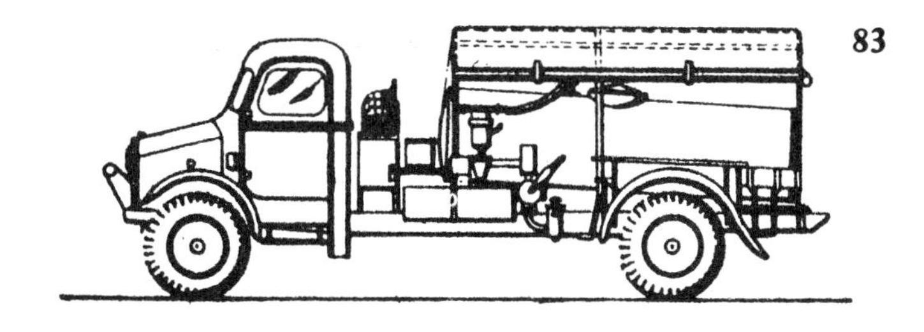

LORRY 3 TON 4 x 2 WATER 350 GALLONS

On Bedford OYC 3 Ton 4 x 2

(CHASSIS DETAILS AS PAGE 37)

Fitted with two hand-operated pumps and duplicate sets of complete filtering equipment. Each pump capable of delivering a complete tankful (350 gallons) in one hour. Filter equipment and delivery hose stored in locker built into rear of tank. Camouflage cover (tilt) carried on detachable superstructure.

CAB:- Enclosed steel panel cab seating two.

OVERALL DIMENSIONS

Length 20' 1" Width 6' 7" Height 7' 7"

WEIGHTS

	UNLADEN	LADEN
F.A.	1 ton 7½ cwt.	2 tons 1 cwt.
R.A.	1 ton 14½ cwt.	3 tons 6 cwt.
Gross	3 tons [illegible]¼ cwt.	5 tons 7 cwt.

STRIPPED HEIGHT:- 7' 0" (Superstructure detached)

LORRY 3 TON 4 x 2 PETROL 800 GALLONS

On Bedford OYC 3 ton 4 x 2

OYC is maker's code name for chassis and cab. Chassis is similar to Bedford OY, details on page 37, except:

Additional cross member is fitted to frame under forward end of tank. Fire screen is fitted behind cab. Exhaust is repositioned to exhaust forward of the fire screen.

TANKS are made by a number of contractors.

General details are:

Two compartments
Fitted with hand operated pump
Two 10 ft. lengths of delivery hose
Camouflage cover mounted on detachable superstructure. (As on Water Tank page 118)

CAB:- Enclosed, steel panelled, seats 2.

OVERALL DIMENSIONS

Length 20' 1" Width 6' 7" Height 7' 7½"

WEIGHTS

	UNLADEN	LADEN
F.A.	1 ton 9½ cwt.	2 tons 1¼ cwt.
R.A.	1 ton 13 cwt.	4 tons 5 cwt.
Gross	3 tons 2½ cwt.	6 tons 6¼ cwt.

STRIPPED SHIPPING HEIGHT:- 7' 0" (Superstructure detached)

LORRY 3-TON 4 x 2 WORKSHOP

(R.A.S.C., H.R.S. Conversion)

ON BEDFORD OY 3-TON 4 x 2
(Chassis Details as page 37)

AND COMMER Q4 3-TON 4 x 2
(Chassis Details as page 38)

This is a standard G.S. body fitted with the following equipment:-

Lathe
Switchboard
Charging Board
7.5 k.w. Dynamo
Dynamo Regulator
Interlocking Switch and Plug
Bench
Electric Drill and Stand
4½" Vice
3½" Vice
Quasi-Arc Welding Plant
Portable Forge
Portable Air Compressor
Portable Benches
Electric Grinder on Stand
3½" Vice and Stand, Portable
Post Drill
Coppersmith Tool Chest
Lathe Starter

Interior Lighting, Tool Boxes, Anvil, etc.

DRIVE TO DYNAMO. Propellor shaft is disconnected and a driving sprocket is bolted to end of cardan shaft.

DIMENSIONS:- As respective G.S. bodies.

LORRY, 3-TON 4 x 2 STORES

On Bedford OY 3-ton 4 x 2 Chassis

(R.A.S.C., H.R.S., CONVERSION)

CHASSIS:- details as page 37.

BODY:- This is a standard G.S. body similar to to that shown on page 43, with the following modifications. Wire mesh covers are fitted to the standard tubular superstructure and above the tailboard at rear. Interior is fitted with full length benches either side.

Steel storage bins are located under the benches, with compartments for small items of equipment above benches.

Writing desk.

Correspondence basket.

DIMENSIONS:- As page 43.

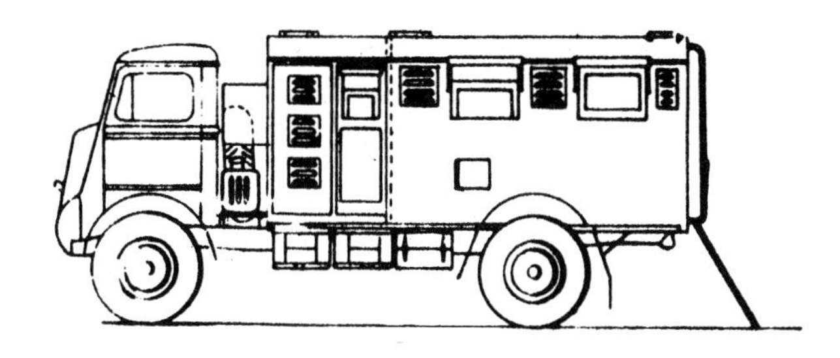

LORRY 3 TON 4 x 4 WIRELESS ON BEDFORD Q.L. CHASSIS PRELIMINARY DETAILS

CHASSIS DETAILS:- As Standard Bedford Q.L. on page 48 except 2 side petrol tanks mounted at side of frame instead of 1 tank behind cab. (Total capacity 32 galls.).

BODY:- House type. Interior fittings to take 7 different sets of equipment:-

1. Cypher Office (No generator)
2. No.12 Set
3. R.C.A. American Sender.
4. H.24 American Sender
5. No.22 High Power Set
6. Signal Office, 3 Teleprinters
7. "P" Vehicle, 6 Receivers and 6 Aerials

SPECIAL FITMENTS:- Generator. W.D. Drawbar Gear.

DIMENSIONS:- Interior (basic, without any Equipment) 12' 8" x 6' 8" x 5' 3" Headroom. Length front Compartment 3' 6". Rear Compartment 9' 0".

Overall: Length 19' 10", Width 7' 6", Height 8' 10".

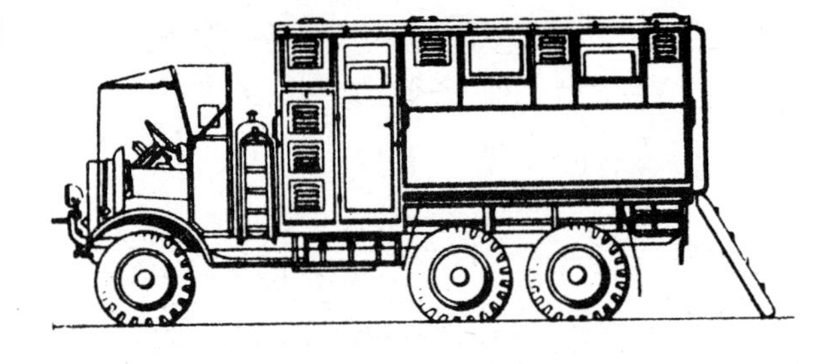

LORRY 3-TON 6 x 4 WIRELESS BODY NO. 7 MK.I

on Thornycroft WOF/DC4 (C.I. engine) 3-ton 6 x 4 Chassis

PRELIMINARY DETAILS:

CHASSIS DETAILS:- Generally as page 66.

BODY:- House Type, divided into 2 compartments.

Wireless Set: H.S.1 Sender
Generator: Stewart 2½ k.v.a. driven from P.T.O.

INTERIOR BODY DIMENSIONS

	Front Compartment	Main Compartment
Length	4' 0"	9' 8½"
Width	6' 10"	6' 10"
Height	6' 9"	5' 9"

OVERALL DIMENSIONS:- Length 22' 6". Width 7' 6" Height 10' 6".

BEDFORD QL 6-TON SEMI-TRAILER G.S. 4 x 4 - 2

Forward Control. Wheelbase Tractor 11' 11"

Semi-trailers made by Glover, Webb & Liversidge and Scottish Motor Traction to common design.

TRACTOR UNIT:

Standard Bedford QLC (Chassis and Cad) as specification on page 48 except:-

1. Rear ends of frame sidemembers cut off. Additional cross members fitted to carry trailer coupling.
2. Braking modified by fitment of vacuum reservoir tank and vacuum gauge in car.
3. Modified rear shock absorbers.

SEMI-TRAILER:

Coupling Gear - Permanently coupled to tractor by Tasker 6/8 ton ball type coupling which allows 20° angle of movement either side and fore and aft.

Brakes - Foot: vacuum servo operated from vacuum tank on tractor unit. Hand: cable operated hand brake lever fitted to trailer.

Tyres - 10.50-20. Spare wheel carried in retractable carrier above trailer axle.

Springs - Semi-elliptic.

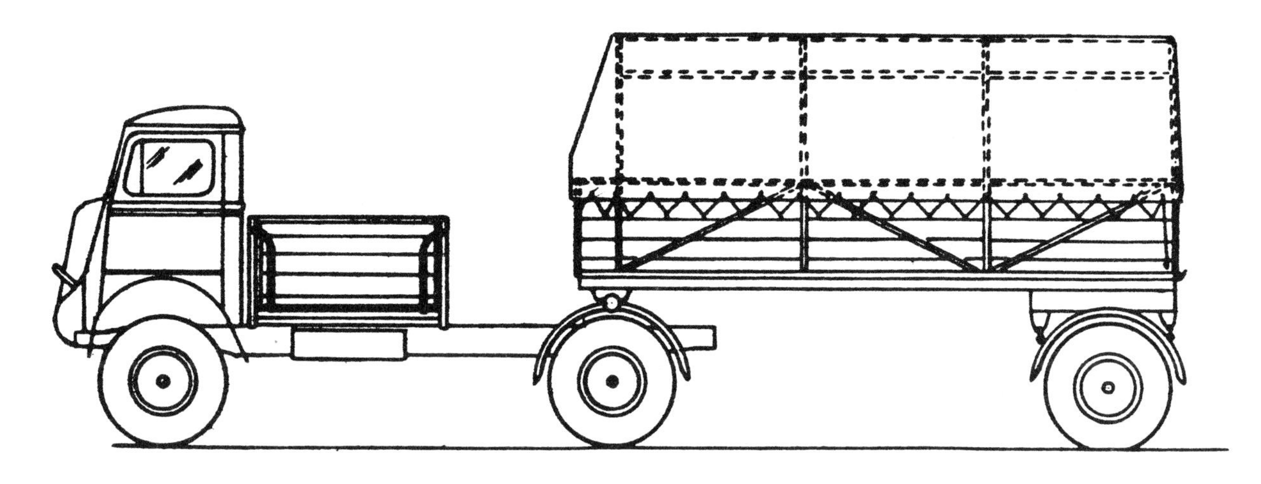

BEDFORD OL SEMI TRAILER G.S. 4 x 4 - 2

BODY

TRACTOR:- Small dropside body designed to carry a load of approximately 1 ton.

SEMI-TRAILER:- Flat floor G.S. body with detachable canvas tilt and bows.

SEMI TRAILER DIMENSIONS

	Overall	Inside Body
Length	17' 8"	16' 11"
Width	7' 6"	6' 10"
Height	10' 8"	6' 0" under tilt 1' 6" sides

Loading height at rear unladen 4' 6"

DIMENSIONS TRACTOR & SEMI TRAILER

Overall Length 30' 6". Width 7' 6"
Height 10' 8"

WEIGHTS

	Unladen	Laden
F.A. Tractor		2 tons 14¼ cwt.
R.A. "		3 tons 18¾ cwt.
Semi Trailer Axle		4 tons 5 cwt.
GROSS	4 tons 18 cwt.	10 tons 18 cwt.

STRIPPED SHIPPING HT.:- (Approx.) 7' 6".

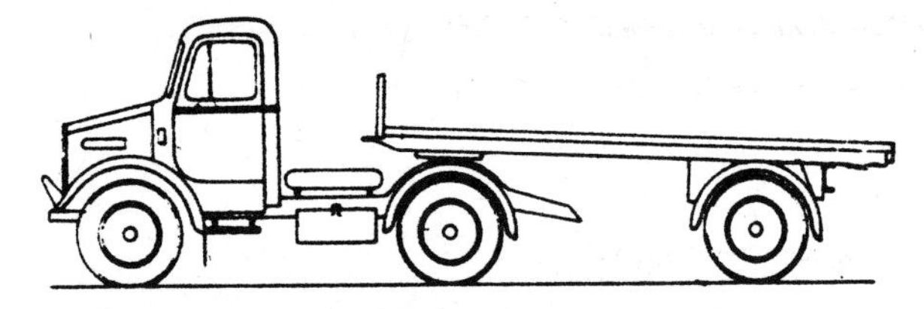

BEDFORD-SCAMMELL 6-TON SEMI TRAILER 4 x 2 - 2

ON BEDFORD OXC CHASSIS

WHEELBASE: Tractor 9' 3" Overall 18' 2"

TRACTOR UNIT:- Generally same as Bedford OX on page 25, except:-

Rear Axle Ratio: 7.4 to 1
Overall Ratios: 53.4 to 1, 25.6 to 1. 12.6 to 1, 7.4 to 1, rev. 52.9 to 1.

*Tyres: 32 x 6 H.D., twin rear. Spare carried behind cab. Petrol Tank: One only, 16 gallons.

Fitted with Scammell Automatic coupling device. To couple; back Tractor into Trailer so that latter rides up ramps and engages securing hook.

To uncouple; apply trailer brakes, operate uncoupling lever and drive away.

SEMI-TRAILER:- Straight frame ball mounted trailer, fitted with jockey wheels. *Tyres 32 x 6 H.D. twin Body, flat platform 15' x 7'.

OVERALL DIMENSIONS:- Length 25' 0", tractor and trailer
Width 7' 6" (Trailer)
Ht. 6' 11".

LADEN WEIGHTS:- F.A.W. 1 ton 5¼ cwt.
Driving A.W. 4 tons 15¼ cwt.
Trailed A.W. 3 tons 14½ cwt.

*Later models are fitted with 10.50-16 tyres, singles all round.

FODEN DG/6/12 10-TON 6-WHEEL WITH SINGLE REAR TYRE EQUIPMENT

6 x 4 Forward Control Wheelbase 15' 8"
Bogie Centres 4' 4"

ENGINE:- Make, Gardner 6 L.W. C.1. 6-cyl. Bore 4¼" Stroke 6". Capacity 8.36 litres. Max. B.H.P. 102 @ 1700 r.p.m. Max. torque lbs./ins. 4164 @ 1100 r.p.m.

GEARBOX:- Main 4 speed & 1 reverse Auxiliary, superlow, ratio 2.16 to 1, on 1st & reverse only. Overall Ratios (High), 46.1 to 1, 22.7 to 1, 12.2 to 1, 7.5 to 1, reverse 46.1 to 1.

REAR AXLES:- Full floating, overhead worm drive. Ratio 7.5. to 1.

SUSPENSION:- Front, semi-elliptic. Rear semi-elliptic free ends.

BRAKES:- Foot, Hydraulic on all wheels. Hand, Mechanical on four rear wheels.

TYRES:- 13.50 - 20 singles all round.

TRACK:- f. 6' 10", r. 6' 3".

FUEL CAPACITY:- 50 galls. Radius 400 miles.

COOLING SYSTEM CAPACITY:- 8 galls.

BODY DETAILS:- G.S. body with fixed sides, detachable canvas tilt and hoops. Internal Dimensions 18' 6" x 6' 11¼" x 6' 0" high.

CAB:- All enclosed. Detachable top.

WEIGHTS (Estimated):- Add 3 cwt. to F.A. weight and 7 cwt. to R.A. weight of Foden model DG/6/10.

STRIPPED SHIPPING HT.:- Approx. 7' 4".

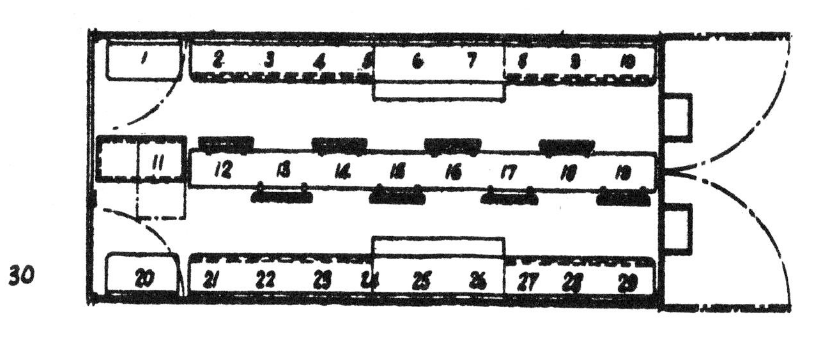

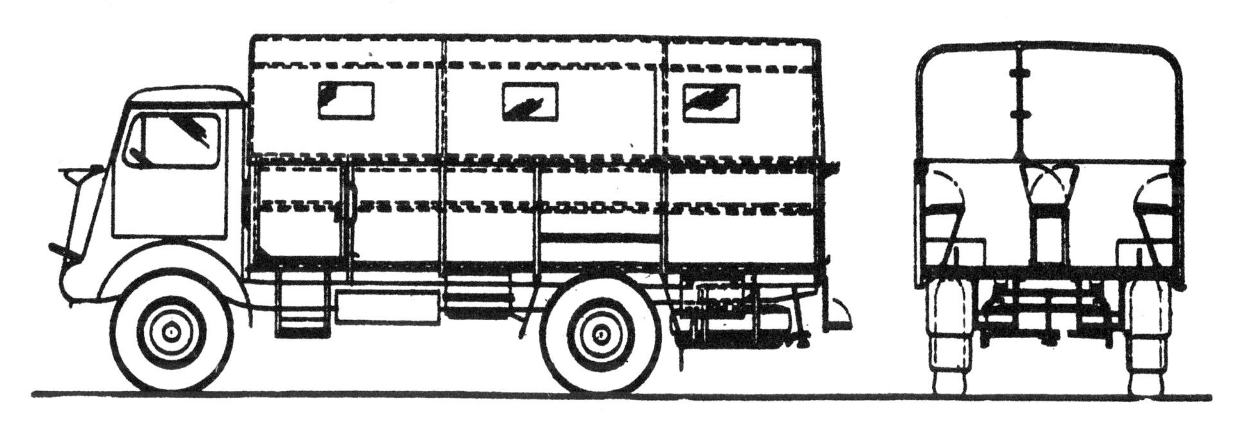

LORRY 3-TON 4 x 4 TROOP CARRYING

DIMENSIONS

	OVERALL	INSIDE BODY
Length	21' 10"	15' 10"
Width	7' 6"	7' 1½"
Height	9' 8"	6' 0"

LADEN WEIGHT

F. A.	2 tons 6	cwt.
R. A.	4 tons 5	cwt.
GROSS	6 tons 11	cwt.

STRIPPED SHIPPING HEIGHT:- 7' 6" (Superstructure and top of cab detached)

LORRY, 3-TON 4 x 4, TROOP CARRYING

ON BEDFORD QL TROOP CARRIER CHASSIS

CHASSIS:- Bedford QL Troop Carrier is similar to the Standard QL chassis shown on page 48, except:-

*1. Chassis frame extension.
2. Two 16 gallon fuel tanks either side in place of single 28 gallon tank behind cab.
*3. Spare tyre carried under frame at rear.
4. Extended exhaust.

(* These modifications carried out by body builder)

BODY:- Accommodation for driver, 30 troops and all their kit; 29 seats in main body, the 30th sits by driver. A.A. hatch forward end of body with machine gun mounting. For conversion to load carrying, the side seats can be folded back flat and the centre assembly can be folded up and stored under floor of body. Load must be carried well forward when used as load carrier.

Detachable tilt and bows

CAB:- All enclosed. Seats 2. Detachable top.

6 POUNDER PORTEE & FIRE LORRY 3-TON 4 x 4 ANTI TANK PORTEE

ON AUSTIN 3-TON 4 x 4 & BEDFORD QL 3-TON 4 x 4

CHASSIS DETAILS:- Austin as page 47, Bedford as page 48, except:-

Tyres 10.50 - 20 Run flat
Two side petrol tanks are fitted.
Capacity 32 gallons total.

GENERAL DESIGN:- Gun can be carried in 3 ways:-

1. Fighting Trim, gun mounted forward to fire off forward, gun trail is split as shown in plan view. (Cab top, sides and wind-screen are also folded down).
2. Fighting Trim, gun mounted rearward to fire off rearward. Vehicle can be driven with gun in both positions (1) and (2).
3. Touring Trim with gun out of action. Provision is made in this trim carrying the gun forward with the gun trail unsplit.

Gun is loaded on to vehicle by 2 hand operated winches. Gun wheel ramps and central gun trail ramps are carried in locker under rear of body.

Cab has detachable top canvas and side-screens and folding windscreen which are collapsed to fire off forward.

Detachable canvas tilt on steel super-structure. Four demountable tip-up crew seats in body. Gun shields (not shown in illustration) are fitted to side of body. Three Ammunition lockers either side and containers in body to take a total of 96 rounds. W.D. Drawbar gear at rear.

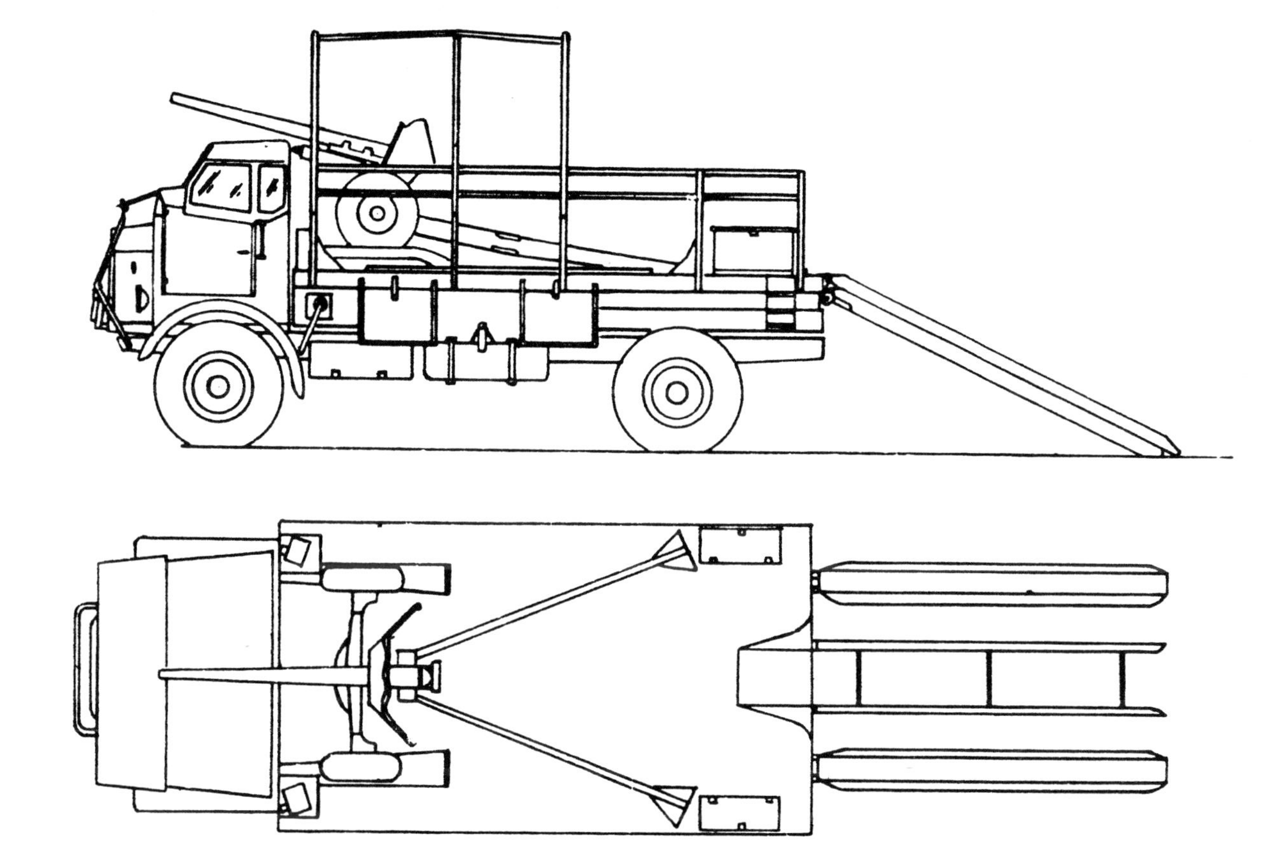

6 POUNDER PORTEE & FIRE LORRY 3-TON 4 x 4 ANTI TANK PORTEE

WEIGHTS (Austin Chassis)*

	Gun Forward	Gun Rearward
F.A.	4 tons 2¾ cwt.	3 tons 10¼ cwt.
R.A.	3 tons 15½ cwt.	4 tons 8 cwt.
Gross	7 tons 18¼ cwt.	

* Bedford as above, but deduct 4-cwt. from F.A., 3-cwt. from R.A. weights.

DIMENSIONS (Both Chassis)

Overall Length 19' 3", Overall Width 8' 4½". Overall Height 10' 2½", Ht., tilt lower position, 9' 6½". Overall length ramp extended 29' 2", Stripped shipping Height (less gun) 7' 5½", Stripped shipping Height gun mounted rear (lowest position) approx. 9' 0".

S.P. BOFORS VEHICLE

On Morris C. 9/B Chassis

AND

S.P. PREDICTOR VEHICLE

On Morris C. 8/P Chassis

PRELIMINARY INFORMATION

These 2 vehicles are designed to work normally as a pair.

The S.P. Bofors can be used however without its companion Predictor vehicle.

Both chassis have been developed from the Morris C.8. F.A.T. 4 x 4 chassis. Many chassis units are the same as those used on the latest Morris C.8. Mk.III chassis described on page 82.

The Bofors C. 9/B chassis has a wheelbase of 10' 0".

The Predictor C.8/P chassis has a wheelbase of 8' 3" - the same wheelbase as the C.8. Mk.III F.A.T. chassis.

There are many special additional fittings; the main items of which are covered in the following 4 pages.

Both vehicles have entirely open cabs or crew compartments fitted with folding deflector windscreens.

MORRIS C.9/B. S.P. BOFORS

4 x 4 Wheelbase 10' 0"

ENGINE:- Make Morris, Petrol, 4 cyl. Bore 100 m.m. Stroke 112 m.m. Capacity 3.5 litres. Max. B.H.P. 70 @ 3,000. Max. torque 1,800 lbs./ins. @ 1,750 r.p.m.

GEARBOX:- 5 speeds and reverse. Overall Ratios 67.7 to 1, 38.2 to 1, 20.4 to 1, 11.7 to 1, 6.57 to 1, r. 67.7 to 1. F. axle drive can be disengaged.

AXLES:- Front and Rear, full floating, spiral bevel. Ratio 6.57 to 1.

SUSPENSION:- F. and r. semi-elliptic with hydraulic shock absorbers.

BRAKES:- Foot, Hydraulic on all wheels. Hand, Mechanical on rear wheels.

TYRES:- 10.50 - 16. Run flat Tyres.

TURNING CIRCLE:- 53' 6".

GROUND CLEARANCES:- F. Axle 9½". Belly 10½".

CAPACITIES:- Fuel 22 galls. Cooling 4½ galls.

SPECIAL FEATURES:- Rear axle can be locked to frame so that gun can be fired off vehicle tyres. Four special jacks permanently attached to vehicle. These are lowered when gun is fired by Predictor control. In addition to manual gun control, a hydraulic control is incorporated electrically operated, power being provided by a 2.75 K.V.A. alternator driven from gearbox P.T.O.
Chassis is fitted with special springing. Channel frame is strengthened by welded flitch plates.

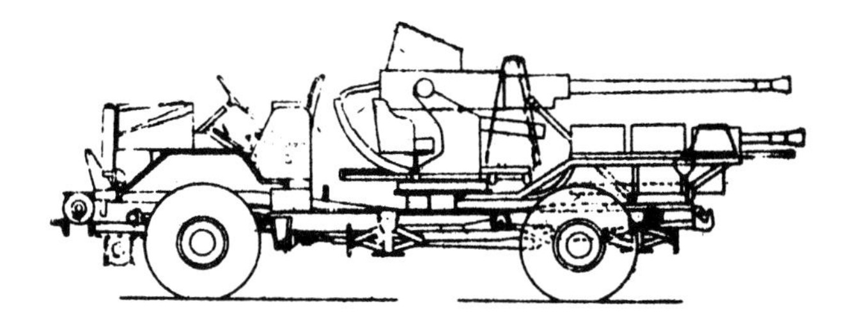

Faint dotted line above bonnet indicates forward position of gun.

S.P. BOFORS VEHICLE

ON MORRIS C. 9/B CHASSIS

CAB:- Seating for driver and a crew of 3 men. Rear of cab open to give easy access to gun platform. Seats can be folded forward. Two single deflector windscreens which can be folded forward. Folding steering wheel. Ammunition: 120 rounds carried. W.D. Drawbar gear at rear.

OVERALL DIMENSIONS

Length 20' 2" (over spare gun barrel)
Width 7' 3" (Jacks in travelling position)
Height 7' 6"

LADEN WEIGHTS

F.A. 2 tons 7 cwt.
R.A. 3 tons 7 cwt.
GROSS 5 tons 14 cwt.

STRIPPED SHIPPING HEIGHT:- As overall height.

MORRIS C.8/P. S.P. PREDICTOR

4 x 4 Wheelbase 8' 3"

ENGINE:- Make Morris, Petrol, 4-cyl. Bore 100 m.m. Stroke 112 m.m. Capacity 3.5 litres. Max. B.H.P. 70 @ 3,000. Max. Torque 1,800 lbs./ins. @ 1,750 r.m.p.

GEARBOX:- 5 speeds and reverse. Overall Ratios, 67.7 to 1, 38.2 to 1, 20.4 to 1, 11.7 to 1, 6.57 to 1, r. 67.7 to 1, F. axle drive can be disengaged.

AXLES:- Front and rear, full floating, spiral bevel. Ratio 6.57 to 1.

SUSPENSION:- F. and r. semi-elliptic with hydraulic shock absorbers.

BRAKES:- Foot, Hydraulic on all wheels. Hand, Mechanical on rear wheels.

TYRES:- 10.50 - 16.

TURNING CIRCLE:- Track. f. and r. 5' 5¼".

CAPACITIES:- Fuel 22 galls. Cooling 4½ galls.

SPECIAL FEATURES:- Three screw jacks permanently fitted, one each side of cab, one at centre rear. They are locked in up position for travelling.

S.P. PREDICTOR VEHICLE

ON MORRIS C. 8/P. CHASSIS

CAB:- Seating for driver and a crew of two. Rear of cab open. Two single deflector windscreens which can be folded forward.

STOWAGE:- Large locker at rear. Small lockers in cab.

AMMUNITION:- Provision for 192 rounds.

SPECIAL FITMENTS:- W.D. Drawbar gear at rear. Two spare wheel carriers.

OVERALL DIMENSIONS

Length	16' 1"
Width	7' 3"
Height	8' 1"

LADEN WEIGHTS

Not available, but known they are less than S.P. Bofors, page 135.

STRIPPED SHIPPING HEIGHT:
As overall ht.

Make	Type		
BEDFORD, ML		4 x 2	Ambulance
AUSTIN, K2		4 x 2	Ambulance
GUY, Quad Ant	30 cwt.	4 x 4	F.A.T.
MORRIS, C8	30 cwt.	4 x 4	F.A.T.
MORRIS, C8	30 cwt.	4 x 4	Anti Tank Portee 2 Pdr.
BEDFORD, QL	3 ton	4 x 4	6 Pdr. Portee
AUSTIN	3 ton	4 x 4	6 Pdr. Portee
A.E.C., Matador		4 x 4	Medium Tractor
SCAMMELL		6 x 4	Heavy Gun Tractor
SCAMMELL SV/2S		6 x 4	Heavy Breakdown
A.E.C., Matador		4 x 4	Armoured Command and Armoured Personnel
A.E.C., Matador		4 x 4	Armoured Demolition

These Wts. and Dimensions replace those shown in text pages

Unladen Weights			Laden Weights			Overall Dimensions			Inside Body Dimensions		
F.A.	R.A.	Gross	F.A.	R.A.	Gross	Length	Width	Height	Length	Width	Height
1. 5. 3	1.14. 0	2.19. 3			3.18. 0	19' 0"	7' 3"	9' 3"	8' 7"	6' 5"	5' 6"
1. 8. 1	1.14. 3	3. 3. 0	1.12. 2	2. 9. 3	4. 2. 1	18' 0"	7' 3"	9' 2"	8' 7"	6' 5"	5' 6"
1.14. 3	2. 1. 2	3.16. 1	1.15. 0	3.10. 0	5. 5. 0	13' 9"	7' 3"	7' 3"	6' 0"	6' 9"	4' 0"
1.14. 3	1.14. 1	3. 9. 0	1.18. 0	2.18. 0	4.16. 0	14' 9"	7' 3"	7' 5"	6'10"	6'11"	4' 0"
2. 3. 1	2. 5. 3	4. 9. 0	2. 5. 3	2.14. 0	4.19. 3	14' 8"	7' 0"	8' 2"	–	–	–
Gun Forward			Gun Rear End								
3.18. 3	3.12. 2	7.11. 1	3. 6. 1	4. 5. 0	7.11. 1	19' 3"	8' 4½"	10' 2½"	–	–	–
4. 2. 3	3.15. 2	7.18. 1	3.10. 1	4. 8. 0	7.18. 1	19' 3"	8' 4½"	10' 2½"	–	–	–
F.A.	R.A.	Gross	F.A.	R.A.	Gross						
3.16. 3	3.13. 2	7.10. 1	4.15. 3	6.19. 1	11.15. 0	20' 9"	7'10½"	10' 2"	14' 6"	7' 4"	5' 6"
2. 9. 2	5.18. 0	8. 7. 2	3.10. 3	8.12. 0	12. 2. 3	20' 3"	8' 6"	9' 9"	9' 8"	8' 3"	5' 0"
3.10. 3	6.17. 1	10. 8. 0	2. 3. 1	11. 9. 0	13.12. 1	20' 3"	8' 6"	9' 5"	–	–	–
5. 4. 0	5. 0. 3	10. 4. 3	5.14. 1	5.16. 3	11.11. 0	20' 0"	8' 6"	9' 5"	–	–	–
5. 8. 0	6. 5. 0	11.13. 0	5.11. 0	6. 5. 0	11.16. 0	20' 0"	8' 6"	9' 5"	–	–	–

DATA BOOK OF

WHEELED VEHICLES

PART 2

IMPORTED VEHICLES

OF ALL TYPES

DATA BOOK OF
WHEELED VEHICLES

Part 2

IMPORTED VEHICLES AND DIVERTED SHIPMENTS

This section of the Data Book contains details of all the main types and makes of Imported Vehicles and Diverted Shipments. The only omissions are of one or two vehicles where the number in service did not justify their inclusion.

As with the British vehicles, these specifications must be treated as typical and do not necessarily apply in all details to every contract relating to individual vehicles.

All data in this book has been translated into British terms and phraseology. Some notes on American terms and their British equivalents are included on page 205, which may be useful when referring to American produced Instruction and Maintenance Manuals.

GENERAL NOTES ON DESIGN OF IMPORTED VEHICLES AND DIVERTED SHIPMENTS

The Imported Vehicles and Diverted Shipments covered in this Data Book can be divided into four main groups according to the source of supply. The group to which a particular model belongs is indicated in brackets on the top outside corner of the chassis specification page. Here are the four main groups with notes on the general design.

GROUP 1, CANADIAN MILITARY PATTERN

The Chevrolets and Fords in this group are built in Canada to Canadian Army specifications which closely follow the appropriate W.D. specifications. Performance, ground clearance, turning circles are similar to W.D. requirements. As soon as possible W.D. requirements in equipment are introduced.

Radius of action aimed at is 200 miles; two side petrol tanks with reserve are fitted on all models (On 8-cwt., tanks are behind cabs). The cabs and front end sheet metal work (except engine cover) are interchangeable on all Chevrolets and Fords irrespective of size except 8-cwt. All cabs have detachable tops (referred to as "split cabs") and folding or detachable windscreens to give a max. shipping height of 6 ft. 6 ins. Right hand drive on all models.

NOTE:- "New type cab and front end has now been produced for both Ford and Chevrolet. Components of new cab are not interchangeable with those of old cab."

W.D. type divided wheels with single rear tyre equipment.

GROUP 2, CANADA

These chassis, built in Canada, are standard commercial types which have been adapted to bring them in line with W.D. specifications.

Cabs are standard commercial coachbuilt type (non-detachable tops). Right hand drive is fitted in every case. Wheel and tyre equipment is as stated in the appropriate specification and includes in a few instances dual rear tyres.

The bodies are standard commercial design modified to Canadian Army requirements and in many instances similar to those in Group I. For instance, the Chevrolet and Ford 3-ton, 4 x 2 Stores and Workshop bodies are similar to the corresponding bodies on the 3-ton, 4 x 4 Canadian Army Pattern chassis.

GROUP 3, U.S.A.

The remarks under Group 2 apply also to these models, which have been built in U.S.A. factories instead of Canadian subsidiaries. Cabs are standard U.S. commercial types. Bodies are British-built to W.D. Design in the majority of cases.

GROUP 4, Ex FRENCH

These contracts were originally placed by the French. Generally the vehicles are standard U.S. commercial types, which were modified to comply with very much less rigid specifications than the standard W.D. Certain models have left hand drive and, where so, this is indicated under the subhead "Steering". Dual rear tyre equipment is also fitted in a number of instances.

NOTE: Some of the MACK EXBX & WHITE 920 vehicles, and all Mack N.R. vehicles, are direct orders from the U.S.A. They are included with the EX FRENCH CONTRACT vehicles as their specifications generally comply with this group.

WHO'S WHO AMONG AMERICAN MANUFACTURERS

These notes show the inter-relationship of the various American manufacturers. Only those firms engaged on W.D. contracts are shown.

GENERAL MOTORS CORPORATION (U.S. Parent Co.)
(Canada: General Motors of Canada)

Manufacturing:
Chevrolet Trucks;
G.M.C. Trucks (made by the Yellow Truck & Coach Co., a subsidiary of General Motors).

FORD MOTOR COMPANY (U.S. Parent Co.)
(Canada: Ford Motor Co. of Canada Ltd.)

Manufacturing:
Ford Trucks.

Note: Imported Ford vehicles are not necessarily interchangeable with British Fords built at Dagenham. For instance, a larger bore engine than the British. Many details of the chassis design are different.

CHRYSLER CORPORATION (U.S. Parent Co.)

Manufacturing:
Dodge Trucks.

THE FOUR WHEEL DRIVE AUTO. CO.

F.W.D. Trucks.

(Not to be confused with the Autocar Co. who market F.W.D. Autocar models).

INTERNATIONAL HARVESTER CO. (W.D. Contracts)
International Trucks.

MACK MANUFACTURING CORP.

Mack Trucks.

WHITE MOTOR CO.

White Trucks.

WHO'S WHO AMONG AMERICAN MANUFACTURERS

CHRYSLER CORPORATION (U.S. Parent Co.)

(In Canada: Chrysler Corporation of Canada Ltd)

Manufacturing:

Dodge Trucks
Fargo Trucks
Plymouth Trucks
Chrysler, De Soto Plymouth cars.

(Note: Some Dodge models are made in the Fargo factories, hence Dodge-Fargo)

(Additional)

Autocar Company
American Bantam (Bantam "Jeep")
Brockway Motor Co. Inc.
Corbitt Company
Diamond T
Federal Motor Company
The Hug Company (Hug Tractors)
Harley Davidson (Motor Cycles)
Indian (Motor Cycles)
Studebaker Corporation
Ward La France
Willys-Overland

Main Trailer Mfgs.

Rogers
Winter Weiss (Transporters to Rogers design

AMERICAN TERMS AND BRITISH EQUIVALENTS

Note: All the data in this book has been translated into British equivalents. These notes, however, are included for use when referring to American produced handbooks, maintenance manuals, etc.

American	British Equivalent
1 U.S. gallon	= .835 Imperial gallons
or 1.2 U.S. gallons	= 1 Imperial gallon
U.S. ton	= short ton (2,000 lbs.)

U.S. vehicle weights are usually expressed in lbs. in the form of "gross vehicle weights". This weight includes chassis, cab, body, equipment, accessories, personnel and payload with full fuel, oil and water.

C.O.E. (Cab over engine)	= Forward control
Booster Brakes	= Vacuum servo assisted; Compressed air servo are sometimes called "Booster".
Fenders	= Mudwings
Hood	= Bonnet
Transmission	= Gear-box
Transmission shift lever	= Gear lever
Octane Selector	= Micrometer adjustment of ignition setting by manual control on distributor
Dump Truck	= Tipper
Steel Cargo Body	= Fixed-sided lorry body, similar to W.D. type G.S.
S.A.E. Rating (in h.p.)	= R.A.C. Rating

AMERICAN TERMS & BRITISH EQUIVALENTS (continued)

American	British Equivalent
	BODY DESIGNATIONS
Sedan	= Saloon
Command Reconnaissance	= An unarmoured open 4-str. with hood (e.g. the car, 5 cwt. 4 x 4, or, as it is generally known, the "Peep," "Jeep", or Blitz Buggy.)
Panel Delivery	= Delivery Van (steel) panelled body
Station Wagon	= Heavy Utility (4 side doors)
Carryall	" " but only 2 side doors
Weapons Carrier	= Similar to W.D. 15 cwt. open truck body. Has folding cab.
Tractor-truck	= The tractor end (motive unit) of an articulated semi-trailer
Fifth wheel	= Turntable on semi-trailer types
Upper fifth wheel	= ½ " " " " itself
Lower " wheel	= ½Turntable on motive unit (tractor truck)
Wrecker	= Breakdown
Stake body	= Flat platform load carrier with removable sides to body. Sides approx. 40" high. No W.D. equivalent. Nearest comparable is commercial type dropside lorry.
Gas Tanker	= Petrol tanker
	On Wrecker (Breakdown) Models
The Boom	= Crane jib
Skid pan	= Tractor Scotch

INDEX BY TYPE

(All Models except Ex French Contract Vehicles)

EX FRENCH CONTRACT VEHICLES

INDEX BY MAKE

(ALL MODELS)

IMPORTANT NOTE

All the data in this section has been prepared on the same lines as the British Vehicle section outlined on page 4, except in one instance:-

SHIPPING NOTES -

"Height cut down" in the British section, is the height of the vehicle (unladen) with the tilt and superstructure in the lowered position.

In this Imported Vehicle Section the height given is the Stripped Shipping Height which represents the height of the vehicle with:-

(1) Canvas Tilt and hoop superstructure detached (where applicable)

and/or

(2) Cab "Split", i.e. top of cab roof detached in the case of models with detachable cab tops.

SECTION I

GENERAL LOAD CARRYING VEHICLES

(Excluding Ex French Contract Vehicles which are shown in Section III)

	Chassis Type and Make	Body Type
8-cwt. 4-wheel	Chevrolet 4 x 2	Personnel
	Dodge 4 x 4	"
	Ford 4 x 2	Wireless
15-cwt. 4 x 2	Chevrolet	G.S.
	Ford	"
30-cwt. 4 x 4	Chevrolet	G.S.
	Ford	"
3-ton 4 x 2	Chevrolet	G.S.
	Ford	"
3-ton 4 x 4	Chevrolet	G.S.
	Ford	"

CHEVROLET 8420 (C.8) 8 CWT. 4 WHEEL

4 x 2 Semi-Forward Control Wheelbase 8'5"

ENGINE. Make, Chevrolet, Petrol 6 cyl. Bore 3½" Stroke 3¾", Capacity 3.54 litres, R.A.C. 29.4 h.p. Max. B.H.P. 78 @ 3200 r.p.m. Max. Torque Lbs/ins. 2016 @ 1100 r.p.m.

CLUTCH. Single dry plate, 10¾".

GEARBOX. 4 speeds & 1 reverse. Overall ratios: 32.9 to 1, 15.8 to 1, 7.76 to 1, 4.55 to 1, rev. 32.5 to 1.

FRONT AXLE. I section.

REAR AXLE. Semi-floating, hypoid gear. Ratio: 4.55 to 1.

SUSPENSION. Semi-elliptic with hydraulic shock absorbers front and rear.

BRAKES. Foot, Hydraulic all wheels, Hand, Mechanical rear wheels.

STEERING. Right Hand Drive. Worm and Sector, ratio: 21 to 1.

WHEELS & TYRES. Two piece divided 13 x 6.50. Tyres, 9.00 - 13.

ELECTRICAL EQUIPMENT. 6 volt.

TURNING CIRCLE. 38' Track, f.4'11", r.5'2-5/8". 5/8".

GROUND CLEARANCES: R.Axle 8".

CAPACITIES. Fuel 24½ gals. Cooling 3-1/8" gals.

PERFORMANCE FIGURES. (Personnel Body Laden)
B.H.P. per Ton 31.1.
Tractive Effort per Ton (100%) 1863 lbs/Ton.
M.P.G.16. Radius 380 miles.

DODGE T.212 3 CWT. FWD. 4-WHEEL

4 x 4 Normal Control Wheelbase 9' 8"

ENGINE: Make, Dodge, Petrol 6 cyl. Bore 3-3/8" Stroke 4-1/16" Capacity 3.57 litres R.A.C. 27.34 B.H.P. 74 @ 3,000 r.p.m. Max torque 2,016 lbs. ins. @ 1,200 r.p.m.

CLUTCH. Single dry plate 10" dia.

GEAR-BOX. 4 speeds and 1 reverse. OVERALL RATIOS: 30.3 to 1, 15.1 to 1, 8.2 to 1, 4.89 to 1, rev. 38.2 to 1.

TRANSFER CASE. Ratio direct.

FRONT AXLE. Full floating, Hypoid bevel. Ratio 4.89 to 1.

REAR AXLE. Full floating. Hypoid bevel. Ratio 4.89 to 1.

SUSPENSION. Semi-elliptic, front and rear with hydraulic shock absorbers.

BRAKES. Foot, Hydraulic on all wheels, Hand, transmission brake.

STEERING. Right Hand Drive. Worm and Sector.

WHEELS and TYRES: W.D. type divided wheels, 16 x 6.00. Tyres, 9.25-16.

ELECTRICAL EQUIPMENT. 6 volt.

TURNING CIRCLE: 49 ft. Track f.4' 11-3/8", r. 5' 1-3/8".

GROUND CLEARANCE. Axles 9-3/8".

CAPACITIES: Fuel 21 gals. Cooling 3¾ gals.

PERFORMANCE FIGURES

B.H.P. per ton 25.9.
TRACTIVE EFFORT per ton (100%) 1390 lbs/Ton.
Max. Speed 55 m.p.h.
M.P.G.11. Radius 230 miles (Estimated).

S.M. 2030. (CANADIAN MILITARY PATTERN)

FORD C011DF (F.8) 8 CWT. 4 WHEEL

4 x 2 Semi-Forward Control Wheelbase 8'5"

ENGINE. Ford Mercury, Petrol V.8. Bore 3.19", Stroke 3.75", Capacity 3.91 Litres, R.A.C. 32.5 h.p. Max. B.H.P. 95 @ 3600 r.p.m. Max. torque 2112 lb./ins. @ 1800 r.p.m.

CLUTCH. 11" single dry plate, truck clutch.

GEARBOX. 4 speeds and 1 reverse. Overall ratios: 26.3 to 1, 12.7 to 1, 6.9 to 1, 4.1 to 1, rev. 32.1 to 1.

FRONT AXLE. I section.

REAR AXLE. Full floating, spiral bevel drive. Ratio: 4.11 to 1.

SUSPENSION. F. & R., Semi-elliptic with hydraulic shock absorbers.

BRAKES. Foot, Hydraulic on all wheels. Hand, Mechanical on rear wheels.

STEERING. Right Hand Drive. Worm and roller, ratio: 18.4 to 1.

WHEELS & TYRES. Two piece wheels, 13 x 6.50. Tyres 9.00 - 13.

ELECTRICAL EQUIPMENT. 6 volt.

TURNING CIRCLE 39' TRACK f. 5'0", r. 4'11".

GROUND CLEARANCES. r. axle 7½".

CAPACITIES. Fuel 24 gals. Cooling 4¾ gals.

PERFORMANCE FIGURES. (Wireless Body Laden)

B.H.P. per Ton 38.
Tractive Effort per Ton (100%) 1404 lbs/Ton.
M.P.G. 15. Radius 360 miles.

8-CWT. 4-WHEEL PERSONNEL BODY

ON CHEVROLET 8420 8-CWT. CHASSIS

(Similar body also fitted to Dodge 8-cwt. 4 x 4 chassis. For weights and dimensions of this see Data Sheet at back).

BODY: Similar in design and equipment to 8-cwt. Morris Personnel on Page 11.

CAB: (Chevrolet) All enclosed. Side screens to doors, seats 2. Split top to cab and folding windscreen.
(Dodge). One-piece commercial type cab.

DIMENSIONS

	OVERALL	INSIDE BODY
Length	13' 5"	4' 9¾"
*Width	6' 3½"	5' 8½"
Height	6' 6"	3' 9½"

* 7' 3" wide over r. vision mirrors.

WEIGHTS

	UNLADEN	LADEN
F.A.W.	1 ton 1½ cwt.	1 ton 4¾ cwt.
R.A.W.	17¾ cwt.	1 ton 5½ cwt.
Cross	1 ton 19¼ cwt.	2 tons 10¼ cwt.

SHIPPING NOTES: Tonnage, standing, 13.7. Stripped Height under 6'6".

8-CWT. 4-WHEEL WIRELESS BODY

ON FORD C011DF 8-CWT. CHASSIS

BODY: Similar in design and equipment to 8-cwt. Humber on page 12.

CAB: As Chevrolet above.
Weights and dimensions. See Data sheet at back.

(CANADIAN MILITARY PATTERN)

CHEVROLET 8421 (C.15) 15 CWT. 4 WHEEL

4 x 2 Semi-Forward Control Wheelbase 8' 5"

ENGINE. Make, Chevrolet, Petrol 6 cyl. Bore 3½" Stroke 3¾", Capacity 3.54 litres, R.A.C. 29.4 h.p. Max. B.H.P. 78 @ 3200. Max. torque lbs./ins. 2016 @ 1100 r.p.m.

CLUTCH. Single plate 10¾" dia.

GEARBOX. 4 speeds and 1 reverse Overall Ratios: 44.6 to 1, 21.4 to 1, 10.5 to 1, 6.16 to 1, rev. 44.1 to 1.

FRONT AXLE. Reverse Elliott I section.

REAR AXLE: Full floating, hypoid gear, Ratio 6.16 to 1.

SUSPENSION. Front and rear, semi-elliptic with hydraulic shock absorbers.

STEERING. Right Hand Drive. Worm and sector, ratio 17 to 1.

WHEELS & TYRES: W.D. Divided wheels, 16 x 6.00, Tyres 9.00 - 16.

ELECTRICAL EQUIPMENT: 6 volt.

TURNING CIRCLE: 40 ft. Track, f.5'7", r. 5'5".

GROUND CLEARANCE: R. Axle 9¼".

CAPACITIES: Fuel 24½ gals. Cooling 3¼ gals.

PERFORMANCE FIGURES (G.S. Laden)

B.H.P. per ton 20.1.
Tractive Effort per ton (100%)
1460 lbs/ton.
M.P.G. 12 Radius 290 miles.

S.M.2002, 2031, 2079, 2267, 2459, 2463, 2485

(CANADIAN MILITARY PATTERN)

FORD C101WF (F-15) 15-CWT. 4-WHEEL

4 x 2 Semi-forward Control Wheelbase 8' 5"

ENGINE:- Ford Mercury. Petrol. V.8. Bore 3.19". Stroke 3.75". Capacity 3.91 litres. R.A.C. 32.5 h.p. Max. B.H.P. 95 @ 3,600 r.p.m., Max. torque 2.112 lbs. ins. @ 1,800 r.p.m.

CLUTCH:- Single dry plate 11" dia. truck clutch.

GEARBOX:- 4 speeds and 1 reverse. Overall Ratios: 42.6 to 1, 20.6 to 1, 11.2 to 1, 6.6 to 1, rev. 52.1 to 1.

FRONT AXLE:- I section.

REAR AXLE:- Full floating, spiral bevel drive, Ratio 6.67 to 1.

SUSPENSION:- Semi-elliptic front and rear with hydraulic shock absorbers.

BRAKES:- Foot, Hydraulic on all wheels. Hand, Mechanical on rear wheels.

STEERING:- Worm and roller. Ratio 18.4 to 1.

WHEELS & TYRES:- W.D. type divided wheels. Tyres 9.00-16.

ELECTRICAL EQUIPMENT:- 6 volt.

TURNING CIRCLE:- 40 ft. TRACK: f.5'7", r.5'5".

GROUND CLEARANCE:- 8¼" axle.

CAPACITIES:- Fuel 24½ gals. Cooling 5-1/8 gals.

PERFORMANCE FIGURES (G.S.Laden)

B.H.P. per ton. 25.8
Tractive effort per ton (100%) 1,482 lbs./Ton.
M.P.G.12. Radius 290 miles.

15-CWT. 4-WHEEL G.S. BODY

ON CHEVROLET 8421 15-CWT. CHASSIS

(Similar body fitted to Ford 15-cwt. chassis. For weights and dimensions, see data sheet at back).

BODY: Standard G.S. body, similar to Morris 15-cwt. on page 18 except

(1) It is slightly wider.
(2) Toolbox is situated between body and cab.

Later models being fitted with detachable hoops and tilt.

CAB: Enclosed, side curtains to doors, seats two. Top of cab split and windscreen folds for shipping.
W.D. Drawbar gear at rear.

DIMENSIONS

	OVERALL	INSIDE BODY
Length	14' 0"	6' 5½"
Width	7' 2"	6' 9½"
Height	6' 4½"	1' 10"
Height with Tilt	7' 6"	4' 6"

WEIGHTS (less tilt)

	UNLADEN (E)	LADEN
F.A.W.	1-ton 4-cwt.	1-ton 7¾ cwt.
R.A.W.	1-ton 5 cwt.	2-tons 9½ cwt.
Gross	2-tons 9-cwt.	3-tons 17¼ cwt.

SHIPPING NOTES: Tonnage standing, 15.8.

Stripped Height under 6'6".

CHEVROLET 8441 (C-30S) 30-CWT. AND 3 TON F.W.D. 4 WHEEL

4 x 4 Semiforward Control Wheelbase 11'2"

ENGINE:- Make, Chevrolet. Petrol, 6-cyl. Bore 3½". Stroke 3¾". Capacity 3.54 litres R.A.C. 29.4 h.p. Max. B.H.P. 78 @ 3,200. Max. torque lbs./ins. 2016 @ 1100 r.p.m.

CLUTCH:- Single dry plate 10¾" dia.

GEARBOX:- 4 speeds and 1 reverse. Overall Ratios (High): 51.4 to 1, 24.7 to 1, 12.1 to 1, 7.16 to 1, rev. 50.8 to 1.

TRANSFER CASE:- Ratios direct and 1.87 to 1. Front wheel drive can be disengaged in High gear.

SUSPENSION:- Semi-elliptic front and rear with hydraulic shock absorbers.

BRAKES:- Foot, Hydraulic on all wheels. Hand, Mechanical on rear wheels.

STEERING:- Ball bearing nut and sector. R.H.D.

WHEELS & TYRES:- W.D. divided wheels. Tyres 10.50 - 16.

ELECTRICAL EQUIPMENT:- 6 volt.

TURNING CIRCLE:- 57'. Track, f. 5'10", r. 5'10½".

GROUND CLEARANCE:- r. axle 8¾".

CAPACITIES:- Fuel 24½ gals. Cooling 3½ gals.

PERFORMANCE FIGURES (G.S. Body Laden):-

B.H.P. per ton 14.0
Tractive effort per ton (100%) 1840 lbs/Ton.
M.P.G. 7.5 Radius 180 miles.

S.M. 2003. (CANADIAN MILITARY PATTERN)

FORD C01QF (F-30) 30-CWT.
F.W.D.

4 x 4 Semi-forward Control. Wheelbase 11'2¼"

ENGINE:- Ford Mercury. Petrol V.8. Bore 3.19". Stroke 3.75". Capacity 3.91 litres. R.A.C. 32.5 h.p. Max. B.H.P. 95 @ 3600 r.p.m. Max. torque 2112 lbs.ins. @ 1800 r.p.m.

CLUTCH:- Single dry plate 11" dia.

GEARBOX: 4 speeds and 1 reverse. Overall Ratios (High):- 46.1 to 1, 22.2 to 1, 12.2 to 1, 7.2 to 1, rev. 56.4 to 1.

TRANSFER CASE:- Ratios, Direct and 1.87 to 1. Front Wheel drive can be disengaged in high gear.

FRONT AXLE:- Drive and steer type. Differential assembly identical with rear axle.

REAR AXLE:- Spiral bevel drive. Ratio 7.2 to 1.

SUSPENSION:- Front and rear, semi-elliptic with hydraulic shock absorbers.

BRAKES:- Foot, Hydraulic on all wheels. Hand, Propeller shaft brake.

STEERING:- Right Hand Drive. Worm and roller.

WHEELS & TYRES:- W.D. divided wheel. Tyres 10.50-16.

ELECTRICAL EQUIPMENT:- 6 volt.

TURNING CIRCLE:- 64 ft. Track, 5'10½".

GROUND CLEARANCE:- 9" r.axle.

CAPACITIES: Fuel 24½ gals. Cooling 5 1/8 gals.

PERFORMANCE FIGURES (G.S. Body Laden):-

B.H.P. per Ton 18.3.
Tractive Effort per Ton (100%) 1980 lbs/Ton.
M.P.G. 8. Radius 195 miles.

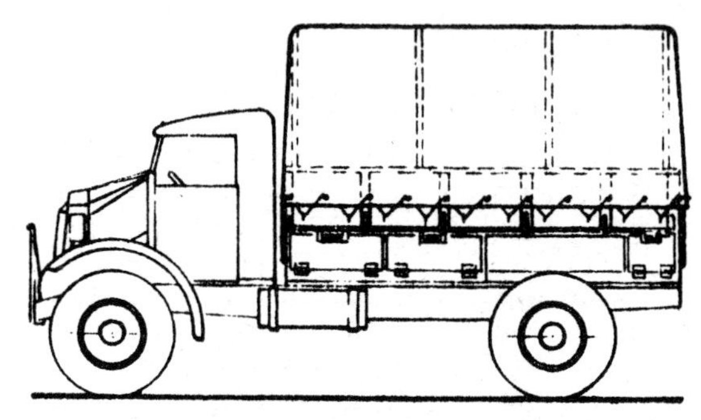

30-CWT. F.W.D. G.S. BODY

ON FORD C01QF CHASSIS

(Similar body also fitted to Chevrolet 30 cwt. 4 x 4 chassis. For weights and dimensions of this see Data sheet at back).

BODY: Standard G.S. (steel cargo) body with well floor, detachable tilt and hoops.

CAB: Enclosed, side curtains to doors, seats two. Split top to cab and folding windscreen for shipping purposes. W.D. Drawbar gear at rear.

DIMENSIONS

	OVERALL	INSIDE BODY
Length	16' 10"	9' 8"
Width	7' 3"	6' 8"
Height	9' 6"	6' 0"

WEIGHTS

	UNLADEN	LADEN
F.A.W.	1-ton 15¾ cwt.	1-ton 18-cwt.
R.A.W.	1-ton 13¼ cwt.	3-tons 9-cwt.
Gross	3-tons 9 cwt.	5-tons 7-cwt.

SHIPPING NOTES: Tonnage standing 28.8. Stripped Height under 6'6".

S.M. 2209, 2263
S.M. 2005, 2006, 2037. (CANADA)

CHEVROLET (15-43 x 2) 3-TON 4 WHEEL

4 x 2 Normal Control. Wheelbase 13' 4"

ENGINE: Make, Chevrolet - Petrol 6 cyl. Bore 3½", Stroke 3¾", Capacity 3.54 litres, R.A.C. 29.4 H.P. Max. B.H.P. 78 @ 3200. Max. torque lbs/ins. 2016 @ 1100 r.p.m.

CLUTCH. Single dry plate, 10¾" dia.

GEARBOX. 4 speeds and 1 reverse.

OVERALL RATIOS. (High) 40.7 to 1, 19.6 to 1, 9.6 to 1, 5.64 to 1, rev. 40.3 to 1.

REAR AXLE. 2 speed double reduction. Spiral bevel drive with helical spir. gears. Ratios 5.64 to 1, and 8.22 to 1.

Note against R. Axle and Brakes "Later Contracts, S.M.2037 and following, 2 speed Axle ratios, 6.33 and 8.31. Vacuum serveo brakes".

FRONT AXLE. Reverse Elliott I beam.

SUSPENSION. Semi-elliptic front and rear.

BRAKES. Foot, Hydraulic on all wheels Hand, Mechanical on propellor shaft.

STEERING. Right Hand Drive. Ball bearing, nut and sector.

WHEELS & TYRES. 16 x 6.00 split disc wheels. Tyres 10.50 - 16.

ELECTRICAL EQUIPMENT. 6 volt.

TURNING CIRCLE. 60 ft. Track, f.5'6-1/16", r.5'9¾".

GROUND CLEARANCE. R. Axle 9¾".

CAPACITIES. Fuel 15 gals. Cooling 3¼ gals.

PERFORMANCE FIGURES. (G.S. Body Laden)

B.H.P. per ton 11.6.
Tractive effort per ton (100%). 1030 lbs/Ton.
*M.P.G. 9.0 Radius 135 miles.

* Estimated.

S.M. 2265, 2293, 2356
S.M. 2004, 2037, 2046 (CANADA)

FORD E.C.O. 98T 3 TON 4 WHEEL

4 x 2 Normal Control Wheelbase 13'2"

ENGINE:- Ford Mercury. Petrol V.8. Bore 3.19" Stroke 3.75". Capacity 3.91 litres. R.A.C. 32.5 h.p. Max. B.H.P. 95 @ 3,600. Max. torque 2112 lbs./ ins. @ 1800 r.p.m.

CLUTCH: Single dry plate 11" dia.

*GEARBOX: 4 speeds and 1 reverse. Overall Ratios (High Axle Ratio): 37.3 to 1, 18.0 to 1, 9.8 to 1, 5.83 to 1, rev. 45.6 to 1.

FRONT AXLE: I beam.

*REAR AXLE: 2 speed. Full floating. Spiral bevel drive. Ratios 5.83 to 1 and 8.11 to 1.

SUSPENSION: Front and rear, semi-elliptic.

*BRAKES: Foot, Hydraulic on all wheels. Hand, Mechanical transmission brake.

STEERING: Right Hand Drive. Worm and roller.

WHEELS & TYRES: W.D. type divided wheels. 16 x 6 ins. Tyres 10.50-16.

ELECTRICAL EQUIPMENT: 6 volt.

TURNING CIRCLE: 58 ft. Track f.5'3" r.5'5".

GROUND CLEARANCE: 10".

CAPACITIES: Fuel 15 gals. Cooling 5 gals.

PERFORMANCE FIGURES: (G.S. Body laden)

B.H.P. per ton 14.5.
Tractive effort per ton (100%) 984 lbs/Ton.
M.P.G. 9. Radius 135 miles.

* Later Contracts, 2-speed axle ratios of 6.33 to 1 and 8.81 to 1. Vacuum servo assisted brakes.

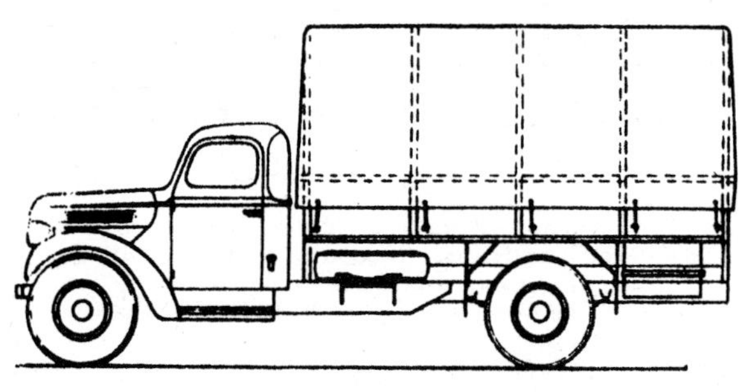

3-TON 4-WHEEL G.S. LORRY

ON FORD E.C.O.98T. 4 X 2 CHASSIS

(Similar body also fitted to Chevrolet 15-43 x 2 3-ton 4 x 2. For weights and dimensions of this, see Data Sheet at back).

BODY:- Steel Cargo (G.S.) body with flat floor. Detachable canvas cover and hoops. Spare Wheel carried under front nearside of body.

CAB:- All Enclosed steel panelled cab. Seats 2. Non-detachable roof. W.D. Drawbar gear at rear.

DIMENSIONS

	OVERALL	INSIDE BODY
Length	21' 1½"	12' 0"
Width	7' 4"	6' 8"
Height	10' 3"	6' 0"

WEIGHTS

	UNLADEN	LADEN
F.A.W.	1 ton 4 cwt.	2 tons 0½ cwt.
R.A.W.	1 ton 17 cwt.	4 tons 11 cwt.
GROSS	3 tons 1 cwt.	6 tons 11½ cwt.

SHIPPING NOTES:- Tonnage Standing 39.0 Stripped Height 7'0".

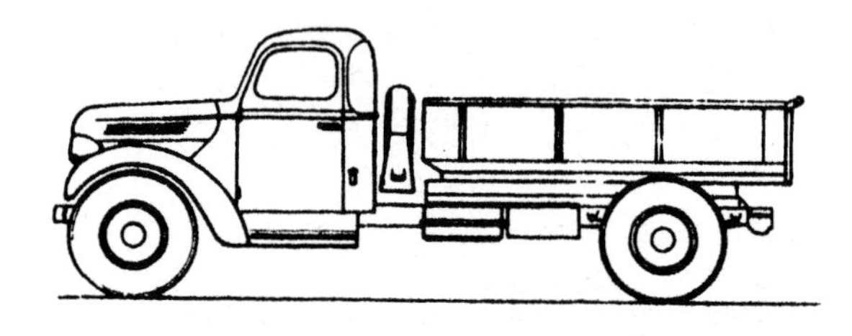

3-TON 4-WHEEL END TIPPER

ON FORD E.C.O.98T 4 x 2 CHASSIS

BODY:- All Steel end tipping body. Fixed sides, hinged tailboard. Hydraulic tipping gear pump (driven from P.T.O. on gearbox and controlled by 2 levers in cab) operates a single cylinder hoist located under floor of body.

CAB:- All enclosed steel panelled commercial type cab. Non-detachable roof. Seats 2.

SPECIAL FITMENTS:- W.D. Drawbar gear at rear.

DIMENSIONS

	OVERALL	INSIDE BODY
Length	18' 10"	9' 0"
Width	7' 4"	6' 9"
Height	7' 0"	1' 8"

WEIGHTS

	UNLADEN	LADEN
F.A.W.	1 ton 7 cwt.	1 ton 16¾ cwt.
R.A.W.	1 ton 18¾ cwt.	4 tons 16¾ cwt.
GROSS	3 tons 5¾ cwt.	6 tons 13½ cwt.

SHIPPING NOTES:- Tonnage Standing 23.4. Stripped Height 7'0".

S.M.2019, 2033, 2037, 2104, 2145, 2264, 2451, 2577

(CANADIAN MILITARY PATTERN)

CHEVROLET 8443 (C - 60L)
3 TON F.W.D.

4 x 4 Semi-forward Control. Wheelbase 13' 2"

ENGINE. Make, Chevrolet, Petrol 6-cyl. Bore 3½", Stroke 3¾", Capacity 3.54 litres. R.A.C. 29.4 h.p. Max. B.H.P. 78 @ 3200. Max. torque lbs/ins. 2016 @ 1100.

CLUTCH. Single dry plate 10¾ dia.

GEARBOX. 4 speeds and 1 reverse. Overall Ratios: (High) 52.0 to 1, 25.0 to 1, 12.3 to 1, 7.2 to 1, rev. 51.5 to 1.

TRANSFER CASE. Ratios: Direct & 1.87 to 1. Front wheel drive can be disengaged in High gear.

FRONT AXLE. Full floating. Spiral bevel drive.

REAR AXLE. Full floating, spiral bevel drive. Ratio: 7.2 to 1.

SUSPENSION. Semi-elliptic, front & rear, with hydraulic shock absorbers.

*BRAKES. Foot, hydraulic on all wheels. Hand, Mechanical on propellor shaft.

STEERING. Right Hand Drive. Ball bearing nut and sector.

WHEELS & TYRES. W.D. divided type wheel 20 x 6. Tyres 10.50 - 20.

ELECTRICAL EQUIPMENT. 6 volt.

TURNING CIRCLE. 67 ft. Track, f. 5' 8½", 4. 5' 9".

GROUND CLEARANCE. 11" at axle housings.

CAPACITIES. Fuel 24½ gals. Cooling 4¼ gals.

PERFORMANCE FIGURES (G.S. BODY LADEN)
B.H.P. per Ton 10.8.
Tractive Effort per Ton (100%) 1420 lbs/Ton.
M.P.G.7. Radius 170 miles.

* Later contracts, vacuum servo assisted.

S.M.2464, 5-6-7, 2266, 2457, 2104, 2145
S.M.2019, 2037. (CANADIAN MILITARY PATTERN)

FORD C0180F (F-60L) 3-TON F.W.D.

4 x 4 Semi-forward Control. Wheelbase 13' 2¼"

ENGINE:- Ford Mercury. Petrol V.8. Bore 3.19". Stroke 3.75". Capacity 3.91 litres. R.A.C. 32.5 h.p. Max. B.H.P. 95 @ 3600 r.p.m. Max. torque 2112 lbs. ins. @ 1800 r.p.m.

CLUTCH:- Single dry plate, 11" dia.

GEARBOX:- 4 speeds and 1 reverse. Overall Ratios (High) 46.1 to 1, 22.2 to 1, 12.2 to 1, 7.2 to 1, rev. 56.4 to 1.

TRANSFER CASE:- Ratios, direct and 1.87 to 1. Front Wheel drive can be disengaged in High gear.

FRONT AXLE:- Differential assembly identical with rear axle.

REAR AXLE:- Spiral bevel drive. Ratio 7.2 to 1.

SUSPENSION:- Front and rear semi-elliptic, with hydraulic shock absorbers.

BRAKES:- Foot, Hydraulic vacuum servo assisted on all wheels. Hand, propeller shaft brake.

STEERING:- Right Hand Drive. Worm and roller.

WHEELS & TYRES:- W.D. type divided wheel, 6.00 x 20. Tyres 10.50 - 20.

ELECTRICAL EQUIPMENT: 6 volt.

TURNING CIRCLE:- 66 ft. Track. f. 5'9", r. 5'9¼".

GROUND CLEARANCE:- 10" r, axle.

CAPACITIES:- Fuel 24½ gals. Cooling 5 1/8 gals.

PERFORMANCE FIGURES (G.S. Body Laden)
B.H.P. per ton 13.4.
Tractive effort per ton (100%) 1328 lbs./Ton.
M.P.G. 7.5 Radius 180 miles.

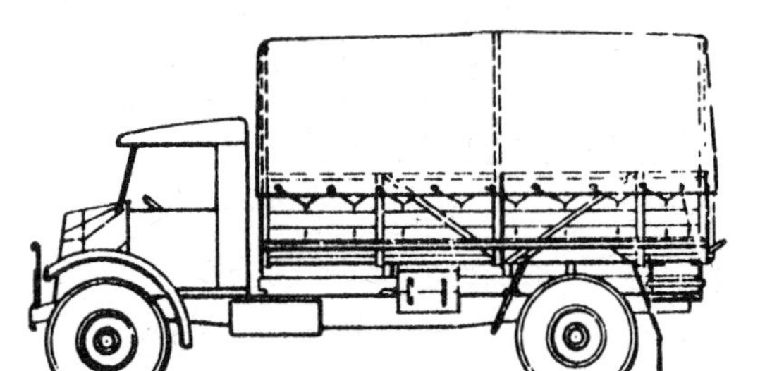

3-TON F.W.D. G.S. BODY

Two types.
1. English body on Ford. 3 ton 4 x 4 chassis (as illustrated).
2. Canadian body on Ford 3 ton 4 x 4 and Chevrolet 4 x 4 (as details at foot).

BODY (English):- Standard G.S. 12'6" body of wooden construction with low wheel arches at rear. Spare wheel carried in body, detachable tilt and hoops.

CAB: (both types):- All enclosed, side curtains to doors, seats two. Split top to cab and folding windscreen.

W.D. Drawbar gear at rear. (Both types).

DIMENSIONS (English Body)

	OVERALL	INSIDE BODY
Length	19' 0"	12' 6"
Width	7' 5"	6' 10"
Height	9' 11"	6' 0"

WEIGHTS (English Body)

	UNLADEN	LADEN
F.A.W.	1 ton 18 cwt.	2 tons 13¾ cwt.
R.A.W.	1 ton 12½ cwt.	4 tons 9 cwt.
Gross	3 tons 10½ cwt.	7 tons 1¾ cwt.

SHIPPING NOTES: Tonnage standing 35.2. Stripped Height 6' 6".

CANADIAN BODY: 12'0" Steel cargo (G.S.) body with low wheel arches, detachable tilt and hoops, spare wheel carried between cab and body. For weights and dimensions see Data sheet at back.

SECTION II

SPECIALISED VEHICLES

(Excluding Ex French Contract Vehicles which are shown in Section III)

	Chassis Type and Make	Body Type
	Ford	Utility (Station Wagon)
3-ton 4 x 2	(Chevrolet	Holmes Wrecker
	(Ford	" "
	(Chevrolet	Stores
	(Ford	"
	(Chevrolet	Workshop
	(Ford	"
3-ton 4 x 4	(Chevrolet	Stores
	(Ford	"
	(Chevrolet	Workshop
	(Ford	"
3-ton 6 x 4	Dodge	Breakdown
Artillery Tractors	(F.W.D.Co.	M.A.T.
	(Chevrolet	F.A.T.
	(Ford	F.A.T.

S.M. 2027, 2044. (CANADA)

FORD C11ADF 4 WHEEL CHASSIS

4 x 2 Normal Control, Wheelbase 9' 6"

ENGINE. Ford Mercury, Petrol V 8 Bore 3.19" Stroke 3.75", Capacity 3.91 litres, R.A.C. 32.5 h.p. Max. B.H.P. 95 @ 3600 r.p.m. Max. torque 2112 lbs/ins. @ 1800 r.p.m.

CLUTCH. Single dry plate, 11" dia.

GEARBOX. 3 speeds and 1 reverse. Overall Ratios: 12.7 to 1, 7.2 to 1, 4.1 to 1, rev. 16.4 to 1.

FRONT AXLE. I beam, 8 cwt. truck type.

REAR AXLE. Full floating, 1 ton tender type, spiral bevel drive, Ratio 4.1 to 1.

SUSPENSION. Front, single transverse, Rear, semi-elliptic.

BRAKES. Foot, Hydraulic on all wheels. Hand, Mechanical on rear wheels.

STEERING. Right Hand Drive. Worm and roller.

WHEELS AND TYRES. W.D. divided wheels 6.50 x 13. Tyres: 9.00 - 13.

ELECTRICAL EQUIPMENT. 6 volt.

TURNING CIRCLE. 39 ft. Track, f, 4' 7½", r. 5' 1".

GROUND CLEARANCES. R. Axle 8". Belly 13".

CAPACITIES: Fuel 14 gals. Cooling 21 qts.

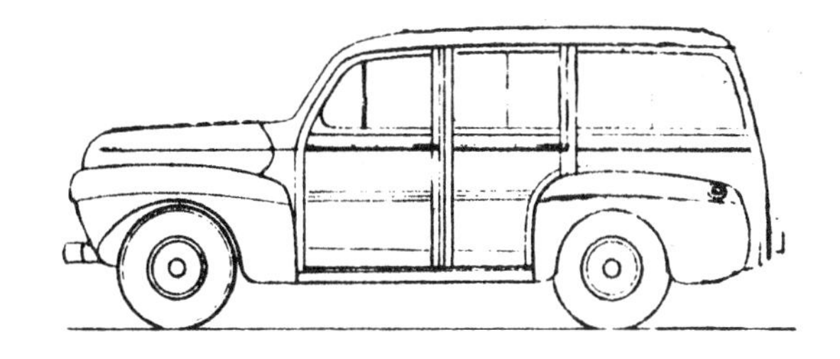

STATION WAGON (HEAVY UTILITY) BODY

ON FORD C11ADF CHASSIS

BODY:- All steel body. Seating for five passengers including driver; two in front and three on the single, straight-across seat behind. Four side doors. Hinged ventilator panes in front door windows, sliding panels in rear door and quarter light windows.

The rear portion of the body is designed for equipment carrying. The rear door is full width, split horizontally and hinged top and bottom so that the lower portion forms a tail board. The windows in this rear section are protected inside by wire mesh in a metal frame.

OVERALL Length 16' 1". Width 6' 5".

UNLADEN WEIGHT:- 1 ton 15 cwt.

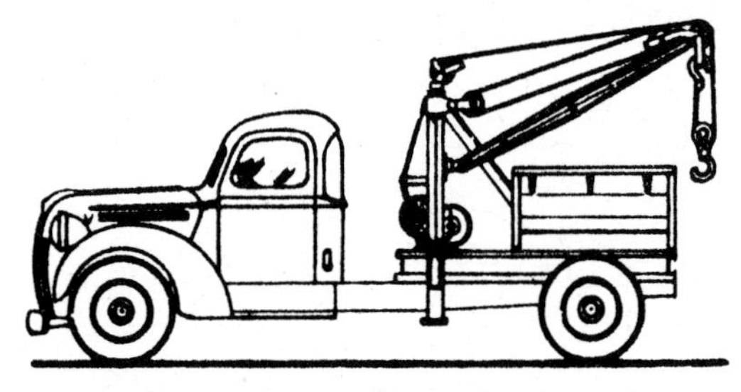

HOLMES WRECKER (BREAKDOWN UNIT)

ON FORD 3-TON 4 x 2 13 2" WHEELBASE CHASSIS

(Similar equipment also fitted to Chevrolet 3-ton 15-43 x 2 13' 4" wheelbase chassis. For weights and dimensions of this, see Data Sheet at back).

WRECKER UNIT has two swinging booms with 2 separate hand powered winches. With booms locked together at rear, a pull off the rear of the vehicle can be made with either or both winches. When booms are swung to side, the line from one winch is used to anchor the unit to the ground (by ground anchor or convenient post or tree) while the other winch is used to lift from the opposite direction. The extension leg under the pulling winch is dropped to the ground to relieve frame of vertical load. Cables can be removed from boom and pull made direct from top of wrecker frame.
V type towing bars at rear of unit.
Each cable 100' 3/8" dia. steel.
Equipment lockers above rear wheels.

OVERALL DIMENSIONS:-

Length 19 ft. Width 7' 1"
Height top Wrecker Frame 8' 2"

UNLADEN WEIGHT:- F.A.W. 1 ton 8 cwt.
R.A.W. 1 ton 18 cwt.
GROSS 3 tons 6 cwt.

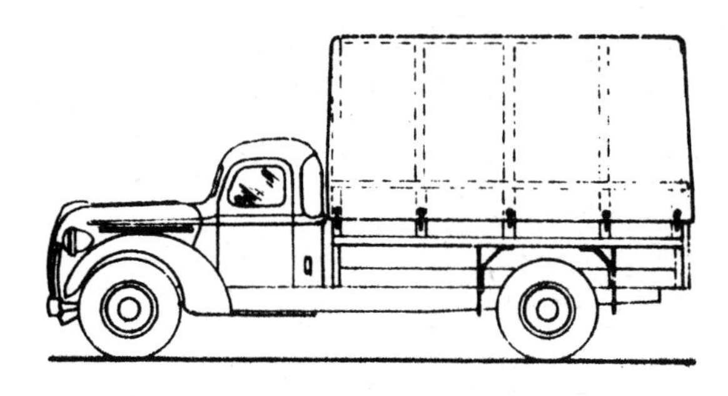

3-TON 4-WHEEL STORES BODY

ON FORD 3-TON 4 x 2 CHASSIS

(Similar body also fitted to Ford 3-ton 4 x 4 Chevrolet 3-ton 4 x 2 and 4 x 4 chassis. For wts. and dimensions of these, see Data Sheet at back).

BODY:- Lower part of body is standard steel Cargo (G.S.) Body with flat floor and hinged tailboard. Superstructure consists of curved steel hoops inter-spaced with wire mesh which entirely enclose body. Divided swing doors above rear tailboard, consisting of light steel framework with wire mesh. Canvas cover to superstructure, as illustrated. Two full length steel counters along either side of body with steel storage bins underneath. Smaller steel bins above counters. Left hand counter has a writing desk, right counter a three tier correspondence basket.

OVERALL DIMENSIONS:- Length 21' 1½". Width 7' 4". Height 10' 3".

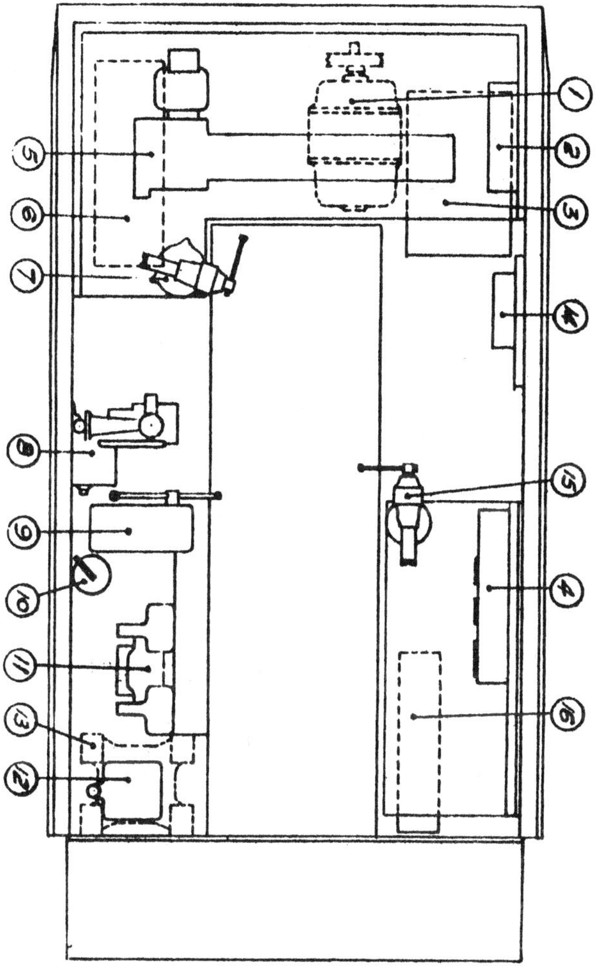

3-TON 4-WHEEL WORKSHOP BODY

(On Chevrolet 3-ton 4 x 2 & 4 x 4, and Ford 3-ton 4 x 2 & 4 x 4)

PRINCIPAL ITEMS OF EQUIPMENT

(Numbered to correspond with plan on facing page).

Forward end of Body, Lathe Bench. Right Hand side, Fitters Bench. Left Hand side, Drill Bench.

(1) 7½ K.W. D.C. Generator driven from P.T.O. of gearbox (4 x 2 models) of transfer case (4 x 4 models).
(2) 110 volt switchboard for above.
(3) 7½ volt battery charger.
(4) Charging panel for above.
(5) 10" Bench lathe.
(6) Air Compressor.
(7) 5" swivel base vise.
(8) Post Mill.
(9) Arbor Press.
(10) 2 qt. blow torch.
(11) Bench grinder.
(12) ½" Black and Decker Electric Drill.
(13) Anvil.
(14) Battery Service Kit.
(15) 5" Plain Vise.
(16) Portable Forge.

Also 3½" plain vise on portable stand, cargo lights, Hand lamps, etc. Detachable canvas covers on steel hoop superstructure.
W.D. Drawbar gear at rear.
Later Types have a 3.k.w. generator with a separate welding generator, in place of 7.k.w. generator.

DIMENSIONS

	OVERALL	INSIDE BODY
Length	21' 1½"	12' 0"
Width	7' 4"	6' 9"
Height	10' 3"	6' 0"

ESTIMATED WEIGHT (Ford 3 ton 4 x 2)

Laden with complete equipment and personnel, 6 tons 16½ cwt.

DODGE VK60 C.O.E. 3-ton 6-wheel

6 x 4 Forward Control. Wheelbase 13' 8"
Bogie Centres 4' 0"

ENGINE:- Make, Dodge T.124. Petrol 6 cyl. Bore 3¾". Stroke 5". Capacity 5.42 litres. R.A.C. 33.7 h.p. Max. B.H.P. 91 @ 2,800 r.p.m. Max. torque lbs/ins. 2760 @ 800 r.p.m. Comp. Ratio 5.2 to 1.

CLUTCH:- Single dry plate 13" dia.

GEARBOX:- 5 speeds and 1 reverse. Overall Ratios (High): 66.1 to 1, 38.2 to 1, 20.9 to 1, 12.9 to 1, 8.73 to 1, rev. 53.2 to 1.

AUX. GEAR in Thornton rear bogie transfer case. Ratios 1.18 to 1 and 2.04 to 1.

REAR AXLES:- Spiral Bevel drive. Ratio 7.4 to 1. Drive to 4 rear wheels.

FRONT AXLE:- Reverse Elliott I beam.

SUSPENSION:- Semi-elliptic front. Inverted tandem semi-elliptic rear.

BRAKES:- Foot, Hydraulic vacuum servo assisted to all wheels. Hand, transmission brake.

STEERING:- Right Hand Drive. Worm and sector.

WHEELS & TYRES:- Wheels, split type. Tyres 10.50 - 20.

ELECTRICAL EQUIPMENT:- 6 volt.

TURNING CIRCLE:- 57½ ft. Track f. 5' 8½". r. 5' 6".

GROUND CLEARANCE:- 9¾" r. axle.

CAPACITIES:- Fuel 37 gals. Cooling 5-7/8 gals.

*PERFORMANCE FIGURES:- (Breakdown Body Laden):-
B.H.P. per ton 9.3.
Tractive effort per ton (100%). 1680 lbs/Ton.
M.P.G. 5.5. Radius 200 miles.

* Estimated Figures.

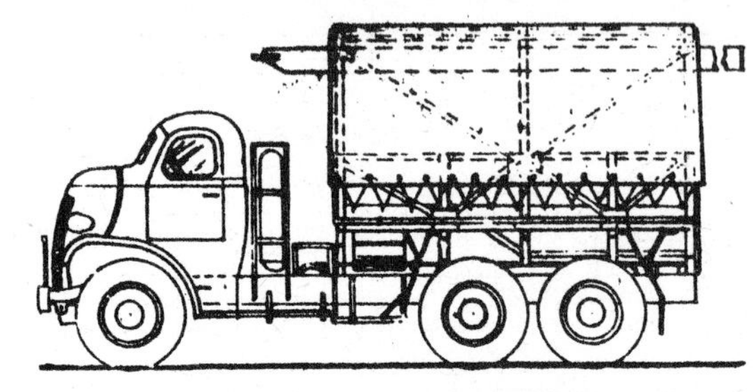

3-TON 6-WHEEL BREAKDOWN

ON DODGE VK.60 CHASSIS

BODY:- British built and identical to that fitted to Leyland, Crossley and Guy 3-ton 6-wheelers (Page 68). Floor partly flat, partly well type. Superstructure supports longitudinal runway with hand-operated travelling block. Runway capacity 2½ tons at 3' from end of body, 1½ tons at 5' and 1 ton at 6'. Height of lift can be increased by dropping front end of runway to floor, max. load then 15 cwts.

CAB:- All enclosed, steel panel cab. Seats 2, roof not detachable.

SPECIAL FITMENTS:- Gar Wood Winch. Model WSK-11-16-401, with 150 ft. steel cable. W.D. Drawbar, gear at rear.

DIMENSIONS

	OVERALL	INSIDE BODY
Length	21' 8"	12' 0"
Width	7' 6"	6' 8½"
Height	11' 3"	* 5' 4½"

* Flat floor to inside runway joist.

WEIGHTS (E)

	UNLADEN	LADEN
GROSS	6 tons 15 cwt.	9 tons 15 cwt.

SHIPPING NOTES:- Tonnage standing 46.1.

S.M.2018, 2021 (U.S.A.)

THE F.W.D.Co. TYPE S.U. C.O.E. TRACTOR CHASSIS

4 x 4 Forward Control. Wheelbase 12' 0"

ENGINE:- Make, Waukesha SRKR. Petrol 6 cyl. Bore 4-5/8". Stroke 5-1/8". Capacity 7.57 litres. R.A.C. 51.3 h.p. Max. B.H.P. 126 @ 2250. Max. torque 4428 @ 800 r.p.m.

CLUTCH:- Single dry plate. 14" dia.

GEARBOX:- 5 speeds and 1 reverse, Overall Ratios: 62.0 to 1, 33.5 to 1, 19.6 to 1, 10.62 to 1, rev. 95.0 to 1. (On types in W.D. Service ex S/M2018 and 2021 the bottom gear of the 5 speed gearbox has been blanked off.)

TRANSFER CASE:- Chain drive from gearbox to transfer case. Transfer Ratio direct. Drive to all four wheels, no provision for disconnecting front wheel drive. Third differential located in transfer case, capable of being locked.

FRONT AXLE:- Full floating with ball and socket universal joints.

REAR AXLE:- Full floating, spiral bevel drive. Ratio 4.3 to 1. Transfer Ratio 2.47 to 1.

SUSPENSION:- Front, semi-elliptic. Rear, semi-elliptic, with helper springs.

BRAKES:- Foot. Hydraulic with compressed air servo on all wheels. Hand, transmission brakes. Warner electric braking attachment also fitted.

STEERING:- Right Hand Drive. Cam and lever.

WHEELS & TYRES:- W.D. 2 piece divided wheel. Tyres 13.50 - 20.

ELECTRICAL EQUIPMENT:- 12 volt.

TURNING CIRCLE:- 76 ft. Track, f.6'1½", r.5'9".

GROUND CLEARANCE:- 13" r. axle.

Fuel 70 gals. Cooling 7½ gals.

PERFORMANCE FIGURES:-

B.H.P. per ton 11.5.
Tractive effort per ton (100%) 1948 lbs/Ton.
M.P.G. 3.3. Radius 230 miles.

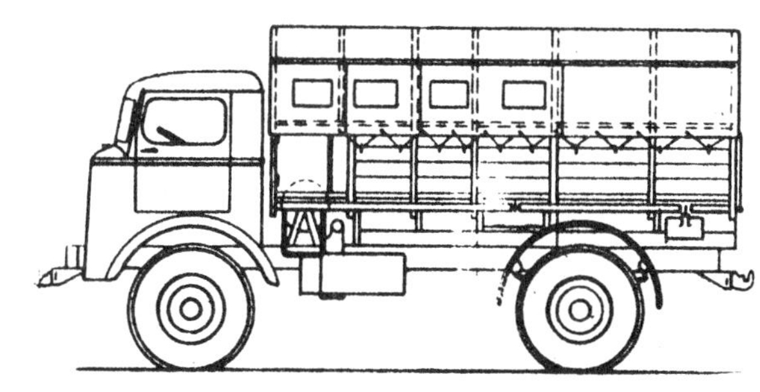

F.W.D. Co. 4-WHEEL MEDIUM TRACTOR

ON F.W.D. Co. S.U. Chassis

BODY:- This British Built body is generally similar to the A.E.C. Matador Tractor on page 87. That is it has a flat floor, detachable hoops and canvas cover, seating for 10 men with lockers and equipment racks. Adjustable shell carrier runners on floor. Four seats can be folded away to take alternative A.A. loading. An exception to page 87 is that two entrance doors are provided, one nearside, one offside.

CAB:- All enclosed, seats two.

SPECIAL FITMENTS:- Winch. W.D. Drawbar gear front and rear.

DIMENSIONS

	OVERALL	INSIDE BODY
Length	22' 1"	14' 5½"
Width	7' 9¼"	7' 0"
Height	10' 2"	5' 3"

WEIGHTS (E)

	UNLADEN	LADEN
F.A.W.	3-tons 16-cwt.	4-tons 10-cwt.
R.A.W.	3-tons 12-cwt.	6-tons 14-cwt.
Gross	7-tons 8-cwt.	11-tons 4-cwt.

SHIPPING NOTES:- Tonnage standing 43.4. Stripped Height 8' 0" (Cab Ht.).

CHEVROLET 8440 (C-GT) F.A.T. CHASSIS

4 x 4 Semi Forward Control. Wheelbase 8' 5"

ENGINE. Make, Chevrolet, Petrol 6 cyl, Bore 3½" Stroke 3¾", Capacity 3.54 litres, R.A.C. 29.4 H.P. Max. B.H.P. 78 @ 3200
Max. torque lbs/ins. 2016 @ 1100 r.p.m.

CLUTCH. Single dry plate 10¾" dia.

GEARBOX. 4 speeds and 1 reverse. Overall ratios (High) 51.4 to 1, 24.7 to 1, 12.1 to 1, 7.16 to 1, rev. 50.8 to 1.

TRANSFER CASE. Ratios, Direct & 1.87 to 1, Front wheel drive can be disengaged in high ratio.

FRONT AXLE. Differential assembly interchangeable with rear axle.

REAR AXLE. Full floating spiral bevel drive, Ratio 7.16 to 1.

SUSPENSION. Front and rear, semi elliptic with hydraulic shock absorbers.

*BRAKES. Foot, Hydraulic on all wheels. Hand, Mechanical on propellor snaft.

STEERING. Right Hand Drive. Worm and roller, Ratio 18.4 to 1.

WHEELS AND TYRES. W.D. Divided wheels, Tyres: 10.50 - 20. R.F.

ELECTRICAL EQUIPMENT. 6 volt.

TURNING CIRCLE. 49 ft. Track f. 5' 10½", r. 5' 9".

GROUND CLEARANCE. 12" rear Axle.

CAPACITIES. Fuel 24½ gals. Cooling 3¼ gals.

PERFORMANCE FIGURES
B.H.P. per ton 14.1
Tractive effort per ton (100%) 1762 lbs/Ton.
M.P.G. 7.5. Radius 180 miles.

* Vacuum servo assisted on later models.

FORD C011QF (F-GT) F.A.T. CHASSIS

4 x 4 Semi Forward Control. Wheelbase 8' 5¼"

ENGINE: Ford, Mercury, Petrol V.8. Bore 3.19". Stroke 3.75" Capacity 3.91 litres R.A.C. 32.5 h.p. Max. B.H.P. 95 @ 3600 r.p.m. Max. torque 2116 lbs/ins. @ 1800 r.p.m.

CLUTCH. Single dry plate 11" dia.

GEARBOX. 4 speeds & 1 reverse.

OVERALL RATIOS (High) 46 to 1, 22.2 to 1, 12.2 to 1, 7.2 to 1, rev. 56.4 to 1.

TRANSFER CASE. Ratios, Direct and 1.87 to 1, F.W. Drive can be disengaged in High Ratio.

FRONT AXLE. Differential assembly identical with rear axle.

REAR AXLE. Spiral bevel drive, Ratio 7.2 to 1.

SUSPENSION. Front and rear, semi-elliptic, with hydraulic shock absorbers.

BRAKES. Foot, Hydraulic vacuum servo assisted. Hand, operates on propellor shaft.

STEERING. Right Hand Drive. Worm and roller.

WHEELS AND TYRES. 6.00 x 20 divided wheels. Tyres: 10.50 - 20 R.F.

ELECTRICAL EQUIPMENT. 6 volt.

TURNING CIRCLE. 46 ft, Track 5' 9" r., 5' 8¼" f.

GR ND CLEARANCE. 10½" r. axle.

CAPACITIES. Fuel 24½ gals. Cooling 5¼ gals.

PERFORMANCE FIGURES
B.H.P. per ton 18.0.
Tractive effort per ton (100%) 1810 lbs/Ton.
M.P.G. 7.0. Radius 170 miles.

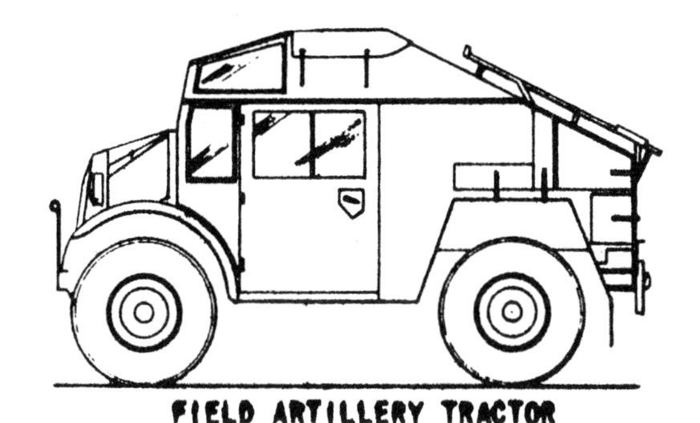

ON FORD CO11QF 4 x 4 CHASSIS

(Similar body also fitted to Chevrolet 8440 chassis. For weights see Data Sheet at back).

BODY:- The general design is similar to that of the Morris F.A.T. (Page 84) All metal gun tractor body with detachable canvas top to roof. Accommodation for driver, commander and four men. Fitments at rear to carry a spare gun wheel.

SPECIAL FITMENTS:- Winch to pull from front or rear of tractor. W.D. type Drawbar gear at rear.

OVERALL DIMENSIONS

Length 14' 9". *Width 8' 6¾". Height 7' 9".

*Over rear vision mirrors.

WEIGHTS

	UNLADEN	LADEN
F.A.W.	1 ton 15 cwt.	2 tons 0½ cwt.
R.A.W.	2 tons 4¼ cwt.	3 tons 5¾ cwt.
GROSS	3 tons 19¼ cwt.	5 tons 6¼ cwt.

SHIPPING NOTES:- Tonnage Standing 24.5.

SECTION III

EX FRENCH CONTRACT VEHICLES

ALL TYPES

Chassis Make and type	Body Type
Chevrolet Thornton 3-ton, 6 x 4	G.S.
Dodge T.203 .30-cwt. 4 x 4	G.S.
Dodge V.K.62 3-ton 4 x 2	G.S.
G.M.C. ACK 353 30-cwt. 4 x 4	G.S.
G.M.C. ACKW 353 3-ton 6 x 6	G.S.
G.M.C. AC.504 3-ton 4 x 2	G.S.
G.M.C. AFWX 354 3-ton 6 x 4	Searchlight
Mack NR 10-ton 6 x 4	
Mack EXBX 18-ton 6 x 4	Transporter
White 920 18-ton 6 x 4	Transporter
White Ruxtall 922 18-ton 6 x 4	Transporter and Recovery Vehicle

NOTE: Some of the MACK EXBX & WHITE 920 vehicles, and all Mack N.R. vehicles, are direct orders from the U.S.A. They are included with the EX FRENCH CONTRACT vehicles as their specifications generally comply with this group.

 S.M. 2011. (EX FRENCH CONTRACT)

CHEVROLET-THORNTON 3 TON 6 WHEEL

6 x 4 NORMAL CONTROL WHEELBASE 14' 6¼"
BOGIE CENTRES 3' 6"

ENGINE. Make - Chevrolet, Petrol 6 cyl, Bore 3½", Stroke 3¾", Capacity 3.54 litres, R.A.C. 29.4 h.p. Max. B.H.P. 81½ @ 3100. Max. torque lbs/ins. 2088 @ 1600 r.p.m.

CLUTCH. Single dry plate. 10¾" dia.

GEARBOX: 4 speeds and 1, reverse. Overall Ratios (High) 52.7 to 1, 26.4 to 1, 12.5 to 1, 7.3 to 1, rev. 52.2 to 1.

AUX. GEAR in Thornton rear bogie Transfer Case. Ratios 1.18 to 1, and 2.04 to 1.

FRONT AXLE. I Section beam.

REAR AXLES. Full floating spiral bevel drive. Ratio 6.17 to 1.

SUSPENSION. Front, semi-elliptic type, Rear, inverted tandem semi-elliptic.

BRAKES. Foot, Hydraulic vacuum servo assisted on all wheels, Hand, Mechanical on rear wheels.

STEERING. Left Hand Drive. Worm and sector.

WHEELS & TYRES. 20 x 7 Disc wheels, single front, dual tandem rear. Tyres 7.50 - 20.

ELECTRICAL EQUIPMENT. 6 volt.

TURNING CIRCLE. 64 ft, Track, f. 4' 11, r. 5' 6½".

GROUND CLEARANCE. 8½" r. spring shackle.

CAPACITIES. Fuel 16 gals, Cooling 3¼ gals.

PERFORMANCE FIGURES. (G.S. Body Laden)
B.H.P. per ton 10.3
Tractive effort (100%) 1410 lbs/Ton.
M.P.G. 7.0. Radius 112 miles.

3 TON 6 WHEEL G.S. BODY

ON CHEVROLET THORNTON 6 x 4 CHASSIS

BODY:- All steel G.S. (Cargo type) body with flat floor and fixed sides. Two longitudinal troop carrying seats either side of body. Detachable canvas tilt and hoops.

CAB:- All enclosed, steel panelled cab. Seats 2. Non-detachable top.

SPECIAL FITMENTS:- Rear pintle hook.

DIMENSIONS

	OVERALL	INSIDE BODY
Length	23' 6"	14' 0"
Width	7' 10"	7' 6"
Height	9' 11"	5' 9"

WEIGHTS

	UNLADEN	LADEN
F.A.W.	1 ton 2¼ cwt.	1 ton 11 cwt.
R.A.W.	3 tons 3¾ cwt.	6 tons 6 cwt.
GROSS	4 tons 6 cwt.	7 tons 17 cwt.

SHIPPING NOTES:- Tonnage standing 44.2
Stripped Height 6' 11" (Superstructure removed).

S.M.2012. (EX FRENCH CONTRACT)

DODGE T.203 30-CWT. F.W.D. 4-WHEEL

4 x 4 Normal Control Wheelbase 13' 4"

ENGINE:- Make, Dodge. Petrol 6-cyl.
Bore 3-3/8" Stroke 4½" Capacity 3.95 litres.
R.A.C. 27.3 h.p. Max. B.H.P. 75 @ 3000.
Max. torque lbs./ins. 2100 @ 1200 r.p.m.

CLUTCH:- Single dry plate, 11" dia.

GEARBOX:- 4 speeds and 1 reverse.
Overall Ratios (High): 42.2 to 1, 20.4 to 1, 11.2 to 1, 6.6 to 1, rev. 51.5 to 1.

TRANSFER CASE:- Ratios, Direct and 1.87 to 1.

FRONT AXLE:- Full floating, spiral bevel, Ratio 6.6 to 1.

REAR AXLE:- Full floating, spiral bevel, Ratio 6.6 to 1.

SUSPENSION:- Semi-elliptic, front and rear, with helper rear. Hydraulic Shock absorbers at front.

BRAKES:- Foot, Hydraulic vacuum servo assisted on all wheels. Hand, Transmission brake.

STEERING:- Left Hand Drive. Worm and sector.

WHEELS & TYRES:- Steel disc wheels, 7" rims. Tyres 7.50 - 20, singles front, twin rear.

ELECTRICAL EQUIPMENT:- 6 volt.

TURNING CIRCLE:- 67 ft. Track f. & r. 5' 5½"

GROUND CLEARANCE:- 9½" under axles.

CAPACITIES:- Fuel 32 gals. Cooling 4½ gals.

PERFORMANCE FIGURES:-

B.H.P. per ton 13.3.
Tractive effort per ton (100%) 1720 lbs./Ton.
M.P.G. 8.0. Radius 255 miles.

30-CWT. 4 WHEEL G.S. BODY

ON DODGE T.203 4 x 4 CHASSIS

BODY:- All steel G.S. (Cargo type) body with flat floor and solid sides 2' 4" high. Removable folding troop carrier seats. Detachable canvas tilt and hoops.

Two spare wheels carried under nearside and offside front of body.

CAB:- All enclosed steel panelled cab. Seats 2. Non-detachable top.

DIMENSIONS

	OVERALL	INSIDE BODY
Length	21' 7"	12' 0"
Width	7' 7"	7' 0"
Height	9' 9"	5' 10"

WEIGHTS

	UNLADEN	LADEN
F.A.W.	1 ton 9 cwt.	1 ton 16 cwt.
R.A.W.	2 tons 2½ cwt.	3 tons 16½ cwt.
GROSS	3 tons 11½ cwt.	5 tons 12¼ cwt.

SHIPPING NOTES:- Tonnage standing 39.8 Stripped height 6' 11" (Superstructure detached).

 S.M. 2017. (EX FRENCH CONTRACT)

DODGE VK.62 3-TON 4-WHEEL

4 x 2 Normal Control. Wheelbase 15' 8"

ENGINE:- Make, Dodge. Petrol 6-cyl. Bore 3¼". Stroke 5". Capacity 5.42 litres R.A.C. 33.7 h.p. Max. B.H.P. 91 @ 2800. Max. Torque lbs./ins. 2880 @ 800 r.p.m. Comp. Ratio 5.85 to 1.

CLUTCH:- Single dry plate 13" dia.

GEARBOX:- 5 speeds and 1 reverse. Overall Ratios: 56.1 to 1, 32.4 to 1, 17.7 to 1, 10.9 to 1, 7.4 to 1, rev. 45.1 to 1.

FRONT AXLE:- Reverse Elliott I beam.

GEAR AXLE:- Full floating, spiral bevel. Ratio 7.4 to 1.

SUSPENSION:- Semi-elliptic front and rear, with helper springs at rear.

BRAKES:- Foot, Hydraulic vacuum servo assisted on all wheels. Hand transmission brake.

STEERING:- Left Hand Drive. Worm and sector.

WHEELS & TYRES:- Steel disc wheels with 8" rims. Tyres 9.00 - 20, twin rear.

ELECTRICAL EQUIPMENT:- 6 volt.

TURNING CIRCLE:- 68 ft. Track, f. 5' 9", r. 5' 10½".

GROUND CLEARANCE:- 9¼" r.axle.

CAPACITIES:- Fuel 42 gals. Cooling 5-7/8 gals.

PERFORMANCE FIGURES (G.S. Laden):-

B.H.P. per ton 11.4.
Tractive effort per Ton (100%) 1078 lbs./Ton.
M.P.G. 7.0. Radius, 290 miles.

3 TON 4 WHEEL G.S. BODY

ON DODGE VK 62 4 x 2 CHASSIS

BODY:- All steel G.S. (Cargo type) body with flat floor and solid sides 2' 4" high. Full length folding removable troop seats along each side. Detachable canvas tilt and hoops.

Spare wheel carried under frame at rear.

CAB:- All steel cab. Seats two. Non-detachable top.

SPECIAL FITMENTS:- U.S. Army type pintle hook at rear.

DIMENSIONS

	OVERALL	INSIDE BODY
Length	25' 1"	15' 0"
Width	7' 8"	7' 0"
Height	10' 0"	5' 6"

WEIGHTS

	UNLADEN	LADEN
F.A.W.	1 ton 12 cwt.	2 tons 0¼ cwt.
R.A.W.	2 tons 14 cwt.	5 tons 19¼ cwt.
GROSS	4 tons 6 cwt.	7 tons 19½ cwt.

SHIPPING NOTES:- Tonnage standing 43.8
Stripped Height (Superstructure detached) 6' 11".

S.M. 2008. (EX FRENCH CONTRACT)

G.M.C. A.C.K.353 30-CWT. F.W.D. FOUR WHEEL

4 x 4 Normal Control Wheelbase 13' 1¼"

ENGINE:- Make G.M.C. 248, 6-cyl. Petrol Bore 3.23/32", Stroke 3.13/16". Capacity 4.07 litres. R.A.C. 33.2 h.p. Max. B.H.P. 77 @ 3000 r.p.m. Max. torque 2340 lbs./ins. @ 1100 r.p.m.

CLUTCH:- Single dry plate, 10¾" dia.

GEARBOX:- 4 speeds and 1 reverse. Overall Ratios: (High) 43.3 to 1, 23.6 to 1, 11.4 to 1, 6.6 to 1, rev. 52.0 to 1.

TRANSFER CASE:- Ratios, Direct and 1.87 to 1. F. Axle drive can be disengaged in high gear.

FRONT AXLE:- Reverse Elliott drive and steer type, same carrier assembly as r. axle.

REAR AXLE:- Full floating spiral bevel. Ratio 6.6 to 1.

SUSPENSION:- Semi-elliptic with hydraulic shock absorbers front and rear. Rear helper springs.

BRAKES:- Foot, hydraulic on all wheels. Hand mechanical on rear wheels.

STEERING:- Ball bearing nut and sector type. L.H. Drive.

WHEELS & TYRES:- Wheels, 20 x 7. Tyres 7.50-20, single front and dual rear. Front hubs designed to take dual wheels.

ELECTRICAL EQUIPMENT:- 6 volt.

TURNING CIRCLE:- 64 ft. TRACK:- f. 5'0-1/8" r. 5' 5¾".

GROUND CLEARANCE:- 9½" r. axle.

CAPACITIES:- Fuel 32 gals. Cooling 4¾ gals.

PERFORMANCE FIGURES (G.S. Laden):-
B.H.P. per ton 14.1.
Tractive effort per ton (100%) 2020 lbs/Ton.
M.P.G. 8.0. Radius 255 miles.

30-CWT. 4 WHEEL G.S. BODY

ON G.M.C. ACK 353 4 x 4 CHASSIS

BODY:- All steel G.S. (Steel Cargo) body with flat floor and fixed sides 2' 6" high. Longitudinal removable troop carrying seats on either side of body. Detachable canvas tilt and hoops.
Spare wheel carried between cab and body.

CAB:- All enclosed, steel panelled cab. Seats two. Non-detachable top.

SPECIAL FITMENTS:- Rear pintle hook.

DIMENSIONS

	OVERALL	INSIDE BODY
Length	21' 4"	11' 0"
Width	7' 6"	7' 0"
Height	9' 9"	5' 9"

WEIGHTS

	UNLADEN	LADEN
F.A.W.	1 ton 7¾ cwt.	1 ton 12¾ cwt.
R.A.W.	1 ton 19½ cwt.	3 tons 16¼ cwt.
GROSS	3 ton 7¼ cwt.	5 tons 9 cwt.

SHIPPING NOTES:- Tonnage standing 38.1. Stripped Height 7' 5" (Superstructure detached).

 S.M. 2039. (EX FRENCH CONTRACT)

G.M.C. ACKW 353 3 TON 6 WHEEL DRIVE

6 x 6 Normal Control Wheelbase 14' 7¾"

ENGINE:- Make G.M.C. 248, 6-cyl. Petrol. Bore 3.23/32". Stroke 3.13/16". Capacity 4.07 litres. R.A.C. 33.2 h.p. Max. B.H.P. 77 @ 3000 r.p.m. Max. torque 2340 lbs. ins. @ 1100 r.p.m.

CLUTCH:- Single dry place, 10¾" dia.

GEARBOX:- 4 speeds and 1 reverse. Overall Ratios: (High) 47.7 to 1, 22.9 to 1, 11.2 to 1, 6.6 to 1, rev. 47.2 to 1.

TRANSFER CASE:- Ratios, High Direct; Low 2.05 to 1. Front axle drive can be disengaged in High Ratio.

FRONT AXLE:- Reverse Elliott driving and steering type, same carrier assembly as r. axle.

REAR AXLES:- Full floating spiral bevel. Ratio 6.6 to 1. Drive to all 4 rear wheels.

SUSPENSION:- Semi-elliptic front, inverted semi-elliptic rear.

BRAKES:- Foot, Hydraulic vacuum servo-assisted on all wheels. Hand, transmission brake.

STEERING:- Ball bearing nut and sector type. L.H. Drive.

TYRES:- 7.00-20, single front, dual rear. Front hubs designed for dual wheels.

ELECTRICAL EQUIPMENT:- 6 volt.

TURNING CIRCLE:- 66'. TRACK:- f. 5' 1"; r. 5'7".

GROUND CLEARANCE:- 9" f. and r. axles.

CAPACITIES:- Fuel 32 gals. Cooling 4¾ gals.

PERFORMANCE FIGURES (G.S. Laden)

B.H.P. per ton 9.8.
Tractive Effort per Ton (100%) 1710 lbs./Ton.

3 TON 6 WHEEL G.S. BODY

ON G.M.C. ACKW 6 x 6 CHASSIS

BODY:- All steel G.S. (Cargo type) body with flat floor and fixed sides 2' 6" high. Longitudinal folding troop carrying seats on either side of body. Detachable canvas tilt and hoops. Spare wheel carried between cab and body.

CAB:- All enclosed, steel panelled cab. Seats two. Non-detachable top.

SPECIAL FITMENTS:- Rear pintle hook.

DIMENSIONS

	OVERALL	INSIDE BODY
Length	22' 0"	12' 0"
Width	7' 10½"	7' 6"
Height	9' 10"	5' 9"

WEIGHTS (E)

	UNLADEN	LADEN
GROSS	4 tons 8 cwt.	7 tons 18 cwt.

SHIPPING NOTES:- Tonnage standing 42.5
Stripped height 7' 6" (Superstructure detached).

G.M.C. A.C.504 3 TON 4 WHEEL

4 x 2 Normal Control Wheelbase 14' 10"

ENGINE:- Make G.M.C. 278, o.h.v., 6-cyl. Petrol. Bore 3 5/8". Stroke 4½". Capacity 4.53 litres. R.A.C. 31.5 h.p. Max. B.H.P.85 @ governed speed of 2800 r.p.m. Max. torque 2640 lbs. ins. @ 1200 r.p.m.

CLUTCH:- Dry plate. 12" diameter.

GEARBOX:- 5 speeds and 1 reverse. Overall ratios: 47.7 to 1, 27.6 to 1, 19.2 to 1, 10.8 to 1, 6.29 to 1, rev. 38.4 to 1.

FRONT AXLE:- Reverse Elliott I beam.

REAR AXLE:- Full floating, spiral bevel. Ratio 6.29 to 1.

SUSPENSION:- Front and rear, semi-elliptic with helper springs at rear.

BRAKES:- Foot, Hydraulic vacuum servo assisted on all wheels. Hand, operates on transmission.

STEERING:- Left Hand Drive. Recirculating ball type, ratio 23.6 to 1.

WHEELS & TYRES:- Wheels, 7.00 x 20. Tyres, 9.00-20, dual rear.

ELECTRICAL EQUIPMENT:- 6 volt.

TURNING CIRCLE:- 66'. Track f. 5' 6", r. 5' 8".

GROUND CLEARANCE:- 9" spare wheel carrier.

CAPACITIES:- Fuel 49 gals. Cooling 4½ gals.

PERFORMANCE FIGURES (G.S. Lorry Laden)

B.H.P. per ton 11.0.
Tractive Effort per ton (100%) 868 lbs./Ton
M.P.G. 8.0. RADIUS 390 miles.

ON G.M.C. A.C.504 4 x 2 CHASSIS

BODY:- All steel G.S. (Cargo type) body with flat floor and fixed sides 2' 6" high. Longitudinal removable troop carrying seats along either side of body. Detachable canvas tilt and hoops.

Spare tyre carried under rear of frame.

CAB:- All enclosed steel panelled cab. Seats 2. Non-detachable top.

SPECIAL FITMENTS:- Rear pintle hook.

DIMENSIONS

	OVERALL	INSIDE BODY
Length	23' 8"	14' 0"
Width	7' 11"	7' 6"
Height	10' 2"	5' 9"

WEIGHTS

	UNLADEN	LADEN
F.A.W.	1 ton 16 cwt.	2 tons 7 cwt.
R.A.W.	2 tons 13 cwt.	5 tons 10 cwt.
Gross	4 tons 9 cwt.	7 tons 17 cwt.

SHIPPING NOTES:- Tonnage standing 47.5.
Stripped Height 7' 5" (Superstructure Removed).

 S.M.2016. (EX FRENCH CONTRACT)

G.M.C. A.F.W. 354 3 TON 6 WHEEL

6 x 4 Forward Control Wheelbase 13' 11"

ENGINE:- Make, G.M.C.248, o.h.v., 6-cyl., Petrol Bore 3.23/32". Stroke 3.13/16". Capacity 4.07 litres. R.A.C. 33.2 h.p. Max. B.H.P. 77 @ 3000 r.p.m. Max. torque 2340 lbs. ins. @ 1100 r.p.m.

CLUTCH:- Single dry plate, 10½" dia.

GEARBOX:- 4 speeds and 1 reverse. Overall Ratios (High) 47.7 to 1, 22.9 to 1, 11.2 to 1, 6.6 to 1, rev. 47.2 to 1.

TRANSFER CASE:- Ratio, High direct, Low 1.87 to 1. Drive by 2 prop. shafts to both r. axles.

FRONT AXLE:- Reverse Elliott. I section beam.

REAR AXLES:- Full floating, spiral bevel. Ratio 6.6 to 1.

SUSPENSION:- Front, semi-elliptic. Rear, inverted semi-elliptic.

BRAKES:- Foot, Hydraulic, servo-assisted on all wheels. Hand, Mechanical at rear of transfer case.

STEERING:- Left Hand Drive. Ball bearing nut and sector, ratio 23.6 to 1.

WHEELS & TYRES:- Wheels 20 x 6. Tyres 7.00-20 single front, dual rear.

ELECTRICAL EQUIPMENT:- 6 volt.

TURNING CIRCLE:- 71 ft. Track: f. 5'0 7/8", r. 5'8".

GROUND CLEARANCE:- R. axle, 9¼".

CAPACITIES:- Fuel 24 gals. Cooling 4½ gals.

PERFORMANCE FIGURES (Searchlight Body Laden):-
B.H.P. per ton 9.1.
Tractive Effort per ton (100%) 1460 lbs./Ton.

3 TON 6 WHEEL SEARCHLIGHT BODY

ON G.M.C. A.F.W. 354 6 x 4 CHASSIS

BODY:- All steel G.S. (Cargo type) body with flat floor and fixed sides. Detachable canvas tilt and hoops.

CAB:- All enclosed, steel panelled cab, forward control type. Standard American "Sleeper" type extension to back of cab, adapted as a searchlight crew compartment. Seating for driver and 1 man on single seats in front, for 3 more men on the single, straight-across rear seat. (60" wide).

SPECIAL FITMENTS:- Rear pintle hook.

DIMENSIONS

	OVERALL	INSIDE BODY
Length	23' 11½"	14' 0"
Width	7' 11"	7' 6"
Height	11' 3"	7' 6"

WEIGHTS

	UNLADEN	LADEN
F.A.W.		
R.A.W.		
GROSS	4 tons 18¼ cwt.	8 tons 9¼ cwt.

SHIPPING NOTES: Tonnage standing 54.0
Stripped Height 8' 4" (Superstructure detached).

ADDITIONAL

CANADIAN

AND U.S.A.

VEHICLES

(U.S. Army Types are included in Part 4 commencing page 401)

DODGE T-222 15-CWT. 4 x 2

Normal Control Wheelbase 10' 8½"

ENGINE:- Make, Dodge, 6-cyl. Petrol. Bore 3.7/16". Stroke 4¼". Capacity 236.6 cu. ins. (3.87 litres). Max. B.H.P. 95 @ 3,600. Max. torque 2.196 lbs. ins. @ 1,200 r.p.m.

CLUTCH:- Single dry plate, 11" dia.

GEARBOX:- 4 speed and 1 reverse. Overall ratios, 40.2 to 1, 19.4 to 1, 10.6 to 1, 6,285 to 1. r. 49.1 to 1.

FRONT AXLE:- Elliott I beam.

REAR AXLE:- (U.S. Supply), Hypoid, ratio 6.285 to 1. Early types have Canadian type, ratio 6.33 to 1, Hypoid.

SUSPENSION:- Semi-elliptic, front and rear.

BRAKES:- Foot, Hydraulic, on all wheels.

STEERING:- Right Hand Drive. (Early types Gemmer 170 later types Gemmer 140)

WHEELS AND TYRES:- 9.00-16 pneumatic, singles.

ELECTRICAL EQUIPMENT:- 6-volt.

TURNING CIRCLES:- 51'6" right, 46'6" left.

GROUND CLEARANCES:- F. axle 11", r. axle 9½".

CAPACITIES:- Fuel 15 galls. Cooling 4 galls.

PERFORMANCE FIGURES (based on estimated wt. G.S.)

B.H.P. per Ton 27.1
Tractive Effort per Ton (100%) 1,475 lbs./Ton.
Maximum Speed 50 m.p.h.
M.P.G. 12. Radius 180 miles.

BODY TYPES OF DODGE T-222
15-CWT. 4 x 2

15-cwt. 4 x 2 G.S.

S.M.2326, 2485, 2459, 2674

Body to the same general design as Chevrolet 15-cwt. 4 x 2 shown on page 216. Fitted with detachable bows and tilt.

Cab, all enclosed commercial type. Detachable top.

OVERALL DIMENSIONS:- 16' 9" x 7' 3" x 7' 6" high.

ESTIMATED WEIGHTS

UNLADEN	LADEN
2 Ton 3 cwt.	3 Ton 10 cwt.

STRIPPED SHIPPING HT.:- (Approx.) 6' 3".

15-CWT. 4 x 2 WATER TANK

S.M. 2548

Capacity 200 gallons (nett.) Design of tank generally similar to Bedford shown on page 22. Pump is driven from P.T.O. as on latest Bedford models.

Estimated Laden Weight 3 Tons 15-Cwt.

DODGE T-110-L 3-TON 4 x 2

Normal Control Wheelbase 13' 4"

ENGINE:- Make, Dodge, 6-cyl. Petrol. Bore 3.7/16", Stroke 4¼" Capacity 236.cu. ins. (3.87 litres). Max. B.H.P. 95 @ 3,600. Max. torque 2,196 lbs. ins. @ 1,200 r.p.m.

CLUTCH:- Single dry plate, 10" dia.

GEARBOX:- 4 Speeds and 1 reverse. Overall Ratios 40.2 to 1, 19.4 to 1, 10.6 to 1, 6.285 to 1; R. 49.1 to 1.

FRONT AXLE:- Elliott I beam.

*REAR AXLE:- Full floating, Hypoid. Ratio 6.285 to 1.

SUSPENSION:- Semi-elliptic, front and rear, with auxiliary rear.

BRAKES:- Foot, Hydraulic vacuum servo assisted on all wheels. Hand transmission.

STEERING:- Right Hand Drive. Worm and Sector.

WHEELS AND TYRES:- Early types, 7.50-20 (34 x 7) 10 ply on 7" rims, dual rear. Later types 10.50-16 singles all round, on 6.00 x 16 divided wheels.

ELECTRICAL EQUIPMENT:- 6-volt.

TURNING CIRCLE:- 57 ft. Track, f. 57 5/8", r. 65.13/16" (duals).

CAPACITIES:- Fuel 15 galls. Cooling 4 galls.

PERFORMANCE FIGURES:-

B.H.P. per ton: 15.8
TRACTIVE EFFORT per ton (100%) 870 lbs. Ton.
Maximum Speed 45 m.p.h.
M.P.G. 8. Radius 120 miles.

*Model 110-L-3 has hypoid axle & twin rear types 110-L-4 has 2 speed axle (6.33 & 8.81) & twin rear tyres. 110-L-5 has 2 speed axle & 10.50-16 single tyres.

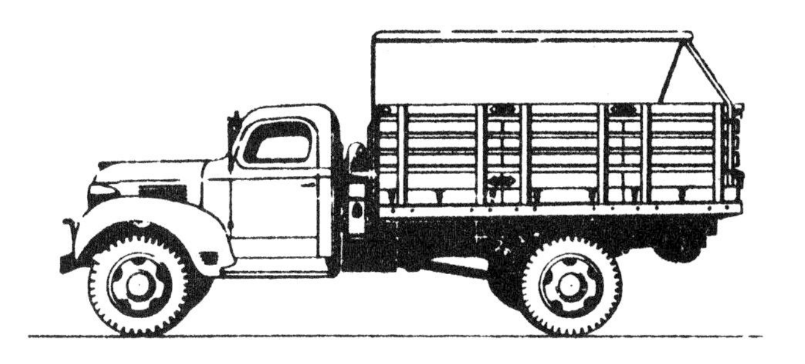

*3-TON 4 x 2 STAKE BODY

ON DODGE T-110L-3 CHASSIS

BODY:- The Stake body consists of a flat platform with detachable racks to the sides, 36" high. The canvas tilt is mounted on a single central ridge pole.

*CAB:- All enclosed. Seats 2.

SPECIAL FITMENTS:- W.D. Drawbar gear at rear.

DIMENSIONS

	OVERALL	INSIDE BODY
Length	20' 9"	* 11' 0"
Width	7' 3"	6' 10"
Height	9' 10"	5' 6"

WEIGHTS (Estimated)

	UNLADEN	LADEN
F.A.	1 ton 4 cwt.	1 ton 12 cwt.
R.A.	1 ton 10 cwt.	4 tons 8 cwt.
GROSS	2 tons 14 cwt.	6 tons 0 cwt.

*STRIPPED SHIPPING HT.:- Approx. 7' 0".
*Dodge T-110L-4 and 5 models are fitted with a 12' G.S. body & tilt similar to Ford on page 222. Cab is also split to reduce shipping height. Effective from and including S.M.2458.

2 PDR. ANTI TANK PORTEE

(Lorry 30-cwt. 4 x 4 A.T. Portee, 2-pdr.)

ON CHEVROLET 8440 (C GT)
4 x 4 Semi-forward control Chassis
Wheelbase 8' 5"

CHASSIS DETAILS:- As Chevrolet F.A.T. Tractor on page 238, except that power winch is not fitted to the A.T. Portee.

BODY DESIGN:- Has been based on that of the Morris C.8. A.T. Portee as shown on page 85.

Gun is hoisted into vehicle by hand operated winches in a similar manner, except that a central gun trail ramp is provided on the Chevrolet.

WEIGHTS

	UNLADEN	LADEN
F.A.	1 ton 17¼ cwt.	2 tons 3½ cwt.
R.A.	2 tons 0¾ cwt.	3 tons 5¾ cwt.
Gross	3 tons 18 cwt.	5 tons 9¼ cwt.

CHEVROLET 4412 3 Ton 4 x 2.

Nominal Rating 1½ Tons.

ACTUAL RATING FOR ROAD WORK 3 TONS

4 x 2 Normal Control Wheelbase 13' 4"

ENGINE:- Make: Chevrolet, 6 Cyl. O.H.V. Petrol. Bore 3½" Stroke 3¾". Capacity 216.5 cu. ins. (3.54 litres) *Max. B.H.P. 90 @ 3,300 r.p.m. Max Torque 2,088 lbs. ins.

CLUTCH:- Single dry plate, 10¾" dia.

GEARBOX:- 4 speed and 1 reverse. Overall ratio 43.5 to 1, 21.5 to 1, 12.1 to 1, 6.17 to 1, r. 43.1 to 1.

FRONT AXLE:- Reverse Elliott.

REAR AXLE:- Full floating. Hypoid. Ratio 6.17 to 1.

SUSPENSION:- Front and rear, semi-elliptic. Auxiliary rear.

BRAKES:- Foot, Hydraulic. Hand, mechanical on rear wheels.

STEERING:- Right hand drive. Recirculating ball type.

WHEELS AND TYRES:- 20 x 6, pierced steel disc. Tyres 7.00-20 (32 x 6) 10 ply, dual rear.

ELECTRICAL:- 6 volt.

TURNING CIRCLE:- 60 ft. Track, f. 4' 10". r. 5' 6".

GROUND CLEARANCES:- Axles, f. 10½"; r. 9"

CAPACITIES:- Fuel 15 gals. Cooling 3 gals.

PERFORMANCE FIGURES:- (3 ton load).

B.H.P. per Ton. 16.9
Tractive Effort per Ton (100%) 1040 lbs. ton.
Max. Speed 45 m.p.h.
M.P.G.8. Radius of Action 120 miles.

SPECIAL NOTE RE

BODY TYPES FOR CHEVROLET 4412 3 TON 4 x 2 CHASSIS

S.M.2387 and S.M.2471 call for the shipment of chassis only, completely knocked down, to the specification shown on the facing page.

Body and Cab are being provided at the vehicle destination.

The following details are generally applicable when a standard cargo or G.S. body is fitted to the chassis.

DIMENSIONS

	OVERALL	INSIDE BODY
Length	21' 2"	12' 0"
Width	7' 3"	6' 10"

WEIGHTS (3 TON LOAD)

	UNLADEN	LADEN
F.A.	1 ton 2 cwt.	1 ton 8 cwt.
R.A.	1 ton 4 cwt.	3 tons 18 cwt.
GROSS	2 tons 6 cwt.	5 tons 6 cwt.

DODGE (FARGO) 3 TON 4 x 2

DODGE MODEL WF. 32 (T.118)
1½ TON NOMINAL CAPACITY
3 TON ACTUAL CAPACITY FOR ROAD WORK ONLY

4 x 2 Normal Control Wheelbase 13' 4"

ENGINE:- Make, Dodge. 6 cyl. Petrol. Bore 3 3/8". Stroke 4½". Capacity 228.1 cu. ins. (3.73 litres) Max. B.H.P. 92 @ 3,000 r.p.m. Max. torque 2,112 @ 1,200 r.p.m.

CLUTCH:- Single dry plate. 9 7/8" dia.

GEARBOX:- 4 speed and reverse. Overall ratios: 40.2 to 1, 19.4 to 1, 10.6 to 1, 6.285 to 1, r. 49.1 to 1.

REAR AXLE:- Full floating, Hypoid gear. Ratio 6.285 to 1.

SUSPENSION:- Semi-elliptic, front and rear, with auxiliary rear.

BRAKES:- Foot, Hydraulic on all wheels. Hand, transmission brake.

STEERING:- Left-Hand Drive. Worm and Sector gear.

WHEELS AND TYRES:- 7.50-20, 8 ply tyres. mud and snow tread, dual rear.

ELECTRICAL EQUIPMENT:- 6 volt.

TURNING CIRCLE:- 55 ft. Track 4' 8"; r. 5' 4".

CAPACITIES:- Fuel 15 galls. Cooling 4 1/8 galls.

PERFORMANCE FIGURES:- (3 ton load)

B.H.P. per ton 17.1
Tractive effort per Ton (100%) 925 lbs. Ton.
Maximum Speed. 45 m.p.h.
M.P.G. 8 Radius 120 miles.

BODY TYPES ON DODGE (FARGO) 3 TON 4 x 2

Two different bodies are being supplied on S.M.2451.

1. Standard Cargo body (G.S.) with tilt and troop seats. Body measures approx. 12' x 7'.
2. Stake Body
   Generally similar to Canadian Dodge Stake body shown on page 267; except tilt and central ridge pole are not fitted.

CAB:- (Both Models). Enclosed steel panel. Seats 2.

DIMENSIONS (Stake body)

	OVERALL	INSIDE BODY
Length	21' 2"	12' 0"
Width	7' 4"	6' 10"
Height	6' 10"	3' 4"

WEIGHTS (Stake, 3-ton load)

	UNLADEN	LADEN
F.A.	1 ton 2 cwt.	1 ton 8 cwt.
R.A.	1 ton 5 cwt.	3 tons 19 cwt.
GROSS	2 tons 7 cwt.	5 tons 7 cwt.

*FORD 3 TON 4 x 2 6-cyl. 158" w.b. CHASSIS

4 x 2 Normal control, Wheelbase 158" (13' 2")

ENGINE:- Make, Ford. SIX Cylinder, Petrol. Bore 3.3". Stroke 4.4". Capacity 226 cu. ins. (3.7 litres) Max. B.H.P. 90 @ 3,300. Max. Torque 2,160 lbs. ins. @ 1,200 r.p.m. R.A.C. rating 26.1 h.p.

CLUTCH:- Single dry plate.

GEARBOX:- 4 speed. Overall Ratios 42.7 to 1, 20.6 to 1, 11.27 to 1, 6.67 to 1, r. 52.1 to 1.

REAR AXLE:- Ratio 6.66 to 1. Full floating, spiral bevel.

SUSPENSION:- Semi-elliptic front and rear, plus auxiliary rear.

FRAME:- Reinforced.

BRAKES:- Foot, Hydraulic. Hand, transmission.

TYRES:- 7.50 - 20, 8 ply, Mud and snow tread. Dual rear.

GROUND CLEARANCE:- 9½" r. axle.

CAPACITIES:- Fuel 15½ galls. Cooling 3½ galls.

* Note. Nominal Rating 1½ tons. Actual rating 3 tons for road work (with special tyre equipment, reinforced frame and auxiliary springs as being supplied on S.M.2451).

FORD 3 TON 4 x 2 6 CYL. 158" w.b. BODY DETAILS

STAKE BODY

BODY:- Ford Stake body is generally similar to the Canadian Dodge Stake Body illustrated on page 267. A plain tarpaulin cover is furnished; there is no tilt supported on central ridge pole as Dodge.

CAB:- Enclosed steel panel, commercial type.

DIMENSIONS

	OVERALL	INSIDE BODY
X Length	21' 0"	12' 0"
Width	7' 3"	6' 10"
Height	7' 2"	X 3' 6"

X (Estimated)

WEIGHTS (Estimated)

	UNLADEN	LADEN
F.A.		1 ton 6 cwt.
R.A.		4 tons 6 cwt.
GROSS	2 tons 12 cwt.	5 tons 12 cwt.

BROCKWAY 146S. 833 gallon TANKER

4 x 2. Normal Control. Wheelbase 14' 2"

ENGINE:- Make, Continental 40 B. 6 cyl. Petrol. Bore 4" Stroke 4 3/8". Capacity 330 c. ins., 5.4 litres. Max. B.H.P. 104 @ 2,800. Max. Torque 2,940 lbs. ins. @ 1,200.

CLUTCH:- Lipe. Single Dry Plate. 13" dia.

GEARBOX:- Fuller. 5-A-33. 5 speed. Overall Ratios: 55.5 to 1, 31.8 to 1, 18.5 to 1, 10.4 to 1. 7.4 to 1, and 54.1 to 1.

FRONT AXLE:- Timken, reverse Elliott, I beam.

REAR AXLE:- Timken 56411-H. Full floating. Spiral bevel. Ratio 7.4 to 1.

SUSPENSION:- Semi-elliptic with auxiliary at rear. Shock absorber f. and r.

BRAKES:- Four wheel hydraulic, vacuum servo assisted. Hand on propellor shaft.

STEERING:- Left Hand Drive. Ross cam and lever.

WHEELS AND TYRES:- Budd steel disc, dual rear. Tyres 8.25 - 20. 1 spare.

TURNING CIRCLE:- 65 ft. Track f. 64½"; r. 69½".

CAPACITIES:- Fuel 64 gals. Cooling 6 gals.

PETROL TANK:- Make Heil. Steel single compartment tank. Capacity 1,000 U.S. gals. = 833 Imperial gals.

LADEN WEIGHT:- Gross weight 7-tons 12-cwt. (estimated).

OVERALL DIMENSIONS:- 20' 5" x 7' 5" x 7' 3" high.

BROCKWAY 166S 1,660 GALLON TANKER

4 x 2. Normal Control. Wheelbase 15' 10"

ENGINE:- Make, Continental 42B. 6 cyl. Petrol. Bore 4 5/16". Stroke 4 5/8". Capacity 405 c. ins. 6.6 litres. Max. B.H.P. 112 @ 2,500. Max. torque 3,660 lbs. @ 1,000.

CLUTCH:- Lipe. Single dry plate. 14" dia.

GEARBOX:- Fuller 5-A-43. 5 speed. Ratios 8.0 to 1, 4.6 to 1, 2.4 to 1, 1.4 to 1, 1.00 to 1, r. 8.0 to 1.

FRONT AXLE:- Make, Shuler. I beam.

REAR AXLE:- Make, Eaton A-9-20,000. 2 speed. Ratios 6.71 and 9.13 to 1.

SUSPENSION:- Semi-elliptic with auxiliary rear springs. Shock absorbers front and rear.

BRAKES:- Westinghouse Air Pressure. Hand propellor shaft.

STEERING:- Left hand drive. Ross cam and lever.

WHEELS AND TYRES:- Budd steel disc. Dual rear. Tyres 11.00-20. 1 spare fitted.

TRACK:- f. 70 3/8"; r. 69½".

ELECTRICAL EQUIPMENT:- 6 volt.

CAPACITIES:- Fuel 64 galls. Cooling 7 galls.

PETROL TANK:- Make, Heil. 2 compartment steel tank. Capacity 2,000 U.S. Galls. = 1,660 Imperial Galls.

LADEN WEIGHT:- 12 tons 10 cwt. (Estimated).

OVERALL DIMENSIONS:- 24' 9" x 8' 0" x 8' 5" high.

WHITE MODEL 1064. 10-TON 6 x 4

Normal Control Wheelbase 16' 8" Bogie Centres 4' 4"

ENGINE:- Make, Cummins H.B. 600. C.I. 6 cyl. Bore 4 7/8". Stroke 6". Capacity 672 cu. ins. (11 litres). Max. B.H.P. 150 @ governed speed of 2,000 r.p.m. Max. torque 6,000 lbs./ins.

CLUTCH:- Brown Lipe, 2 plate, dry disc. 14" dia.

GEARBOX:- 5 speed and 1 reverse. Overall ratios, 64.4 to 1, 35.5 to 1, 17.7 to 1, 10.27 to 1, overdrive 6.88 to 1, R. 83.7 to 1.

FRONT AXLE:- Make, White, Reverse Elliott type.

REAR AXLES:- Full floating, double reduction, ratio 7.33 to 1.

SUSPENSION:- Front Semi-elliptic. Rear inverted semi-elliptic Timken SD-353 - W Constant Parallelogram bogie.

BRAKES:- Foot, Westinghouse, Compressed Air. Hand, transmission. Trailer air brake connections (2 line) at rear.

STEERING:- Left Hand Drive. Ross twin cam and lever.

WHEELS AND TYRES:- Budd. Tyres, front 11.00 -24 (12 ply). Rear, 14.00-20 (16 ply). Singles all round.

ELECTRICAL:- Starter 24 v. Generator 12 v. Lighting 6 v.

TRACK:- 73½" front, 72¼" rear.

CAPACITIES:- Fuel 125 galls.

PERFORMANCE FIGURES:- (10 long ton load). B.H.P. per ton 7.8. Tractive effort per ton (100%) 894 lbs./ton. Max. Speed, direct 26 m.p.h. overdrive 39 m.p.h. Fuel Consumption (estimated) 6 m.p.g. Radius 750 miles.

WHITE 1064 10 TON G.S.

BODY:- All steel cargo, equipped with bows and tilt. Folding troop seats along sides.

CAB:- All enclosed, seats 3. Full opening windscreen. Carryall platform on roof of cab with 7" sides.

SPECIAL FITMENTS:- Rear pintle hook. Front tow hooks.
1 spare tyre and wheel 11.00-24
" " " " 14.00-20

DIMENSIONS

	OVERALL	INSIDE BODY
Length	27' 3" *	15' 0"
Width	8' 0"	7' 4"
Height	10' 3"	5' 6"

* Estimated.

WEIGHTS

	UNLADEN	LADEN
F.A.		3 tons 16 cwt.
R.A.		13 tons 8 cwt.
GROSS	9 tons 4 cwt.	19 tons 4 cwt.

T.T.2/1. S.M.2036, 2075, 2150, 2307, 2432
U.S.A.

MACK NR 10-TON 6 x 4

Normal Control. Wheelbase 16' 8½"
Bogie centres 4' 7"

ENGINE:- Make, Mack ED. C.1 6 cyl. Bore 4-3/8". Stroke 5¼". Capacity 519 cu. ins. (9.0 litres). Max. B.H.P. 131 @ 2,000 r.p.m. Max. torque 4,572 lbs. ins. @ 1,300 r.p.m.

CLUTCH:- Single dry plate 15" dia.

GEARBOX:- 5 speed + auxiliary. Auxiliary ratios direct and 1.38 to 1. Overall Ratios (High) 60.8 to 1, 34.4 to 1, 17.3 to 1, 9.02 to 1, overdrive 7.08 to 1, r. 61.3 to 1.

FRONT AXLE:- Reverse Elliott.

REAR AXLES:- Double reduction. Ratio 9.02 to 1.

*SUSPENSION:- Semi-elliptic front with shock absorbers. Rear, inverted semi-elliptic.

BRAKES:- Foot, Westinghouse air. Hand transmission.

STEERING:- Left hand drive. Mack Archimoid.

*TYRES:- Front 11.00-24; Rear 14.00-20 singles. 2 spares carried, one both sizes.

ELECTRICAL:- Starter 24 volt. Lighting 12 volt.

TURNING CIRCLE:- 82 ft. Track, f. 6' 7½"; r. 6' 3¼".

GROUND CLEARANCE: 13"

CAPACITIES:- Fuel 125 galls. Cooling 10¾ galls.

PERFORMANCE FIGURES:- (Cargo truck laden)

B.H.P. per ton 6.9
Tractive effort per ton (100%) 925 lbs./ton.
Maximum speed 35 m.p.h.
M.P.G. 6 Radius 750 miles.

* For Mack N.R. Tank Transporter, which differs in these and other details, see page 511.

BODY DETAILS

(For Tank Transporter, see page 511)

BODY:- G.S. body with flat floor and tilt carried on detachable bows.

CAB:- All enclosed, steel panelled type with insulated roof.

SPECIAL FITMENTS:- Drawbar gear at rear.

DIMENSIONS

	OVERALL	INSIDE BODY
Length	26' 10"	15' 0"
Width	8' 7"	8' 0"
Height	10' 6"	6' 0"

WEIGHTS (Estimated)

	UNLADEN	LADEN
F.A.	3 tons 15 cwt.	4 tons 11 cwt.
R.A.	5 tons 11 cwt.	14 tons 11 cwt.
GROSS	9 tons 6 cwt.	19 tons 2 cwt.

SHIPPING HEIGHT:- Superstructure detached 8' 5".

MACK LMSW 6 x 4 HEAVY BREAKDOWN

Normal Control Wheelbase 13' 10½"
Bogie Centres 4' 7"

ENGINE:- Make, Mack EP. Six cylinder. Petrol. Bore 4⅜". Stroke 5⅜". Capacity 611 cu.ins. (10 litres) Max. B.H.P. 160 @ governed speed of 2,100. Max. torque 5544 lbs./ins. @ 1,000 r.p.m.

CLUTCH:- Single dry plate.

GEARBOX:- 5 speed main gearbox with 2 speed. Auxiliary. Auxiliary ratios, direct & 1.38 to 1. Overall ratios (High) 49.4 to 1, 28.0 to 1, 15.8 to 1, 9.0 to 1, overdrive 6.94 to 1, reverse 50.0 to 1.

FRONT AXLE:- Reverse Elliott.

REAR AXLES:- Double reduction, ratio 9.02 to 1. One differential in each driving axle, plus Mack power divider between axles.

SUSPENSION:- Front, semi-elliptic with shock absorbers. Rear, inverted semi-elliptic.

BRAKES:- Foot, Westinghouse compressed air on all wheels. Hand, transmission. Two line air trailer brake connections are being fitted to all vehicles.

STEERING:- Left hand drive. Mack Archimoid.

TYRES:- 14.00-20 singles all round.

ELECTRICAL:- 12 V. lighting and starting.

TURNING CIRCLE:- 73' Track, f. 6' 11"; r. 6' 3¼".

GROUND CLEARANCES:- F. Axle 15". R. Axle 14¼".

CAPACITIES:- Fuel 75 galls. Cooling 10 galls.

PERFORMANCE FIGURES:- B.H.P. per ton: 12.8
Tractive effort per ton: (100%) 1,385 lbs./ton.
Maximum speed: 35 m.p.h.
Petrol Consumption: 3.5 m.p.g. Radius 250 miles.

HEAVY BREAKDOWN on MACK LMSW 6 x 4

BODY:- Gar Wood, all steel, with equipment lockers.

CAB:- All enclosed, steel panelled type.

RECOVERY EQUIPMENT:- Three winches are provided.

1. Power, Main Winch for pulling from rear.
2. Power, Crane Winch for hoisting (Capacity 10,000 lbs. 3 ft. from end of chassis) (Both above, make Garwood Model 401).
3. Hand operated, Boom Winch for extending and retracting the Crane Boom (Jib).

W. D. Drawbar gear front and rear.

OVERALL DIMENSIONS

Length 24' 6" Width 8' 3" Height 9' 10" (jib in low position)

WEIGHTS

	UNLADEN	LADEN
F.A.	4 tons 10½ cwt.	4 tons 13½ cwt.
R.A.	6 tons 9½ cwt.	7 tons 16½ cwt.
GROSS	11 tons 0 cwt.	12 tons 10 cwt.

BROCKWAY 10-TON SEMI TRAILER G.S.

Brockway 156 Tractor with Trailmobile Semi Trailer

4 x 2 - 2 Normal Control. Tractor Wheelbase 12' 6".

ENGINE:- Make, Continental 42-B. 6 cyl. Petrol. Bore 4 5/16", Stroke 4 5/8", Capacity 405 c.ins., 6.6 litres. Max. B.H.P. 112 @ 2,500. Max. Torque 3,660 lbs. ins. @ 1,000 v.p.m.

CLUTCH:- Single dry plate. 14" dia.

GEARBOX:- Fuller 5A-43 4 speed. Ratios 8.03 to 1, 4.61 to 1, 2.46 to 1, 1.41 to 1, 1.00 to 1, and 8.00 to 1.

FRONT AXLE:- Timken. I beam.

REAR AXLE:- Eaton A-9-2000. Two speed, spiral bevel. Full floating. Ratios 6.71 to 1 and 9.13 to 1.

SUSPENSION:- Semi-elliptic front and rear, with auxiliary rear.

BRAKES:- Westinghouse Air Pressure. Hand Propellor Shaft.

STEERING:- Left Hand Drive. Ross Cam and lever.

TYRES:- 10.00-20. 12 ply, dual rear, tractor and semi trailer.

ELECTRICAL EQUIPMENT. 6 volt.

TURNING CIRCLE:- 58 ft. Track 64¾" f; 71¾" r.

CAPACITIES:- Fuel 38¼ galls. Cooling 5.8 galls.

SEMI-TRAILER:-

BODY:- Inside:- length 15' by 8' width x 2' fixed sides. Tyres: as Tractor. Brakes: Air Pressure.

LADEN WEIGHTS:- F. axle Tractor 2-tons 1-cwt. R. Axle Tractor 6-tons 6½-cwt. Semi Trailer Axle 6-tons 0-cwt.

MACK EHT 7 TON SEMI-TRAILER G.S.

Mack EHT Tractor with Mack S.T. 20 Semi-Trailer
4 x 2-2 Normal Control Wheelbase Tractor 11' 9"

ENGINE:- Make, Mack EN.354. Petrol 6 cyl. Bore 3 7/8" Stroke 5". Capacity 354 ins., 5.8 litres. Max B.H.P. 112 @ 2,700. Max. torque 3048 lbs. ins. @ 1,200.

CLUTCH:- Single dry plate. 11 7/8" dia.

GEARBOX:- 5 speed. Mack TR.31. Overall ratios: 68.6 to 1, 37.4 to 1, 22.1 to 1, 12.4 to 1, 8.59 to 1, r. 69.7 to 1.

FRONT AXLE:- Reverse Elliott.

REAR AXLE:- Mack R.A.44. Dual Reduction. Ratio 8.59.

SUSPENSION:- Front and rear semi-elliptic.

BRAKES:- Compressed air brakes with hand control valve for trailer.

STEERING:- Left Hand Drive. Ratio 18 to 1.

TYRES:- 9.00 - 20, dual rear.

ELECTRICAL EQUIPMENT:- 6 volt.

CAPACITIES:- Fuel 83 gals. Cooling 6 gals.

SEMI-TRAILER:- 18' x 7' 6" body with bows and tilt. Air brakes. Dual rear tyres 9.00-20 and two spare.

WEIGHTS (Estimated):- Laden F.A. 2 tons. R.A. 4 tons 10 cwt. Trailer A 5 tons.

INTERNATIONAL KR.10 6 TON G.S.

4 x 2 Normal Control Wheelbase 16' 5"

ENGINE:- Make, International FBC.361 6 cyl. Petrol Bore 4 1/8" Stroke 4½". Capacity 360 c. ins. 5.9 litres Max B.H.P. 94 @ 2700. Max torque 3120 lbs. ins.

CLUTCH:- Single dry plate 11 7/8" dia.

GEARBOX:- 5 speed Overall ratios: 72.6 to 1, 4.38 to 1, 22.2 to 1, 12.5 to 1, 9.03 to 1, r. 72.0 to 1.

FRONT AXLE:- Make, Eaton I beam.

REAR AXLE:- Full floating. Spiral bevel. Double reduction. Ratio 9.03 to 1.

SUSPENSION:- Semi-elliptic front and rear.

BRAKES:- Hydraulic on all wheels, vacuum servo assisted. Hand, mechanical on propellor shaft.

STEERING:- Left hand drive. Cam and twin lever.

WHEELS AND TYRES:- Spoke steel wheels, dual rear. Tyres 11.00-20, 14 ply. 2 spares.

ELECTRICAL EQUIPMENT:- 6 volt.

TURNING CIRCLE:- 68' Track front and rear 71½".

GROUND CLEARANCE:- R. Axle 10¼" Belly 13½".

CAPACITIES:- Fuel 43 gals. Cooling 5¼ gals.

BODY:- Cargo. 15' x 8' x 2' sides.

OVERALL DIMENSIONS:- Length 25' 9". Width 8' Height 7' 9".

LADEN WEIGHTS (ESTIMATED):- F.A. 2 tons 14 cwt. R.A. 7 tons 6 cwt.

INTERNATIONAL KR 11 7 TON G.S.

4 x 2 Normal Control Wheelbase 16' 5"

ENGINE:- Make, International FBC. 401 6 cyl. Petrol. Bore 4 1/8" Stroke 5". Capacity 400.9 c. ins., 6.57 litres. Max B.H.P. 114 @ 2,600. Max torque 3,684 lbs. ins. @ 800 r.p.m.

CLUTCH:- Single dry plate 13 7/8" dia.

*GEARBOX:- 5 speed. Overall ratios; 62.8 to 1, 32.1 to 1, 17.1 to 1, 9.03 to 1, overdrive 7.42 to 1; r. 62.6 to 1.

FRONT AXLE:- Reverse Elliott. I beam.

*REAR AXLE:- Full floating. Spiral bevel. Double reduction. Ratio 9.03 to 1 or 8.05 to 1 (15 vehicles).

SUSPENSION:- Semi-elliptic front and rear.

BRAKES:- Compressed air on all wheels. Hand, on propellor shaft.

STEERING:- Left hand drive. Cam and twin lever.

WHEELS AND TYRES:- Spoke steel. Dual rear. Tyres 12.00-20 14 ply.

ELECTRICAL:- 6 volt.

TURNING CIRCLE:- 69' Track f. 74½", r. 75½".

GROUND CLEARANCE:- 10" r. axle.

CAPACITIES:- Fuel 43 gals. Cooling 5¼ gals.

BODY:- Cargo 15' x 8' x 2' sides.

OVERALL DIMENSIONS:- Length 25' 9" Width 8' Height 7' 9".

LADEN WEIGHTS (ESTIMATED):- F. Axle 3 tons 6 cwt. R. Axle 7 tons 18 cwt.

TRAILER 10 TON 6 WHEEL LOW LOADING

Make: Rogers H-10-L-S-4

GENERAL DESCRIPTION

Flat platform low loading trailer with gooseneck (kickup) in front. 6 wheels each with single tyres arranged:-
Two wheels on single front axle (sprung)
Two wheels on each of the oscillating rear axles which are arranged in line.

BRAKES:- Hand brake only (by handwheel on front gooseneck section) operating on rear wheels.

TYRES:- Front 10.00-20, 12 ply.
Rear 9.00-13, 12 ply.

DIMENSIONS

Wheelbase 21' 2".
Track, over outer pair rear wheels, 7' 1½".
Overall length including drawbar, 30' 2".
Overall width 8' 0" Overall Height 6' 3".
Available loading space (flat platform) 16' 5" x 8' 0".
(Rear portion of platform is cut away to permit oscillation of rear wheels.
Platform from rear of gooseneck to rear of trailer actually measures 19' 6".
Loading Height 2' 10".

UNLADEN WEIGHT:- F.A. 1 ton 12 cwt.
R.A. 1 ton 11 cwt.

TRAILER 20 TON 6 WHEEL LOW LOADING

Make: Rogers H-30-L-F-1

GENERAL DESCRIPTION

Flat platform, low loading trailer with gooseneck (kickup) in front.
6 wheels each with twin tyres arranged:-
Two wheels on single front axle (sprung).
Two wheels on each of the oscillating rear axles which are arranged in line.

BRAKES

Westinghouse compressed air 2 pipe line system with reservoir on trailer.

TYRES

10.00-15, 14 ply.

LOADING RAMPS:- Two carried on platform.

DIMENSIONS

Wheelbase 23' 0".
Track rear (across outside pairs of duals) 8' 0".
Overall length, front of drawbar to rear of platform, 31' 9".
Overall width 10' 0". Overall Ht. 5' 9½".
Available loading space (flat platform) 17' 0" x 10' 0".
(Rear portion of platform is cut away to permit oscillation of rear wheels.
Platform from rear of gooseneck to rear of trailer actually measures 20' 3")
Loading Height 2' 11½".

UNLADEN WEIGHTS:- F.A. 2 tons 12½ cwt.
R.A. 3 tons 0 cwt.

DATA BOOK OF
WHEELED VEHICLES

PART 3

CARS, UTILITIES, TRAILERS, MOTOR CYCLES

(BRITISH MANUFACTURE)

This section of the Data Book contains details of all the main types and makes of Cars, Utilities, Trailers and Motor Cycles, excluding only imported types.

These specifications must be treated as being typical and do not necessarily apply in all details to every contract relating to individual vehicles.

DATA BOOK OF "B" AND R.A.S.C. VEHICLES

PART 3

SECTION I

CARS AND UTILITIES

ENGINE:- Make, Austin. 4-cyl. S.V. Bore 2.235" Stroke 3.5" Capacity 900 c.c. R.A.C. rating 7.99 h.p. B.H.P. 23.5 @ 4000 r.p.m.

GEARBOX:- 4 speed. Overall Ratios, 21.6 to 1, 13.1 to 1, 8.25 to 1, 5.37 to 1, Rev. 27.8 to 1, Synchro-Mesh for 2nd, 3rd and top.

REAR AXLE:- ¾ floating, spiral bevel, ratio 5.37 to 1.

SUSPENSION:- Front and rear semi-elliptic with Luvax shock absorbers.

BRAKES:- Hand and foot, Mechanical, Girling wedge and roller on all 4 wheels.

WHEELS AND TYRES:- 16 x 3.00 disc wheels. Tyres 5.00-16.

WHEELBASE:- 7' 5". TRACK f. & r. 3' 9¾".

TURNING CIRCLE:- 37'.

GROUND CLEARANCE:- 7 5/8".

TANK CAPACITY:- 6 gals. Radius 240 miles.

COOLING SYSTEM CAPACITY:- 1 5/8 gals.

ELECTRICAL EQUIPMENT:- 6 volt.

BODY

Open 2-seater body with adjustable bucket seats, folding hood and windscreen. Parcel space at back of seats. Integral frame and floor construction.

OVERALL DIMENSIONS:- Length 12' 4". Width 4' 7". Height 5' 2". Height cut down 4' 2".

WEIGHTS

UNLADEN	LADEN
13½ cwt.	17 cwt.

ENGINE:- Make, Ford. V.8 Bore 3.06". Stroke 3.75". Capacity 3.6 litres. R.A.C. rating 30 h.p. Max. B.H.P. 85 @ 3600 Max. torque lbs./ins. 1800 @ 2000.

GEARBOX:- 3 speed and 1 Reverse. Overall ratios, 16.06 to 1, 8.65 to 1, 4.55 to 1, Rev. 18.2 to 1.

REAR AXLE:- ¾ floating, spiral bevel, Ratio 4.55 to 1.

SUSPENSION:- Front and rear, single transverse.

BRAKES:- Foot, Mechanical on all wheels. Hand, Mechanical on all wheels.

WHEELS AND TYRES:- Steel disc wheels with 9.00-13 tyres. Model W.O.A.1.A. fitted 6.50-16.

WHEELBASE:- 9' 0¼". TRACK f. 4' 9½", r. 5' 0½".

TURNING CIRCLE:- 43' 6".

GROUND CLEARANCE:- 9" at r. axle. 7 3/8" r. brake rod yoke ends.

TANK CAPACITY:- 12½ gals. Radius 210 miles.

COOLING SYSTEM CAPACITY:- 4¾ gals.

ELECTRICAL EQUIPMENT:- 6 volt.

BODY

Modified standard saloon body, with special increased clearance at the front and rear. Single straight across front seat, full width rear seat. Spare wheel carried on lid of luggage boot. Modified radiator grille and bonnet. Roof luggage rail with covers.

OVERALL DIMENSION:- Length 13' 11½", width 6' 3½", Height 6' 4".

WEIGHTS

UNLADEN	LADEN
27¾ cwt.	35 cwt.

HUMBER SNIPE SALOON

ENGINE:- Make, Humber. 6-cyl. Bore 85 m.m. stroke 120 m.m. Capacity 4.08 litres. R.A.C. rating 26.9 h.p. Max. B.H.P. 85 @ 3400. Max. torque 2160 lbs./ins. @ 1300 r.p.m.

GEARBOX:- 4 speed and 1 reverse. Overall Ratios: 15.1 to 1, 10.1 to 1, 5.9 to 1, 4.09 to 1, rev. 15.1 to 1.

REAR AXLE:- Semi-floating, spiral bevel, Ratio 4.09 to 1.

SUSPENSION:- Front, Independent, transverse semi-elliptic. Rear, semi-elliptic. Hydraulic Shock Absorbers front and rear

BRAKES:- Foot, Hydraulic on all wheels. Hand Mechanical on rear wheels.

TYRES:- 9.00-13.

WHEELBASE:- 9' 6". TRACK f. 4' 7¾" r. 4' 9½".

TURNING CIRCLE:- 41 ft.

GROUND CLEARANCE:- 9½" f. spring.

TANK CAPACITY:- 14 gals. Radius 210 miles.

COOLING SYSTEM CAPACITY:- 3¾ gals.

ELECTRICAL EQUIPMENT:- 12 volt.

BODY

Standard 5-seater Saloon body. Folding map table fitted between front and rear seats. Spare wheel carried in luggage boot at rear. Roof luggage rail with waterproof cover.

OVERALL DIMENSIONS:- Length 15' 0". Width 5' 10". Height 5' 11½".

WEIGHTS

	UNLADEN	LADEN
F.A.W.	15¾ cwt.	17½ cwt.
R.A.W.	17¼ cwt.	1 ton 2¼ cwt.
GROSS	1 ton 13 cwt.	1 ton 19¾ cwt.

HUMBER SNIPE TOURER

Chassis. Specification as Humber Snipe Saloon.

BODY:- 5 seater tourer with folding hood and detachable side screens.

OVERALL DIMENSIONS:- Length 15' 0". Width 5' 10". Height 5' 7 3/8".

HUMBER PULLMAN LIMOUSINE

Chassis generally as Humber Snipe Saloon, except:-

WHEELBASE:- 10' 7½". TRACK f. 4' 7 7/8", R. 5' 1".
TYRES:- 7.00-16.

TURNING CIRCLE:- 45 ft.

GROUND CLEARANCE:- 7 7/8".

TANK CAPACITY:- 15 gals. Radius 200 miles.

BODY

7-seater. Limousine Body, seating 2 in front, 2 on occasional seats which can be folded away when not required, 3 on rear seat. Division fitted with sliding windows separating front seats from rear compartment. Spare wheel carried in luggage boot at rear. Roof luggage rail with waterproof cover.

OVERALL DIMENSIONS:- Length 16' 5". Width 6' 1". Height 6' 2½".

WEIGHTS (E)

	UNLADEN	LADEN
F.A.W.	16 cwt.	
R.A.W.	18½ cwt.	
GROSS	1 ton 14½ cwt.	2 tons 3½ cwt.

AUSTIN 10 H.P. LIGHT UTILITY

ENGINE:- Make, Austin 4 cyl. Bore 2.62". Stroke 3.5" Capacity 1.23 litres. R.A.C. rating 10.98 h.p. Max. B.H.P. 29.5 @ 4000 r.p.m. Max. torque 580 lbs./ins. @ 2250 r.p.m.

GEARBOX:- 4 speed and reverse. Overall Ratios 27.6 to 1, 16.7 to 1, 10.5 to 1, 6.14 to 1, rev. 35.5 to 1.

REAR AXLE:- ¾ floating, spiral bevel. Ratio 6.14 to 1.

SUSPENSION:- Semi-elliptic with hydraulic shock absorbers, front and rear.

BRAKES:- Foot, Mechanical on all wheels. Hand, Mechanical on all wheels.

WHEELS AND TYRES:- Spoked disc wheels with 6.00-16 tyres.

WHEELBASE:- 7' 10¾. TRACK, f. 4' 0½", r. 4' 3½".

TURNING CIRCLE:- 36 ft.

GROUND CLEARANCE:- 8½ r. axle.

TANK CAPACITY:- 8½ gals. Radius 230 miles.

COOLING SYSTEM CAPACITY:- 2½ gals.

ELECTRICAL EQUIPMENT:- 12 volt.

BODY

General design as Hillman Light Utility on page 311.

OVERALL DIMENSIONS:- Length 13' 1". Width 5' 3¼". Hight 6' 5½".

WEIGHTS

	UNLADEN	LADEN
F.A.W.	9½ cwt.	10 cwt.
R.A.W.	10¼ cwt.	17 cwt.
GROSS	19¾ cwt.	1 ton 7 cwt.

MORRIS 10 H.P. LIGHT UTILITY

ENGINE:- Make, Morris 4 cyl. Bore 63.5 m.m. Stroke 90 m.m. R.A.C. rating 10 h.p. Capacity 1.14 litres. Max. B.H.P. 37.2 @ 4600 r.p.m. Max. torque 628 lbs/ins. @ 1800 r.p.m.

GEARBOX:- 4 speed and 1 reverse. Overall Ratios 28.7 to 1, 16.9 to 1, 11.3 to 1, 6.33 to 1, Rev. 22.3 to 1.

REAR AXLE:- ¾ floating. Ratio 6.33 to 1.

SUSPENSION:- Semi-elliptic with hydraulic shock Absorbers, front and rear.

BRAKES:- Foot, Hydraulic on all wheels, Hand, Mechancial on rear wheels.

WHEELS AND TYRES:- Disc wheels with 6.00-16 tyres.

WHEEL BASE:- 7'10" Track f and r. 4'3"

TURNING CIRCLE:- 40 ft.

GROUND CLEARANCE:- 7½" sump.

TANK CAPACITY:- 8 1/3 gals. Radius 210 miles.

COOLING SYSTEM CAPACITY:- 1 5/8 gals.

ELECTRICAL EQUIPMENT:- 12 volt.

BODY

General design as Hillman Light Utility on page 311.

OVERALL DIMENSIONS:- Length 13'5½", Width 5'0", Height 6'4".

WEIGHTS

	UNLADEN	LADEN
F.A.W.	10½ cwt.	10¼ cwt.
R.A.W.	11¼ cwt.	1 ton 1½ cwt.
Gross	1 ton 1¾ cwt.	1 ton 11¾ cwt.

HILLMAN 10 H.P. LIGHT UTILITY

ENGINE:- Make, Hillman 4 cyl. Bore 63 m.m. Stroke 95 m.m. cap. 1.18 litres. R.A.C. rating 9.8 h.p. Max. B.H.P. 30 @ 4100 r.p.m. Max. torque 580 lbs/ins. @ 2100 r.p.m.

GEARBOX:- 4 speeds and 1 reverse. Overall Ratios 23.4 to 1, 16.2 to 1, 9.8 to 1, 6.57 to 1, Rev. 31.2 to 1.

REAR AXLE:- Semi-floating, spiral bevel, Ratio 6.57 to 1. Later contract, 5.37 to 1.

SUSPENSION:- Semi-elliptic with hydraulic shock absorbers front and rear.

BRAKES:- Foot, Mechanical on all Wheels. Hand, Mechanical on all wheels.

WHEELS AND TYRES:- Pressed Steel disc, 16 x 4.00 Tyres, 6.00-16. Later contracts 500-16".

WHEELBASE:- 7'8". TRACK f.3'11½" r.4'2½".

TURNING CIRCLE:- 37 ft.

GROUND CLEARANCE:- 7½" at front axle.

TANK CAPACITY:- 8 gals. Radius 220 miles.

COOLING SYSTEM CAPACITY:- 2 gals..

ELECTRICAL EQUIPMENT:- 12 volt.

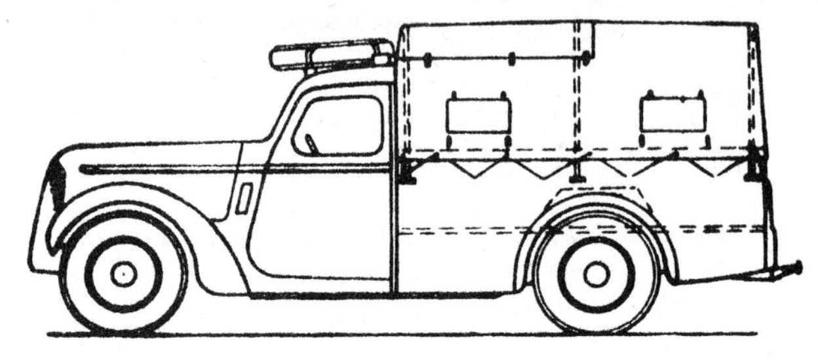

LIGHT UTILITY BODY ON HILLMAN CHASSIS

(Similar bodies also fitted to Austin and Morris Light Utility chassis)

BODY:- Fixed sides, drop tailboard. Detachable canvas tilt carried on 3 hoop sticks. Opening sections to tilt at forward end on roof and either side. Hoopsticks when not in use are carried bunched together in sockets at forward end. Separate front seats for driver and passenger, driver's adjustable, passengers folds forward to give access to rear. Behind are 2 separate seats which can be folded forward to leave a flat floor except for combined wheel arches and lockers.

CAB:- All enclosed. Spare wheel carried on roof. Back of cab open to main body.

DIMENSIONS

	OVERALL	INSIDE BODY
Length	12'7"	5'0"
Width	5'3"	4'8"
Height	6'3"	1'6" sides

WEIGHTS

	UNLADEN	LADEN
F.A.W.	10¼ cwt.	12 cwt.
R.A.W.	10¾ cwt.	17 cwt.
Gross	1 ton 1 cwt.	1 ton 9 cwt.

FORD W.O.A.2. HEAVY UTILITY

ENGINE:- Make, Ford V.8. Bore 3.96". Stroke 3.75". Capacity 3.6 litres. R.A.C. rating 30 h.p. Max. B.H.P.85 at 3600. Max torque lbs/ins. 1800 @ 2000 r.p.m.

GEARBOX:- 3 speed and 1 reverse. Overall ratios 16.06 to 1, 8.65 to 1, 4.55 to 1, Rev. 18.2 to 1.

REAR AXLE:- ¾ floating, spiral bevel, ratio 4.55 to 1.

SUSPENSION:- Front, and rear, single transverse. Hydraulic shock absorbers front and rear.

BRAKES:- Foot, Mechanical on all wheels. Hand, Mechanical on all wheels.

WHEELS AND TYRES:- Steel disc wheels with 9.00-13 tyres.

WHEELBASE:- 9'0¼". TRACK f.4'9½", r.5'0½".

TURNING CIRCLE:- 43'6".

GROUND CLEARANCE:- 9" r. axle 8½" track rod.

TANK CAPACITY:- 12½ gals. Radius 175 miles.

COOLING SYSTEM:- 4¾ gals.

ELECTRICAL EQUIPMENT:- 6 volt.

BODY

6-Seater Steel Utility body with 4 side doors and full width rear doors split horizontally, 2 Bucket seats in front (driver's adjustable), 2 Bucket seats in middle of body and 2 tip-up seats in rear corners. Folding table behind front seats. Middle seats fold forward for goods carrying leaving a space 54" wide x 57" long x 45" high. Sliding panel in centre of roof. Spare wheel carried outside at rear.

OVERALL DIMENSIONS:- Length 14'5", Width 6'3½", Height 5'10".

WEIGHTS (E)

UNLADEN	LADEN
31¼ cwt.	42¾ cwt.

HUMBER F.W.D. HEAVY UTILITY (4 x 4)

ENGINE:- Make, Humber 6-cyl. Bore 85 m.m. Stroke 120 m.m. Capacity 4.08 litres. R.A.C. rating 26.9 h.p. Max. B.H.P. 85 @ 3400. Max. torque 2160 lbs/ins. @ 1300 r.p.m.

GEARBOX:- 4 speed forward and 1 reverse. Transfer case Ratios, direct and 1.48 to 1. Overall Ratios (High), 18.1 to 1, 12.1 to 1, 7.1 to 1, 4.88 to 1, rev. 18.1 to 1. Four wheel drive on low ratio only.

AXLES:- Semi floating, spiral bevel drive, ratio 4.88 to 1.

SUSPENSION:- Front, Independent, transverse semi-elliptic. Rear, semi-elliptic. Hydraulic shock absorbers front and rear.

BRAKES:- Foot, Hydraulic on all wheels. Hand, mechanical on rear wheels.

WHEELS AND TYRES:- Steel disc wheels with 9.25-16.

WHEELBASE:- 9'3¾". Track 5'0¾". Turning circ. 45'.

GROUND CLEARANCE:- 9 1/8"-f. spring.

TANK CAPACITY:- 16 gals. Radius 200 miles.

COOLING SYSTEM CAPACITY:- 4¾ gals.

ELECTRICAL EQUIPMENT:- 12 volt.

BODY

Steel panelled 6-seater Utility body. Arrangement of seats is similar to the Humber 4 x 2 Utility, but otherwise body varies in many details. Front wings, radiator, grille and bonnet are identical to Humber 8-cwt. 4 x 4 chassis.

OVERALL DIMENSIONS:- Length 14'1". Width 6'2". Height 6'5".

LADEN WEIGHT

2 tons 18 cwt.

HUMBER SNIPE HEAVY UTILITY (4 x 2)

ENGINE:- Make, Humber 6-cyl. Bore 85 m.m. Stroke 120 m.m. Capacity 4.08 litres. R.A.C. rating 26.9 h.p. Max. B.H.P. 85 @ 3400. Max. torque 2160 lbs/ins. @ 1300 r.p.m.

GEARBOX:- 4 Speeds forward and 1 Reverse. Overall Ratios 18.1 to 1, 12.1 to 1, 7.16 to 1, 4.89 to 1, Rev. 18.1 to 1.

REAR AXLE:- Semi floating, spiral bevel, ratio 4.89 to 1.

SUSPENSION:- Front, Independent, transverse semi-elliptic spring. Rear, semi-elliptic. Hydraulic shock absorbers front and rear.

BRAKES:- Foot, Hydraulic on all wheels. Hand Mechanical on rear wheels.

WHEELS AND TYRES:- Steel disc wheels with 9.00-13 tyres.

WHEELBASE:- 9'6". TRACK f.5'2", r.5'1½".

TURNING CIRCLE:- 43'.

GROUND CLEARANCE:- 8 7/8" f. suspension. 5¼" silencer.

TANK CAPACITY:- 13 gals. Radius 200 miles.

COOLING SYSTEM CAPACITY:- 3½ gals.

ELECTRICAL EQUIPMENT:- 12 volt.

HEAVY UTILITY BODY ON HUMBER SNIPE (4 x 2) CHASSIS

Steel panelled 6-seater Utility body, with 4 side doors and full width rear door split horizontally to form tailboard. 2 adjustable bucket front seats, 2 intermediate seats which, with cushions detached, can be folded flat into floor, 2 tip-up rear seats. Map table fitted behind front seats. Spare wheel carried on back of tailboard.

OVERALL DIMENSIONS:- Length 15'4". Width 6'1½" Height 6'1".

WEIGHTS

	UNLADEN	LADEN
F.A.W.	16 cwt.	16¾ cwt.
R.A.W.	18½ cwt.	1 ton 7¾ cwt.
Gross	1 ton 14½ cwt.	2 tons 4½ cwt.

DATA BOOK OF WHEELED VEHICLES

PART 3

SECTION 2

TRAILERS

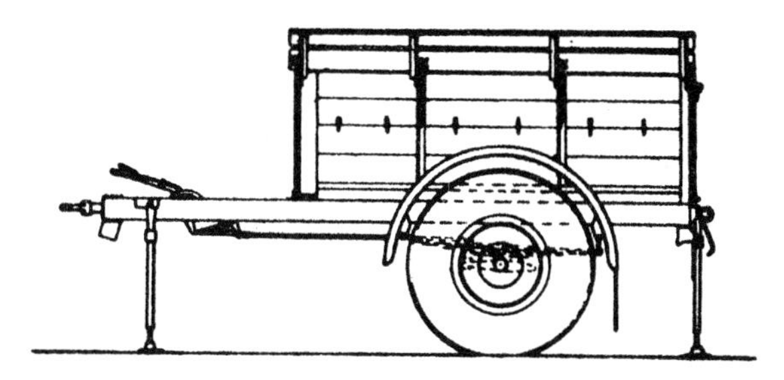

15-cwt. 2-WHLD. LOAD CARRYING

SUSPENSION:- Semi-elliptic.

BRAKES:- Internal expanding, operated by draw-eye overrunning device. Also separate hand lever.

WHEELS:- Detachable steel disc, 10 stud fitting.

TYRES:- 9.00 - 20.

SUPPORTING LEGS:- Adjustable for length fitted front and rear.

BODY:- 15-cwt. G.S. rave-sided body.

DIMENSIONS

TRACK:- 5'6".
TOWING EYE TO AXLE:- 7'9".
HT. OF TOWING EYE:- 2'8".

	OVERALL	INSIDE BODY
Length	11'6".	6'7½"
Width	6'6".	4'2"
Height	5'5".	1'11"

Width over top of raves 6'4".

WEIGHTS

Unladen 14¾ cwt. Laden 1 ton 9¼ cwt.

SHIPPING TONNAGE:- 10.1.

15-CWT. AIR LINE POLE CARRYING

SUSPENSION:- Semi-elliptic.

BRAKES:- Internal expanding, automatically operated on overrun. Hand parking lever.

TYRES:- 9.00 - 20.

SUPPORTING LEGS:- Jack Screw Type, 1 front, and 2 rear.

BODY:- Platform. Fixed sides and front. Detachable rear tailboard. Locker at rear.

DIMENSIONS

TRACK 5'6".

TOWING EYE TO AXLE:- 13'0".

GROUND CLEARANCE:- 16".

	OVERALL	INSIDE BODY
Length	20'10½"	16'0"
Width	6'3"	4'6"
Height	6'6"	2'0"

WEIGHTS

Unladen 19¾ cwt.

SHIPPING TONNAGE:- 21.2

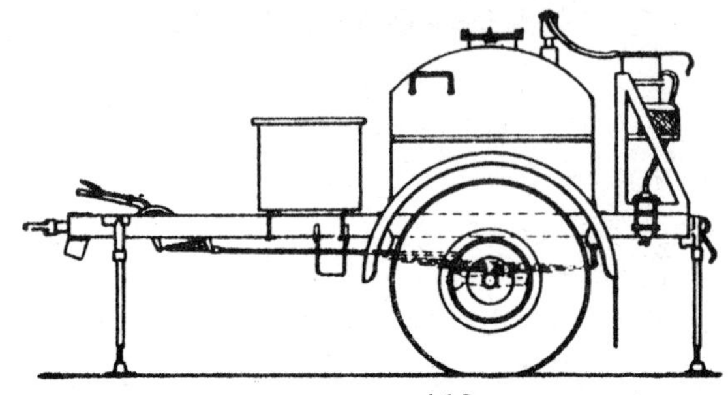

15 cwt. 2-WHLD. 180 GALLON WATER TANK

SUSPENSION: Semi-elliptic

BRAKES:- Internal expanding, operated by draw-eye over-running device. Separate hand lever.

WHEELS:- Detachable steel disc, 10 stud fitting.

TYRES:- 9.00 - 20.

SUPPORTING LEGS:- Adjustable, fitted front and rear.

WATER TANK EQUIPMENT:- 180 gallon tank, pumps, Patterson "Stellar" filters, hose, floats and all fittings.

DIMENSIONS

TRACK:- 5'6".
TOWING EYE TO AXLE:- 7'9".
GROUND CLEARANCE:- 17¼".
OVERALL LENGTH:- 11'8".
OVERALL WIDTH:- 6'6".
OVERALL HEIGHT:- 6'2".

WEIGHTS

Unladen 19 cwt. Laden 1 ton 15½ cwt.

SHIPPING TONNAGE:- 11.67.

2-WHEEL DENTAL TRAILER

(Made by Light Steel
Sectional Constructions, Ltd.)

SUSPENSION:- Semi-elliptic.

BRAKES:- Internal expanding, operated on overrun. Hand parking brake.

WHEELS:- Detachable pressed steel. 5 stud.

TYRES:- 6.50-17.

SUPPORTING LEGS:- Four scissor pattern, brace operated, attached to sub-members of frame at corners.

BODY:- Caravan type (Eccles) body. Divided into reception, office and surgery compartments. Fitted with dental chair, lathe, instruments, etc. Carrier fitted at front of trailer for Tent, Poles and Pegs.

DIMENSIONS

TRACK: 5'5".

TOWING EYE TO AXLE:- 11'7½".

GROUND CLEARANCE:- 8½", r. spring U bolts.

	OVERALL	INSIDE BODY
Length	19'2"	14'0"
Width	6'8½"	6'0"
Height	8'6"	6'3½"

WEIGHT

Unladen, 1 ton 5½ cwt.

With vehicle equipment, 1 ton 8½ cwt. (E).

SHIPPING TONNAGE:- 27.2.

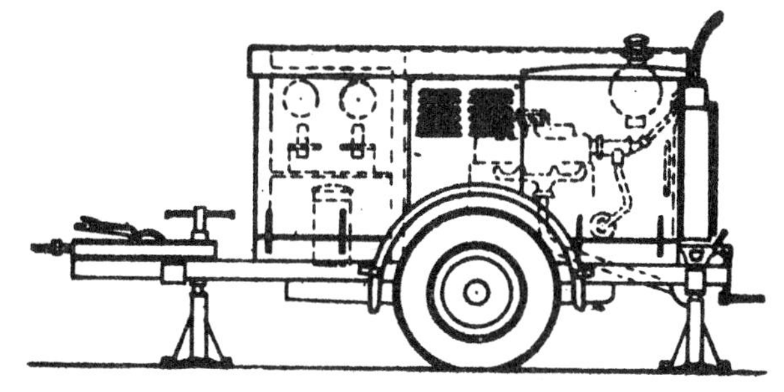

2 WHLD. LIGHTWEIGHT 18 K.W. GENERATING SET

(Make, Bristol Tramways & Carriage Co.)

SUSPENSION:- Solid Axle.

BRAKES:- Internal expanding, operated on overrun. Hand parking lever.

WHEELS:- Detachable steel disc, 10 stud fitting.

TYRES:- 9.00 - 20.

SUPPORTING LEGS:- One adjustable at front, two hinged at rear, with "elephants feet".

EQUIPMENT:- 18 K.W. Generator Set by Norris, Henty & Gardner.

SUPERSTRUCTURE:- Sheet Metal. Semi-permanent top. Readily detachable side panels.

DIMENSIONS

TRACK:- 5'2".
TOWING EYE TO AXLE:- 6'11".
HEIGHT OF TOWING EYE:- 2'5".
OVERALL LENGTH:- 11'4".
OVERALL WIDTH:- 6'6".
OVERALL HEIGHTS:- 6'6".

WEIGHT (E)

With full equipment, 2 tons 2 cwt.

SHIPPING TONNAGE. 11.9.

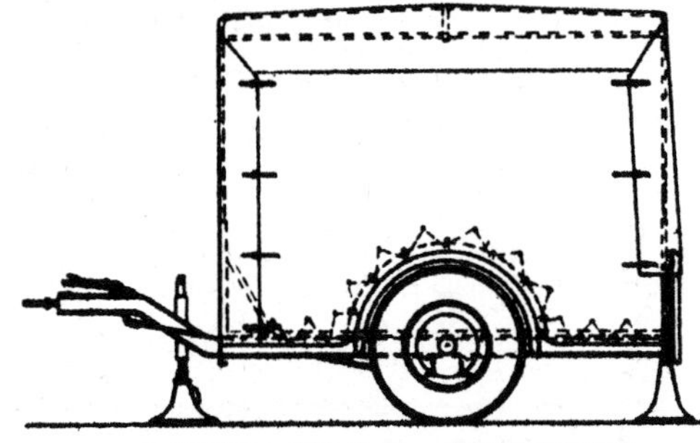

2-WHLD. 22 K.W. GENERATING SET (No. I Mk. II)

(Make Tasker)

SUSPENSION:- Solid Axle.

BRAKES:- Internal expanding, operated on over-run. Hand parking lever.

WHEELS:- Disc, 7.33 x 20.

TYRES:- 10.50 - 20.

SUPPORTING JACKS:- Fitted front and rear.

EQUIPMENT:- Lister 4 cyl. C.I. Engine, Maudslay 22 k.w. 100 volt D.C. generator. Lister switchboard.

SUPERSTRUCTURE:- Tubular supports with canvas tilt and side curtains.

DIMENSIONS

TRACK:- 6'5".
TOWING EYE TO AXLE:- 9'0".
OVERALL LENGTH:- 14'4".
OVERALL WIDTH:- 7'5".
OVERALL HEIGHT:- 9'0".

WEIGHT

Laden with full equipment, 3 tons 6 cwt.

SHIPPING TONNAGE:- 24.1.

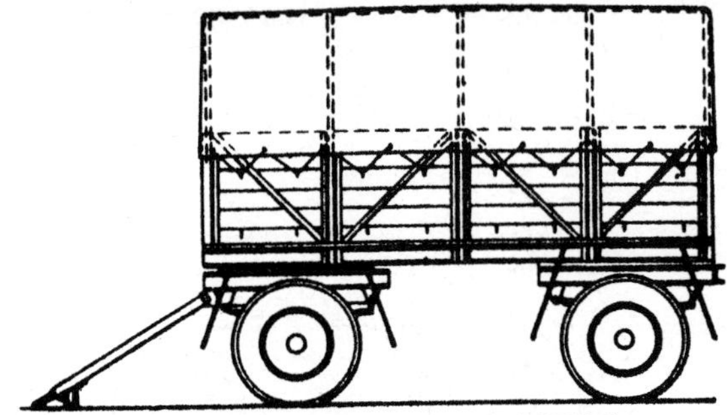

4/5 TON 4-WHL. LOAD CARRYING

(Made by British Trailer Co.)

SUSPENSION:- Semi-elliptic.

BRAKES:- Internal expanding on rear wheels only, operated by tow-bar overrunning device. Parking brake hand lever.

TURNTABLE:- 4' diameter. Tow bar, "Vee" Type fitted standard W.D. draw-eye.

WHEELS:- Detachable steel disc, 8 stud fitting. Rim, 20" x 7.33 F.B.

TYRES:- 10.50 - 20.

BODY:- G.S., fitted with detachable hoops and canvas cover.

DIMENSIONS

WHEELBASE:- 9'0" TRACK:- 5'11"
TOWING EYE TO FRONT AXLE:- 8'0"

	OVERALL	INSIDE BODY
Length	20'0"	13'10"
Width	7'6"	6'10"
Height	10'8"	6' 0"

WEIGHTS

Unladen 2 tons 15½ cwt.

SHIPPING NOTES:- Tonnage standing 34.0, Cut down 28.9. Height Cut Down 9'1". Stripped Shipping Height 7'6". Super-Structure detached.

2 TON 4-WHL. 150 cm. PROJECTOR

SUSPENSION:- Helical, front. Semi-elliptic, rear.

BRAKES:- Manually operated, front wheel drums, operated on overrun.

STEERING:- Through drawbar.

WHEELS:- 5.00 x 16 disc.

TYRES:- 9.00-16.

SUPPORTING LEGS:- Front, through swivelling arms on cross member. Rear, at centre of rear cross member.

DIMENSIONS

WHEELBASE:- 10'4½". TRACK f. 5'0¾", r. 6'7¼".

TOWING EYE TO F. AXLE. 5' 0½".

OVERALL LENGTH drawbar extended 18'4", drawbar folded back 14'10".

OVERALL WIDTH:- 7' 6".

OVERALL HEIGHT with projector loaded, 10' 6".

WEIGHTS:- Unladen 1 ton 5½ cwt.
Laden 3 tons 4¾ cwt.

SHIPPING NOTES (projector loaded):-
Tonnage standing 29.19.

4-WHEELED MOTOR BOAT CARRYING (Made by Brockhouse)

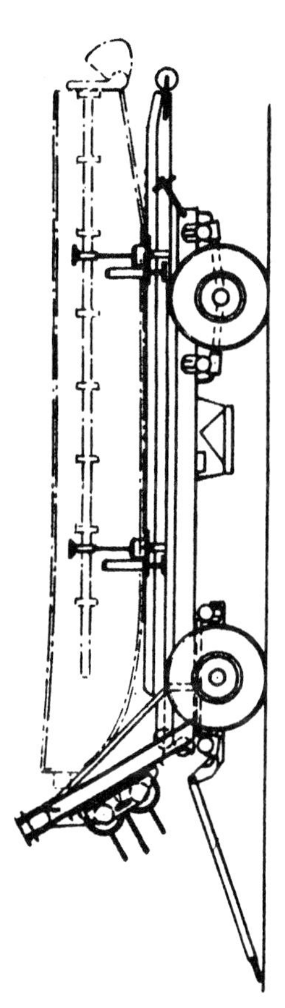

GENERAL:- Straight frame trailer fitted with movable platform which can be moved to rear by wire ropes and winch to form ramp for loading purposes.

PLATFORM:- Formed of 3 lines of rollers with 4 vertical rubber faced guide rollers.

SUSPENSION:- Torsion bar, independent springing.

STEERING:- By turntable.

WHEELS:- Detachable steel disc, divided type.

TYRES:- 10.50 - 13.

BRAKES:- Internal expanding, Lockheed hydraulic type on all wheels, operated by overrunning gear on drawbar. Separate hand parking lever.

DIMENSIONS

WHEELBASE:- 10' 5" (laden)
TRACK:- 6'6".

TOWING EYE TO F. AXLE:- 8' 3".

	OVERALL	PLATFORM
Length	25'0"	16'10"
Width	7'4½"	4'4"
Height	7'6"	

WEIGHTS:- Unladen 2 tons 10 cwt.
Laden 4 tons 10 cwt.

4-wheel FOLDING BOAT EQUIPMENT

(Made by Projectile & Engineering Co. Ltd.)

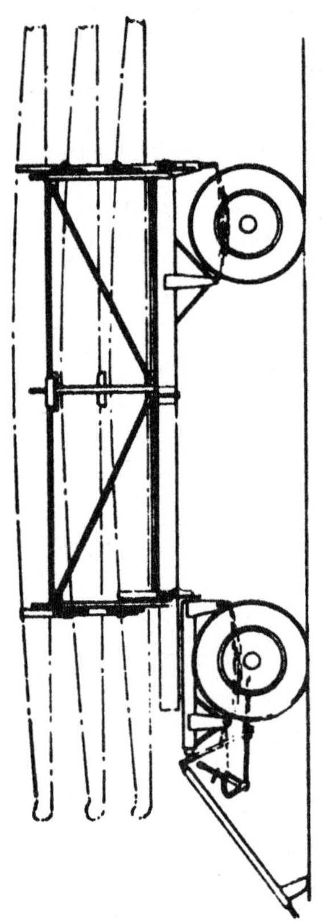

SUSPENSION:- Semi-elliptic.

BRAKES:- Internal expanding, on front wheels only. Automatically operated on overrun.

STEERING:- Turntable, unrestricted.

WHEELS:- Steel disc, 10 stud.

SPARE WHEEL:- Mounted over rear wheels.

TYRES:- 9.00-20.

SUPERSTRUCTURE:- 3 deck, tubular steel welded.

DIMENSIONS

WHEELBASE:- 12' 0".

TRACK:- 5' 7".

TOWING EYE TO F. AXLE:- 8' 0".

OVERALL

Length 22' 1".
Width 6' 5½".
Height 7' 8½".

OVERALL WITH LOAD

28'0".

OVERALL length Drawbar folded back 17' 10".

SHIPPING NOTES:- Tonnage Standing 22.19.

PART 3

SECTION 3

MOTOR CYCLES

MOTOR CYCLES, SOLO

ARIEL 350 C.C. O.H.V.

ENGINE:- Make, Ariel, Single Cylinder, O.H.V. Bore 72 m.m. Stroke 85 m.m. Capacity 346 c.c. Max. B.H.P. 12.8 @ 4500 r.p.m.

CLUTCH:- 3 plate.

GEARBOX:- 4 speed, Burman. Overall Ratios: 18.1 to 1, 12.0 to 1, 8.7 to 1, 5.75 to 1.

TRANSMISSION:- Primary, ½" x .305" x 80 links. Secondary, 5/8" x 3/8" x 92 links.

BRAKES:- Internal expanding with 6 1/2" drum front 7" drum rear.

TYRES:- 3.25 - 19.

WHEELBASE:- 4' 7".

GROUND CLEARANCE:- 5 1/2", sump guard.

TURNING CIRCLES:- Left L. 14' 5". Right L. 14' 10".

TANK CAPACITY:- Petrol, 2 5/8 gals. Oil ½ gal.

OVERALL DIMENSIONS

Length 7' 0". Width 2' 5½". Height 3' 6".

UNLADEN WEIGHT

With full tanks 376 lbs.

PERFORMANCE FIGURES

STANDING ¼ MILE:- 42.9 m.p.h.

AVERAGE SPEED UP 1 in 10, Standing Start:- 38.2 m.p.h.

PETROL CONSUMPTION:- 80 m.p.g. (Road Circuit).

RADIUS OF ACTION:- 210 miles.

B.S.A. M.20 500 c.c. SIDE VALVE

ENGINE:- Made B.S.A. Single cyl. Side Valve, Bore 82 m.m. Stroke 94 m.m. Capacity 496 c.c. Develops 12 B.H.P. @ 4100 r.p.m.

CLUTCH:- Multi plate, Ferodo rings.

GEARBOX:- 4 speed, foot operated. Overall ratios, 15.8 to 1, 10.9 to 1, 7.0 to 1, 5.3 to 1.

TRANSMISSION:- Primary, ½" x .305" x 69 links. Secondary, 5/8" x ¼" x 95 links.

BRAKES:- Front and rear, two shoe, fixed fulcrum with 7" drums.

TYRES:- 3.25 - 19.

WHEELBASE:- 4' 8".

GROUND CLEARANCE:- 4 5/8".

TURNING CIRCLES:- Left lock 13' 6". Right lock 14' 1".

TANK CAPACITY:- Petrol 3½ gals. Oil 5 pints.

OVERALL DIMENSIONS

Length 7' 2". Width 2' 5". Height 3' 3".

UNLADEN WEIGHT

With full tanks - 392 lbs.

PERFORMANCE

STANDING ¼ MILE:- 35.2 m.p.h.

AVERAGE SPEED UP 1 in 10 (Standing Start):- 31.8 m.p.h.

PETROL CONSUMPTION:- 50 m.p.g. (Road Circuit).

RADIUS OF ACTION:- 175 miles.

B.S.A. 350 c.c. O.H.V.
B.30 LIGHTWEIGHT

ENGINE:- Make, B.S.A. Single cylinder O.H.V. Bore 71 m.m. Stroke 88 m.m. Capacity 348 c.c. Max. B.H.P. 14.96 @ 4850 r.p.m.

CLUTCH:- Four plate, three with Ferodo inserts, one with cork insert.

GEARBOX:- 4 speed, foot operated. Overall Ratios: 17.1 to 1, 12.0 to 1, 7.85 to 1, 6.03 to 1.

TRANSMISSION:- Primary, ½" x .305" x 69 links. Secondary, 5/8" x ¼" x 95 links.

BRAKES:- Internal expanding, 7" drums. Hand operated front wheel. Foot operated rear wheel.

TYRES:- 3.25 - 19.

WHEELBASE:- 4' 5".

GROUND CLEARANCE:- 6", sump plate.

TURNING CIRCLES:- Left and Right Lock 13' 8¼".

TANK CAPACITY:- Petrol 2 1/4 gals. Oil 5 pints.

OVERALL DIMENSIONS

Length 6'7½". Width 2'4½". Height 3'0½".

UNLADEN WEIGHT

With full tanks - 339 lbs.

PERFORMANCE

STANDING ¼ MILE:- 43.6 m.p.h.

AVERAGE SPEED UP 1 in 10, Standing Start, 41.8 m.p.h.

PETROL CONSUMPTION:- 80 m.p.g. (Road Circuit).

RADIUS OF ACTION:- 180 miles.

ROYAL ENFIELD, 350 c.c. S.V.

ENGINE:- Make, Enfield, Single cylinder Side Valve Bore 70 m.m. Stroke 90 m.m. Capacity 346 c.c. Max. B.H.P. 9½ @ 4,500 r.p.m.

CLUTCH:- Four plate, Ferodo inserts.

GEARBOX:- 4 speed foot operated. Overall Ratios: 17.6 to 1, 10.7 to 1, 8.05 to 1, 5.96 to 1.

TRANSMISSION:- Primary ½" x .305" x 74 links. Secondary ½" x .305" x 102 links.

BRAKES:- Internal expanding, 6" drums. Hand operating front, foot rear wheel.

TYRES:- 3.25 - 19.

WHEELBASE:- 4' 5".

GROUND CLEARANCE:- 4¼".

TANK CAPACITY:- Petrol 3 gals. Oil 3 pints. (in engine sump).

OVERALL DIMENSIONS

Length 7' 1". Width 2' 4½". Height 3' 2".

UNLADEN WEIGHT

With full tanks - 353 lbs.

PERFORMANCE FIGURES

STANDING ¼ MILE:- 40.7 m.p.h.

AVERAGE SPEED UP 1 in 10 (Standing Start) 36.0 m.p.h.

PETROL CONSUMPTION:- 64.5 m.p.g. (Road Circuit).

RADIUS OF ACTION:- 190 miles.

ROYAL ENFIELD, 350 c.c. O.H.V.

ENGINE:- Make, Enfield, Single cylinder O.H.V. Bore 70 m.m. Stroke 90 m.m. Capacity 346 c.c. Max. B.H.P. 13.5 @ 5000 r.p.m.

CLUTCH:- Four plate, Ferodo inserts.

GEARBOX:- 4 speed. Overall Ratios: 18.6 to 1, 11.2 to 1, 7.9 to 1, 5.65 to 1.

TRANSMISSION:- Primary ½" x .305" x 75 links. Secondary 5/8" x 3/8" x 91 links.

BRAKES:- Internal expanding, 6" drum front, 7" drum rear.

TYRES:- 3.25 - 19.

WHEELBASE:- 4' 6".

GROUND CLEARANCE:- 4½", rear mudguard.

TANK CAPACITY:- Petrol 2¾ gals. Oil 6 pints. (in engine Sump.)

OVERALL DIMENSIONS

Length 7' 1". Width 2' 4". Height 3' 4".

UNLADEN WEIGHT

With full tanks - 367 lbs.

PERFORMANCE

STANDING ¼ MILE:- 41.8 m.p.h.

AVERAGE SPEED UP 1 in 10 (Standing Start) 36.3 m.p.h.

PETROL CONSUMPTION:- 72.72 m.p.g. (Road Circuit).

RADIUS OF ACTION:- 200 miles.

MATCHLESS G.3. 350 C.C. O.H.V.

ENGINE:- Make, Matchless, Single Cylinder, O.H.V. Bore 69 m.m. Stroke 93 m.m. Capacity 347 c.c. Max. B.H.P. 16 @ 5600 r.p.m.

CLUTCH:- Multi plate, Ferodo inserts.

GEARBOX:- 4 speed, foot operated. Overall Ratios:- 15.6 to 1, 10.3 to 1, 7.4 to 1, 5.8 to 1.

TRANSMISSION:- Primary, ½" x .305" x 66 links. Secondary, 5/8" x 3/8" x 91 links.

BRAKES:- Internal expanding, 7" drums.

TYRES:- 3.25 - 19.

WHEELBASE:- 4' 6".

GROUND CLEARANCE:- 4½".

TANK CAPACITY:- Petrol 3 gals. Oil ½ gal.

OVERALL DIMENSIONS

Length 6'10½". Width 2'6½". Height 3'6¾".

UNLADEN WEIGHT

With full tanks - 375 lbs.

PERFORMANCE FIGURES

STANDING ¼ MILE:- 38.3 m.p.h.

AVERAGE SPEED UP 1 in 10 (Standing Start):- 34.5 m.p.h.

PETROL CONSUMPTION:- 59.2 m.p.g. (Road Circuit).

RADIUS OF ACTION:- 175 miles.

MATCHLESS LIGHTWEIGHT G.3/L 350 C.C. O.H.V.

(with Teledraulic Front Forks)

ENGINE:- Make, Matchless, Single Cylinder, O.H.V. Bore 69 m.m. Stroke 93 m.m. Capacity 347 c.c. Max. B.H.P. 16 @ 5600 r.p.m.

CLUTCH:- Multi plate, Ferodo inserts.

GEARBOX:- 4 speed, foot operated. Overall Ratios: 15.6 to 1, 10.3 to 1, 7.5 to 1, 5.8 to 1.

TRANSMISSION:- Primary, ½" x .305" x 66 links. Secondary, 5/8" x 3/8" x 91 links.

BRAKES:- Internal expanding, 7" drums.

TYRES:- 3.25 - 19.

WHEELBASE:- 4' 3".

GROUND CLEARANCE:- 6¼".

TURNING CIRCLES:- Rightlock 15' 10". Left lock 15' 0".

TANK CAPACITY:- Petrol 3 gals. Oil ½ gal.

OVERALL DIMENSIONS

Length 6'10½". Width 2'6½". Height.

UNLADEN WEIGHT

With full tanks - 328 lbs.

PERFORMANCE FIGURES

STANDING ¼ MILE: 43.3 m.p.h.

AVERAGE SPEED UP 1 in 10 (Standing Start): 40.2 m.p.h.

PETROL CONSUMPTION:- 76.1 m.p.g. (Road Circuit).

RADIUS OF ACTION: 225 miles.

NORTON 16H 500 C.C. SIDE VALVE

ENGINE:- Make, Norton, Single cylinder. Side Valve, Bore 79 m.m. Stroke 100 m.m. Capacity 490 c.c. Max. B.H.P. 12 @ 4800 r.p.m.

CLUTCH:- Multi plate. Ferodo linings.

GEARBOX:- 4 speed, foot operated. Overall Ratios: 15.7 to 1, 9.35 to 1, 6.39 to 1, 5.28 to 1.

TRANSMISSION:- Primary, ½" x .305" x 74 links. Secondary, 5/8" x ¼" x 91 links.

BRAKES:- Internal expanding, 7" drums.

TYRES:- 3.25 - 19.

WHEELBASE:- 4' 6".

GROUND CLEARANCE:- 5½".

TURNING CIRCLES:- Left lock 14' 1", Right lock 14' 5½".

TANK CAPACITY:- Petrol 3½ gals. Oil 3 pints.

OVERALL DIMENSIONS

Length 7' 1½". Width 2' 6". Height 3'2¾".

UNLADEN WEIGHT

With full tanks - 388 lbs

PERFORMANCE FIGURES

STANDING ¼ MILE:- 37.2 m.p.h.

AVERAGE SPEED UP 1 in 10 (Standing Start): 31.6 m.p.h.

PETROL CONSUMPTION:- 50 m.p.g. (Road Circuit).

RADIUS OF ACTION:- 175 miles.

NORTON 633 c.c. WITH SIDECAR WHEEL DRIVE

ENGINE:- Make, Norton, Single cylinder, Side valve, Bore 82 m.m. Stroke 120 m.m. Capacity 633 c.c. Max. B.H.P. 14.5 @ 4000 r.p.m.

CLUTCH:- Multi plate. Ferodo inserts.

GEARBOX:- 4 speed, foot operated. Overall Ratios: 23.6 to 1, 15.3 to 1, 9.4 to 1, 6.4 to 1.

TRANSMISSION:- Primary, ½" x .305" x 74 links. Secondary, 5/8" x ¼" x 95 links.

SIDE CAR WHEEL DRIVE:- From a shaft incorporating a dog clutch and Hardy Spicer coupling, engaged by a lever which can be operated either by driver or passenger.

BRAKES:- Internal expanding, 7" drums.

TYRES:- 4.00 - 18. Spare wheel carried on back of sidecar.

WHEELBASE:- 4'6" Track 3'9½".

GROUND CLEARANCE:- 5¼".

TANK CAPACITY:- Petrol 3½ gals. Oil 4 pts.

SIDECAR

Hardwood body, folding screen. Clips fitted to body between machine and sidecar to carry a Bren gun. Ammunition box at back of sidecar. Pillion seat for a third passenger.

OVERALL DIMENSIONS

Length 7'2". Width 5'6½". Height 3'10½" (Cut down) 3'3¾".

WEIGHTS:- Unladen 679 lbs.
Laden 1232 lbs.

PERFORMANCE FIGURES (with sidecar wheel Drive and load of 3 men)

STANDING ¼ MILE, 30.6 m.p.h.

AVERAGE SPEED UP 1 in 10 (Standing Start) 20.9 m.p.h.

RADIUS OF ACTION .107 miles
PETROL CONSUMPTION: 30 m.p.g.

TRIUMPH 350 C.C. TYPE 3 S.W.

ENGINE:- Make, Triumph. Single Cyl. Side Valve. Bore 70 m.m. Stroke 83 m.m. Capacity 343 c.c. Max. B.H.P. 10 @ 4800 r.p.m.

CLUTCH:- Multi plate, cork inserts.

GEARBOX:- 4 speed, foot operated. Overall ratios - 18.7 to 1, 14.0 to 1, 8.84 to 1, 6.1 to 1.

TRANSMISSION:- Primary, ½" x .305" x 74 links. Secondary, 5/8" x .375" x 90 links.

BRAKES:- Internal expanding 7" drums. Hand operating on front wheel. Foot operating on rear wheel.

TYRES:- 3.25 - 19.

WHEELBASE:- 4' 4½".

GROUND CLEARANCE:- 4".

TANK CAPACITY:- Petrol 3 1/8 gals. Oil 6 pints.

OVERALL DIMENSIONS

Length 6' 9". Width 2' 6". Height 3' 4".

UNLADEN WEIGHT

With full tanks - 355 lbs.

VELOCETTE 350 C.C.
O.H.V.

ENGINE:- Make, Velocette, Single cylinder, O.H.V. Bore 68 m.m. Stroke 96 m.m. Capacity 349 c.c. Max. B.H.P. 14.9 @ 5,500.

CLUTCH:- Multi plate with Ferodo inserts.

GEARBOX:- 4 speed, foot operated. Overall Ratios: 16.7 to 1, 9.58 to 1, 7.28 to 1, 5.47 to 1.

TRANSMISSION:- Primary, ½" x .305" x 75 links secondary, 5/8" x 3/8" x 107 links.

BRAKES:- Internal expanding with 6" drums.

TYRES:- 3.25 - 19.

WHEELBASE:- 4' 4¼".

GROUND CLEARANCE:- 5".

TURNING CIRCLES:- Right and left lock 12' 6".

TANK CAPACITIES:- Petrol 2½ gals. Oil ½ gal.

OVERALL DIMENSIONS

Length 7' 1". Width 2' 3½". Height 3' 1½".

UNLADEN WEIGHT

With full tanks - 340 lbs.

PERFORMANCE FIGURES

STANDING ¼ MILE:- 41.7 m.p.h.

AVERAGE SPEED UP 1 in 10 (Standing Start): 38.8 m.p.h.

AVERAGE PETROL CONSUMPTION:- 65.6 m.p.g. (Road Circuit).

RADIUS OF ACTION:- 165 miles.

DATA BOOK OF

WHEELED VEHICLES

PART 3

SECTION 4

LIGHT RECONNAISSANCE CARS

BRITISH TYPES

Page No.

U.S.A. TYPES

GENERAL NOTES

BRITISH TYPES

HUMBER Mk. I (Formerly known as Humber Ironside)

A lightly armed, armoured light reconnaissance car carrying 2 Bren Guns and a crew of 3. Fitted with a No. 11 Wireless Set. The roof is open, a canvas cover being provided for weather protection.

Mounted on a rear wheel drive chassis generally similar to the Humber 8-cwt. on page 7, except that it is fitted with 9.25-16 Runflat tyres.

Production of this model has now ceased.

HUMBER Mk. II

An improved edition of the Mk. I, fitted with an armoured roof and a machine gun turret.

Armament consists of a Bren Gun, Boys A. T. Rifle and a smoke discharger. No. 11 Wireless Set is fitted to certain vehicles in place of the A. T. Rifle.

Chassis generally similar to Humber 8-cwt. on page 7, except that it is fitted with 9.25-16 Runflat tyres.

This model is being superseded by the Mk. III below.

HUMBER Mk. III

Generally as the Humber Mk. II, except that it is mounted on a four wheel drive chassis. Chassis generally similar to Humber F. W. D. 8-cwt. on page 9.

MORRIS Mk. I

Armament similar to the Humber Mk. III, but fitted with slightly heavier armour plate. Mounted on a special rear-engined, rear wheel drive chassis.

CAR, LIGHT RECONNAISSANCE
HUMBER MK. II

GENERAL DESIGN:- A lightly armed, armoured light reconnaissance car with a machine gun turret. The engine is mounted in the normal forward position and the drive is through the rear wheels only.

ARMOUR PLATE:- IT 100. Basis, 14 mm. front, 8 mm. sides.

Front,	10 mm.	Roof, 7 mm.
Sides,	7 mm. and 9 mm.	Turret, 6 mm.
Rear,	7 mm.	

TURRET:- 1 man, all-round traverse, manually operated.

ARMAMENT AND AMMUNITION:- Bren gun in turret on an anti-aircraft mounting which can also be used for ground fire. Ammunition, 1100 rounds in 30 round magazines.

Boys A. T. rifle (except when wireless set carried, see below), fired through Commander's shuttered slot in front plate. Smoke discharger mounted on front plate, 6 generators for discharger.

WIRELESS:- No. 11 Wireless Set and accessories carried on certain vehicles in place of the boys A. T. rifle.

CREW:- 3, Driver, Commander and Gunner. Driver and Commander seated side by side. Gynner behind in turret.

DOORS:- Two side entrance doors.

OBSERVATION:- Driver, through Triplex look-out 2 spare Triplex blocks carried. Look-out carried on a hinged B. P. panel which can be raised when improved vision is required. Side observation slot.

(Continued on facing spread)

CAR LIGHT RECONNAISSANCE
HUMBER Mk.II (continued)

This view of the Humber Mk.II shows the radiator shutters (controlled by the driver) in the open position.

COMMANDER:- through shuttered slot in front plate, also side observation slot. Commander's seat can be used in 2 positions. Low for firing A.T. rifle through shuttered slot in front. High for observation through hinged roof aperture. Slot in rear plate.

CHASSIS

Generally the same as Humber 8-cwt. chassis on page 7 of this Data Book, except:-

TYRES:- 9.25-16, Runflat.

GROUND CLEARANCE:- 10 3/8".

DIMENSIONS

Wheelbase 9' 6" Track f. 4' 11" 1/8"; r. 4' 11 5/8". Overall length 14' 4" overall width 6' 2". Overall Height 6' 11".

WEIGHTS

* Fully laden with crew and all equipment.

F.A.W. 1 ton 5¼ cwt.
R.A.W. 1 ton 14¾ cwt.
GROSS. 3 tons 0¼ cwt.

* Above weights with A.T. rifle. Approximately same if wireless set is substituted.

PERFORMANCE FIGURES

B.H.P. per Ton 28.3

TRACTIVE EFFORT per Ton (100%) 835 lbs/Ton.

M.P.G. (Estimated) 12 m.p.g. Radius of Action 190 miles.

CAR, LIGHT RECONNAISSANCE
HUMBER Mk. III

GENERAL DESIGN:- Generally similar to the Humber Mk. II, except:-

(1) Modifications to turret design to take Mk. II Bren gun.

(2) It is mounted on the Humber 8 cwt. Four Wheel Drive chassis.

ARMOUR PLATE:- IT 100. Basis, 14 mm. front, 8 mm. sides.

Front, 10 mm.
Sides, 7 mm. and 9 mm.
Rear, 7 mm.
Roof, 7 mm.
Turret, 6 mm.

TURRET:- 1 man, all-round traverse, manually operated.

ARMAMENT AND AMMUNITION:- Bren gun in turret on an anti-aircraft mounting which can also be used for ground fire Ammunition, 1120 rounds carried in 4 100-round drums, balance in 30-round magazines.

Boys A.T. rifle (except when wireless set carried, see below), fired through Commander's shuttered slot in front plate.

Smoke discharger mounted on front plate. 6 generators for discharger.

WIRELESS:- No. 11 Wireless set and accessories carried on certain vehicles in place of the Boys A.T. rifle.

CREW:- 3, Driver, Commander and Gunner.

Driver and Commander seated side by side. Gunner behind in turret.

DOORS:- Two side entrance doors.

OBSERVATION:- Driver, through Triplex look-out,

(Continued on facing page)

HUMBER MK. III (continued)

2 spare Triplex blocks carried. Look-out carried on a hinged B.P. panel which can be raised when improved vision is required. Side observation slot.

Commander, through shuttered slot in front plate, also side observation slot. Commander's seat can be used in 2 positions. Low for firing A.T. rifle through shuttered slot in front. High for observation through hinged roof aperture.

Slot in rear plate.

CHASSIS

In all main details similar to the Humber 8 cwt. F.W.D. chassis on page 9 of this Data Book.

DIMENSIONS

Wheelbase 9' 3¾". Track 5' 1".
Overall length 14' 4". Overall width 6' 2".
Overall height 7' 1".

WEIGHT

Fully laden with crew and all equipment.

F.A.W. 1 ton 8½ cwt.
R.A.W. 1 ton 16 cwt.
GROSS 3 tons 4½ cwt.

PERFORMANCE FIGURES

B.H.P. per Ton: 27.3.

TRACTIVE EFFORT per ton (100%) 1,135 lbs./Ton.

M.P.G. (Estimated) 11 m.p.g. Radius 175 miles.

CAR LIGHT RECONNAISSANCE
MORRIS Mk. I

GENERAL DESIGN:- A lightly armed, armoured light reconnaissance car mounted on a special rear-engined rear wheel drive chassis. Generally similar in armament to the Humber Mk. III, except it has the advantages that the A. T. rifle may also be fired to the rear, and that the smoke discharger, being mounted on the turret, can be fired in any direction.

ARMOUR PLATE:- IT 100. Basis, 14 mm. front, 8 mm. sides.

Front, 14 mm.	Rear, 8 mm.
Sides, 8 mm.	Roof, 8 mm.

TURRET:- 1 man, all-round traverse, manually operated.

ARMAMENT AND AMMUNITION:- Bren gun in turret on anti-aircraft mounting which can also be used for ground fire. Ammunition, 1,090 rounds carried in 4 100-round drums, remainder in 30 round magazines.

Boys A. T. rifle (except when Wireless Set carried, see below) fired front or rear through shuttered slits in commanders hatch.

Smoke Discharger, mounted on side of turret, with 6 smoke generators.

WIRELESS:- No. 11 Wireless Set and accessories carried on certain vehicles in place of the Boys A. T. rifle.

DOORS:- Two side entrance doors.

CREW:- 3, Driver, Commander and Gunner.

Driver centrally positioned well forward over the front axle.

(Continued on facing spread)

CAR LIGHT RECONNAISSANCE
MORRIS Mk. I
CHASSIS SPECIFICATION

ENGINE:- Mounted at rear. Make, Morris EK. Petrol 4 cyl. Bore 100 mm. Stroke 112 mm. Capacity 3.52 litres. R.A.C. rating 24.8. h.p. Max. B.H.P. 71 @ 3,100 r.p.m. governed to 2850 r.p.m. Max. torque 1812 lbs/ins @ 1600 r.p.m.

GEARBOX:- Drive taken forward to transfer box and then back to conventional rear axle. 4 speed gearbox.

Ratios, 5.98 to 1, 3.37 to 1, 1.83 to 1, 1.00 to 1. reverse, 7.97 to 1.

TRANSFER BOX:- Two speed. Ratios, High .77 to 1, Low 1.18 to 1.

REAR AXLE:- ¾ floating, spiral bevel. Ratio 6.58 to 1.

SUSPENSION:- Independent front wheel springing, semi-elliptic rear. Hydraulic shock absorbers.

BRAKES:- Foot. Hydraulic on all wheels. Hand, Mechanical on rear wheels.

TYRES:- 9.25-16, Runflat.

ELECTRICAL EQUIPMENT:- 12 volt.

TURNING CIRCLE:- 36 ft.

WHEELBASE:- 8' 2". TRACK, f. 5' 0 5/8", r. 5' 3 1/8".

GROUND CLEARANCE:- 7½" rear.

CAPACITIES:- Fuel, 14 gallons, Cooling 4 1/8 gals.

MORRIS Mk.I (continued)

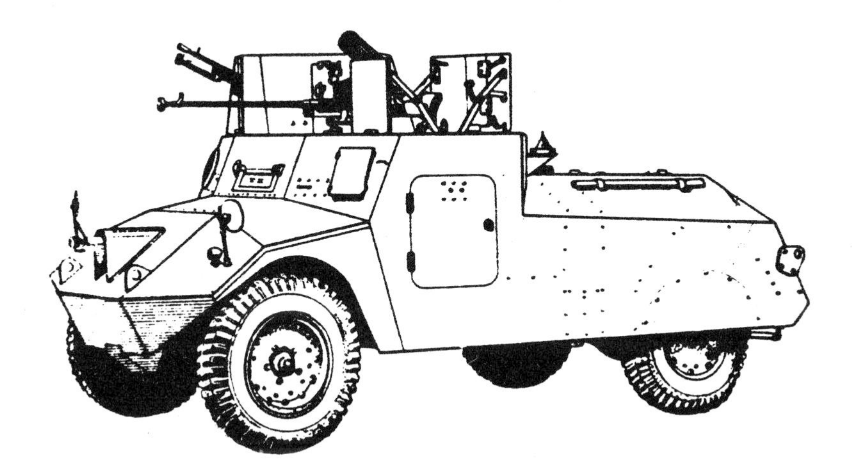

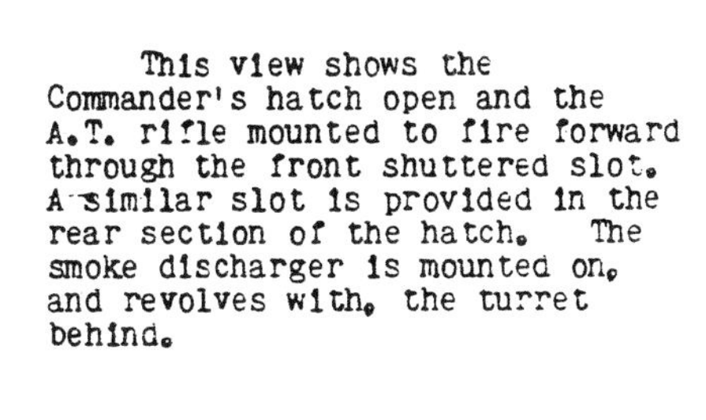

This view shows the Commander's hatch open and the A.T. rifle mounted to fire forward through the front shuttered slot. A similar slot is provided in the rear section of the hatch. The smoke discharger is mounted on, and revolves with, the turret behind.

Commander (nearside) and Gunner (offside) sit side by side.

OBSERVATION:- Driver, through Triplex look-out carried on hinged B.P. panel. Panel can be fully opened when full vision is required. Either side of the panel are two smaller observation slots which face slightly to the left and right respectively.

Commander, provided with a 2-piece observation hatch. Shuttered slots for firing A.T. rifle provided in both halves of this hatch, facing forwards and rearwards respectively.

OVERALL DIMENSIONS

Length 12' 10" width 6' 8"
Height 6' 0".

WEIGHT:- Fully laden with crew and all equipment.

F.A.W.	1 ton 12 cwt.
R.A.W.	2 tons 2 cwt.
GROSS	3 tons 14 cwt.

PERFORMANCE FIGURES

B.H.P. per Ton 19.1.
Tractive Effort per Ton (100%) 1464 lbs/Ton.
M.P.G. 10.5 Radius 145 miles.

GENERAL NOTES

U.S.A. TYPES

¼ TON LIGHT RECONNAISSANCE CARS (BLITZ BUGGIES)

Brief specifications and data are given in the following pages for two makes under this heading:- the American Bantam and the Ford.

Both makes are generally similar.

A similar type is being produced by Willys.

Details concerning this model will follow. An armoured type is also being developed on the Willys chassis. Details to follow.

TRUCK, 1 TON PERSONNEL

(White M.3.A.1 Armoured Scout Car)

Specifications and data are included on page 364 of the White M.3.A.1. Truck 1 ton Personnel (W.D. nomenclature), known in the U.S.A. as the White Armoured Scout car.

White M.3.A.1 Nomenclature

is now

"Truck 15-cwt. Personnel"

22/12/41

AMERICAN BANTAM MODEL BRC. ¼ TON LIGHT RECONNAISSANCE CAR (BLITZ BUGGY)

4 x 4 Wheelbase 6' 7"

ENGINE:- Make, Continental BY.4112. Petrol. 4 cylinder. Side Valve. Bore 3.3/16". Stroke 3½". Capacity 111.7" (1.83 litres) R.A.C. rating 16.25 h.p. Max. B.H.P. 45 @ 3500 r.p.m. Max. Torque 1032 lbs./ins. @ 1800 r.p.m.

GEARBOX:- Make, Warner. 3 speed forward and 1 reverse. Ratios. 2.93 to 1, 1.72 to 1, 1.0 to 1, rev. 3.9 to 1.

TRANSFER BOX:- Ratios, direct and 1.97 to 1. Front wheel drive can be disengaged.

AXLES:- Full floating. Hypoid gear. Ratio 4.88 to 1.

SUSPENSION:- Front and rear, semi-elliptic with Gabriel shock absorbers.

BRAKES:- Foot, Hydraulic on all wheels. Hand, Transmission brake mounted rear of transfer case.

WHEELS:- Kelsey 24562 Disc. 16 x 6.00. 5 stud Drop centre.

TYRES:- 5.50 - 16 or 6.00 - 16, 4 ply.

ELECTRICAL EQUIPMENT:- 6 volt.

TURNING CIRCLE:- 28'. TRACK, 3' 11½".

GROUND CLEARANCES:- 8½" f. & r. Belly 9½".

CAPACITIES:- Fuel 8 1/3 gals. Cooling 8 1/3 qts.

PERFORMANCE FIGURES:- (5-cwt. load).

B.H.P. per Ton:- 38.3.

TRACTIVE EFFORT per Ton (100%) 1,840 lbs./ton.

MAX SPEED:- 55 m.p.h.

AMERICAN BANTAM $\frac{1}{4}$ TON LIGHT RECONNAISSANCE CAR

This drawing shows the side curtains erected in position over the cut-away section in the body sides.

AMERICAN BANTAM BODY DETAILS

BODY: Two separate seats in front, which can be folded forward to give access to the rear seat. Divided type windscreen, each section hinged to open. Complete windscreen assembly can be folded forward to lie flat on the bonnet.

Weather protection afforded by detachable hood, supported at rear on a single vertical bow and attached at front to windscreen. Side curtains provided to fit over cut-away section of body sides. Safety straps fitted above cut-away section. Spare tyre carried at rear. Rear pintle hook.

OVERALL DIMENSIONS

Length 10' 6". Width 4' 6". Height 5' 11½" (Hood raised)

WEIGHTS

	Unladen	Laden
F.A.W.	8¾ cwt.	9½ cwt.
R.A.W.	9½ cwt.	14 cwt.
GROSS	18¼ cwt.	23½ cwt.

FORD ¼ TON LIGHT RECONNAISSANCE CAR
(BLITZ BUGGY)

4 x 4 Wheelbase 6' 8"

ENGINE:- Make, Ford G.P. Petrol, 4 cylinder, Side valve. Bore 3.3/16". Stroke 3¾". Capacity 119.5" (1.95 litres). R.A.C. Rating 16.28 h.p. Max. B.H.P. 45 @ 3600 r.p.m. Max. Torque 1020 lbs./ins. @ 1,700 r.p.m.

GEARBOX:- 3 speeds forward and reverse. Ratios 3.12 to 1, 1.85 to 1, 1.00 to 1, rev. 3.74 to 1.

TRANSFER BOX:- Ratios, Direct and 1.97 to 1. Front wheel drive can be disengaged.

AXLES:- Full floating, Hypoid Gear. Ratio 4.88 to 1.

SUSPENSION:- Front and rear, semi-elliptic with Houdaille hydraulic shock absorbers.

BRAKES:- Foot, Hydraulic on all wheels. Hand, Transmission brake mounted on transfer case.

WHEELS:- Kelsey 24562 Disc. 16 x 4. Drop centre.

TYRES:- 5.50 - 16, or 6.00 - 16, 4 ply.

ELECTRICAL EQUIPMENT:- 6 volt.

TURNING CIRCLE:- 37 ft. TRACK: f. & r. 3' 11½".

GROUND CLEARANCES:- 8" axles, 9½" belly.

CAPACITIES:- Fuel 8.1/3 gals. Cooling 8 1/3 qts.

FORD ¼ TON LIGHT RECONNAISSANCE CAR
(BLITZ BUGGY)

BODY DETAILS

BODY:- Two separate seats in front, which can be folded forward to give access to the rear seat. Divided type windscreen, each section hinged to open. Complete windscreen assembly can be folded forward to lie flat on the bonnet.

Weather protection afforded by detachable hood, supported at rear on a single vertical bow and attached at front to windscreen. Side curtains provided to fit over cutaway section of the body sides. Safety straps fitted above cut away section. Spare tyre carried at rear. Rear pintle hook.

OVERALL DIMENSIONS:

Length 10' 9". Width 5' 2". Height 5' 11".

LADEN WEIGHTS (5 cwt. Load)

F. A. W.	10¼ cwt.
R. A. W.	13 cwt.
TOTAL	1 ton 3¼ cwt.

PERFORMANCE FIGURES:- (5 cwt. load)

B.H.P. per ton: 38.8.

TRACTIVE EFFORT:- per ton (100%) 1948 lbs./ins.

MAX SPEED:- 55 m.p.h.

PETROL CONSUMPTION:- 20 m.p.g. road circuit towing 2-pdr. A/T gun. Radius of action, 165 miles.

This drawing shows the armoured folding shields raised in position above the side doors.

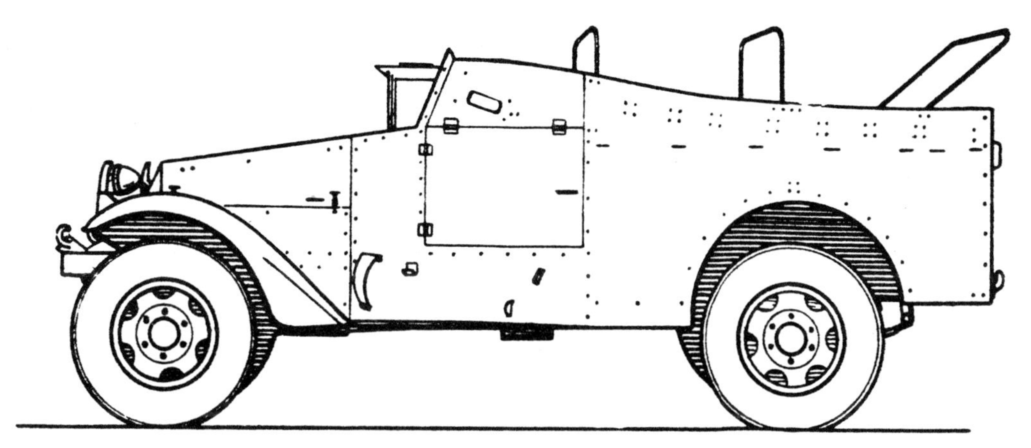

WHITE M.3.A.1. TRUCK 1 TON PERSONNEL BODY DETAILS

Armoured. Hull ¼ inch armoured plate with ½ inch front screen.

Divided type windscreen with hinged armour-plate flap which, when lowered, covers entire screen. Vision for driver and front seat passenger then restricted to 2 slots each measuring 8" x 3½". Two windscreen wipers.

Folding armoured shields are fitted above the 2 side doors. When raised these shields extend to the top of the windscreen, vision slots are provided in each shield.

Seating for driver and 7 men arranged in four pairs.

Pair 1, Driver and front passenger on separate seats.

Pair 2, Back to back, facing either side.

Pair 3, Side by side, facing rear.

Pair 4, Back to back, facing either side.

Weather protection afforded by canvas hood erected on three detachable bow irons.

Shutters fitted to radiator, which can be operated from inside the vehicle.

Two towing hooks at front, spring drawbar gear at rear.

OVERALL DIMENSIONS

Length 18' 3". Width 6' 5½". Height 6' 1".

WEIGHTS

	*UNLADEN	LADEN
F.A.W.	1 ton 19¼ cwt.	2 tons 1¼ cwt.
R.A.W.	2 tons 1¼ cwt.	2 tons 19¼ cwt.
GROSS	4 tons 0½ cwt.	5 tons 0½ cwt.

*Including 2 men and full fuel tanks.

WHITE M.3.A.1. TRUCK 1 ton PERSONNEL (ARMOURED SCOUT CAR)

4 x 4   Normal Control.   Wheelbase 11' 0"

ENGINE:- Make, Hercules JXD. Petrol. 6 cylinder Bore 4". Stroke 4¼". Capacity 5.24 litres. R.A.C. rating 38.4 h.p. Max. B.H.P. 110 @ 3000 r.p.m. Max. Torque 2892 lbs./ins. @ 1100 r.p.m.

GEARBOX:- 4 speed and 1 reverse. Ratios. 5.00 to 1, 3.07 to 1, 1.71 to 1, 1.00 to 1. Reverse 5.83 to 1.

TRANSFER CASE:- 2-speed. Ratios, direct and 1.87 to 1. No provision for disconnecting front wheel drive.

AXLES:- Full floating, spiral bevel, Ratio. 5.14 to 1.

SUSPENSION:- Front and rear, semi-elliptic with hydraulic shock absorbers.

BRAKES:- Foot, Hydraulic vacuum servo assisted. Hand, propeller shaft brake.

TYRES:- Early models 8.25 - 20 U.S.A. bullet proof tyres. Later models 9.00 - 20 Runflat.

ELECTRICAL EQUIPMENT:- 12 volt.

*TURNING CIRCLE:- 60 ft. TRACK: 5' 6".

*GROUND CLEARANCES:- Axles 10". Belly 15".

CAPACITIES:- Fuel 25 gals. Cooling 4 gals.

PERFORMANCE FIGURES:-

B.H.P. per ton 21.3

*Tractive effort per ton (100%) 1770 lbs./Ton.

Petrol Consumption 7.0 m.p.g.
Radius of action 175 miles.

* With 8.25 - 20 U.S.A. Tyres.

CAR 5-CWT. 4 x 4

("Jeep")

WILLYS MODEL M.B. & FORD MODEL G.P.W.

These two models were supplied against later orders. They are built to the same design and it is understood that all parts are interchangeable between Willys M.B. and Ford G.P.W.

The design is generally similar to that of the American Bantam (illustrated on page 361) and the Ford G.P. model shown on pages 362 and 363.

The main differences are:

A larger engine, 60 B.H.P. for the Willys and Ford G.P.W. as against 45 B.H.P. for the earlier Ford G.P. model.

A different gearbox with higher 1st speed, 2.66 to 1 as against 3.12 to 1.

Larger petrol tank - 12 Imp. Gallons.

Tyre equipment 6.00 - 16, this is the oversize of the 5.50 - 16 supplied on Bantam and Ford G.P. models.

WILLYS M.B. CAR 5-CWT. 4 x 4

(U.S. Nomenclature, "¼ ton 4 x 4" also known as "Peep", "Jeep" or "Blitz Buggy".)

FORD G.P.W. CAR 5-CWT. 4 x 4

(Built to similar design as Willys M.B., with interchangeable parts. This Ford G.P.W. differs from Ford G.P. Model shown on page 362.)

Wheelbase: 80" (6' 8")

ENGINE:- Make, Willys. Petrol 4 cylinder Bore 3 1/8" Stroke 4 3/8". Capacity 134.2 cu. ins. (2.19 litres). B.H.P. 60 @ 3600 v.p.m. Max. torque 1260 lbs. ins. @ 2000 v.p.m.

GEARBOX:- Make, Warner T.84J. 3 speeds and 1 reverse. Ratios, 2.66 to 1, 1.56 to 1, 1.00 to 1, r. 3.55 to 1.

TRANSFER CASE:- Make, Spicer 18 2-speed. Ratios, Direct and 1.97 to 1. F. Axle drive can be disengaged.

AXLES:- Full floating. Hypoid gear. Ratio 4.88 to 1.

SUSPENSION:- Front and rear, semi-elliptic with hydraulic shock absorbers.

BRAKES:- Foot, Hydraulic on all wheels. Hand, transmission brake.

STEERING:- Left Hand Drive. Ross cam and lever.

WHEELS:- Kelsey 16 x 4.00 drop centre. (16 x 4.50 for combat wheels, divided type.)

TYRES:- 16 - 6.00.

ELECTRICAL EQUIPMENT:- 6 volt.

TURNING CIRCLE:- 38 ft. Track 4' 0¼".

GROUND CLEARANCES:- F. Axle 8½". Transfer case 9½".

CAPACITIES:- Fuel 12½ galls. Cooling 2¼ galls. Performance Figures: See facing page.

WILLYS M.B. AND FORD G.P.W. CAR 5-CWT. 4 x 4

BODY:- General design similar to American Bantam illustrated on page 361.

Two separate seats in front which can be folded forward to give access to the rear seat. Divided type windscreen, each section hinged to open. Complete windscreen assembly can be folded forward to lie flat on bonnet. Weather protection by folding hood, supported at rear on a single folding bow, and attached at front to windscreen. Side curtains to fit over cutaway sections of the body sides. Safety straps fitted above cutaway section. Spare wheel carried at rear. Rear pintle hook.

DIMENSIONS

Overall Length	11' 0"
Overall Width	5' 4"
Overall Width, handling grips on outer side of body detached	5' 0"
Overall Height, hood up	6' 0"
" " hood folded down	4' 4"

WEIGHTS

	UNLADEN	LADEN (800 lbs.)
F.A.	10¾ cwt.	12 cwt.
R.A.	9¼ cwt.	15½ cwt.
Gross	1 ton 0 cwt.	1 ton 7½ cwt.

Stripped shipping Ht.: 4' 4" (Top of steering wheel)

PERFORMANCE FIGURES

B.H.P. per ton: 44.4
Tractive effort per ton (100%) 1795 lbs./ton.
Max. Speed 56 m.p.h.
M.P.G. (Road) 20.
Radius of Action 250 miles.

WHEELED VEHICLES

PART 4

U.S. ARMY TYPES

EXCLUDING U.S.A. CAR 5-CWT. 4 x 4 (JEEP, AND WHITE SCOUT CAR WHICH ARE INCLUDED WITH LIGHT RECONNAIS-SANCE CARS.

PAGES 359 - 368

U.S. ARMY TYPES
SPECIAL NOTES

1. This section includes mainly the multi-axle drive and other types specially developed for the U.S. Army. The Car, 5-cwt. 4 x 4 (Jeep, Peep or Blitz Buggy) and the Truck, 15-cwt. 4 x 4 Personnel (White Scout Car) are included in the Light Reconnaissance Car Section, pages 359 - 368.

2. U.S. Commercial and Modified Commercial types (some of which may be used by the U.S. Army) are included in Part 2, Pages 201 onwards.

3. WEIGHTS

All U.S. Army Types are rated in Short Tons (2,000 lbs. to a ton). All weights throughout this Data Book are given in Long Tons (2,240 lbs. to a Ton).

To preserve continuity, all Unladen and Laden Weights in this section are given in Long tons, performance Figures are also calculated for Long Tons.

But the actual rating of the vehicle at the top of each page is still the U.S. Army Short Ton rating.

4. LIQUID MEASURE.

U.S. GALLON = .833 of an Imperial Gallon.

All figures are translated to Imperial Gallons throughout this book.

For list of American Terms and British equivalents see pages 205 and 205A.

CHEVROLET 5 PASSENGER SEDAN

CAR, 4 SEATER 4 x 2

Model B.G. Makers Designation 1941/1503
Wheelbase 9' 8"

ENGINE:- Make, Chevrolet A.A. 6-cyl. O.H.V. Petrol. Bore 3½". Stroke 3¾", Capacity 216.5 C.ins., 3.54 litres. Max. B.H.P. 83 @ 3200. Max. Torque 2088 lbs. ins.

CLUTCH:- Single dry plate (Chev. 1941 Pass.)

GEARBOX:- 3 speed. Synchro Mesh. Vacuum assisted gear shift mechanism. Overall ratios, 12.1 to 1, 6.9 to 1, 4.11 to 1, r. 12.1 to 1.

FRONT AXLE:- Chev. 1941 Pass. Knee Action.

REAR AXLE:- Semi floating. Hypoid gear. Ratio 4.11 to 1.

SUSPENSION:- Front, Independent front wheel springing. Rear, Semi-elliptic. Hydraulic shock absorbers, f. and r.

BRAKES:- Foot, Hydraulic on all wheels. Hand, mechanical on rear wheels.

STEERING:- Left hand drive. Worm and sector.

WHEELS & TYRES:- Spoke disc. 16 x 4.00. Tyres 6.00 - 16.

ELECTRICAL EQUIPMENT:- 6 volt.

GROUND CLEARANCE:- Axles, f. 9¼", r. 8½".

CAPACITIES:- Fuel 13 galls. Cooling 3 galls.

BODY:- Conventional 4-door all-steel saloon - no sliding roof - luggage boot at rear.

OVERALL DIMENSIONS:- Length 16' 4". Width 6' 0¾". Ht. 5' 8".

WEIGHTS:- Unladen 1 ton 9 cwt. Laden 1 ton 16½ cwt.

GENERAL NOTES
DODGE ½-TON 4 x 4 T.215 AND DODGE 3/4-TON 4 x 4 T.214

The Dodge ½-Ton 4 x 4 is now an obsolescent type; it has been replaced by the improved 3/4-Ton Model T.214.

1/2-Ton 4 x 4 models have been supplied against early S.M. Nos. It is understood that 3/4-ton models will be supplied against later S.M. Nos.

Abridged specifications for both types are given on the following two pages. The main improvements in the design of the 3/4-ton model are:-

1. Lower Silhouette.
2. Shorter Wheelbases (as shown below)
3. Larger tyres (9.00-16 runflat)
4. Wider bodies.

Dodge Chassis types are denoted by the T. No. series. Body types are indicated by the Sales Symbol W.C. series as follows:-

(Note wheelbases are shown in brackets).

BODY TYPE	1/2-TON T.215	¾-TON T.214
Weapons Carrier	-	W.C.51 (98")
" " with winch	-	W.C.52 "
Command Reconnaissance.	W.C.23(116")	W.C.56 "
" " with winch	-	W.C.57 "
Radio	-	W.C.58 "
Ambulance	W.C.27(123")	W.C.54(121")
Carryall	W.C.26(116")	W.C.53(114")
Radio Panel Delivery	W.C.42(116")	-
Bucket Seat Pick Up	W.C.21(116")	-

DODGE T.215 ½-TON 4 x 4

Normal Control. Wheelbase 9' 8"
(Ambulance Chassis 10'3")

ENGINE:- Make, Dodge, 6-cyl. L. Head. Petrol. Bore 3¼". Stroke 4 5/8". Capacity 230 c.ins. 3.76 litres. Max. B.H.P. 92 @ governed speed of 3200 r.p.m. Max. torque 2160 lbs./ins. at 1200.

CLUTCH:- Single dry plate. 10" dia.

GEARBOX:- 4 speed and reverse. Overall ratios: 31.3 to 1, 15.1 to 1, 8.26 to 1, 4.89 to 1.

TRANSFER CASE:- Single speed. Radio direct. Front axle drive can be disengaged.

FRONT AXLE:- Full Floating. Hypoid. R.4.89.

REAR AXLE:- Full Floating. Hypoid. R.4.89.

SUSPENSION:- Front and rear, semi-elliptic with hydraulic shock absorbers.

BRAKES:- Foot, Hydraulic on all wheels. Hand, mechanical on transmission.

STEERING:- Left Hand Drive. Worm and sector.

WHEELS AND TYRES:- 16 x 5.50. Singles. Tyres 7.50-16.

TURNING CIRCLE:- 51 ft.

ELECTRICAL EQUIPMENT:- 6 volt.

CAPACITIES:- Fuel 21 galls. Cooling 3½ galls.

PERFORMANCE FIGURES (Pick Up)
B.H.P. per Ton 35.4. Tractive Effort per Ton (100%) 1810 lbs./Ton. Maximum Speed 55 m.p.h. M.P.G. 12. Radius 250 miles.

DODGE T.214 3/4-TON 4 x 4

Normal Control. Wheelbase 8' 2" except
9' 6" Carryall Chassis
10' 1" Ambulance "

Specification generally similar to Dodge T.215 on facing page, except:-

OVERALL RATIOS:- 37.3 to 1, 18.0 to 1, 9.85 to 1, 5.83 to 1, r. 45.6 to 1.

TRANSFER CASE:- Same internal parts, different housing.

AXLES:- Ratio 5.83 to 1. Larger constant velocity joints.

SUSPENSION:- Heavier springs.

WHEELS AND TYRES:- 16 x 6.50 split type, 9.00-16.

TURNING CIRCLES:- 43' 4" (8' 2" w.b.); 49' 0" (9' 6" w.b.) 52' 6" (10' 1" w.b.)

TRACK:- 5' 4½" front and rear.

ELECTRICAL EQUIPMENT:- 12 volt on Command and Carryall types 6 volt others.

CAPACITY:- Fuel 25 galls.

PERFORMANCE FIGURES:- (Weapons Carrier)
B.H.P. per ton 29.5. Tractive Effort per ton (100%) 1,570 lbs./Ton. Maximum Speed 55 m.p.h. M.P.G. 11. Radius 275 miles.

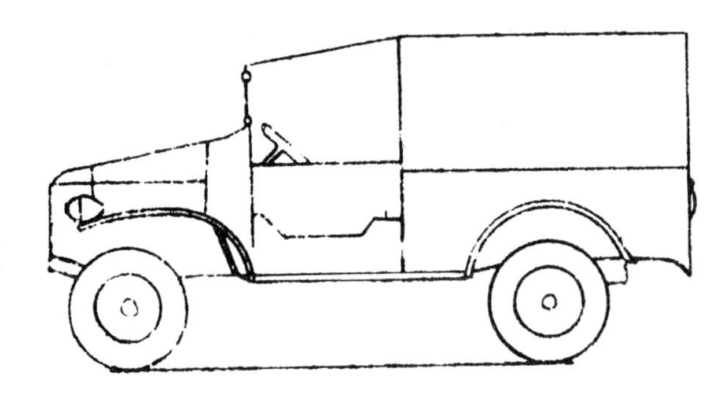

S.M.2225

DODGE 1/2-TON 4 x 4 WC. 21
BUCKET SEAT PICKUP

Fixed sided all-steel body with a drop tail board. Troop seats fitted in body.

Open canopy type cab. Windscreen can be folded flat onto bonnet. Rear pintle hook. Spare wheel carried in recess on side.

DIMENSIONS

	OVERALL	INSIDE BODY
Length	15' 1"	6' 6"
Width	6' 2"	4' 0¼"
Height	7' 4"	4' 2"

WEIGHTS

	UNLADEN	LADEN
F.A.W.		1 Ton 2 Cwt.
R.A.W.		1 Ton 10 Cwt.
GROSS	2 Ton 0¼ Cwt.	2 Ton 12 Cwt.

Winch shown above is not fitted on S.M.2224.

S.M.2224.

DODGE 1/2-TON 4 x 4 WC. 23
COMMAND RECONNAISSANCE

(NOTE: 3/4-Ton 4 x 4 Command Car W.C.56 has a similar body)

Open 4 seater all steel body with folding hood and detachable side curtains. Windscreen can be folded flat onto bonnet. Map case and chart body fitted behind front seat. Trunk compartment at rear with tailboard hung on chains. Rear pintle hook. Spare wheel carried in recess on side. Drawing shows winch, not fitted on S.M.2224.

OVERALL DIMENSIONS

(½-Ton) Length 14' 11½" Width 6'2"
Height 6' 11 3/8"
(¾-Ton) Length 13' 8½" Width 6' 5¾"
Height 6' 9½"

WEIGHTS (½ Ton)

	UNLADEN	LADEN
F.A.W.		1 ton 2 cwt.
R.A.W.		1 ton 11 cwt.
GROSS	2 ton 1¾ cwt.	2 ton 13 cwt.

Laden wt. 3/4 Tons, F.A. 1 Ton 5 cwt.
R.A. 1 Ton 16 cwt.

S.M.2129

DODGE ½-TON 4 x 4 AMBULANCE WC. 27

Designed to carry 4 stretcher or 7 sitting patients together with driver and attendant. Doors either side at front, double doors at rear. Two longitudinal seats in rear compartment for sitting patients which can be folded up when stretchers (U.S. Army type) are carried. Ventilator and 2 lights in roof. Hot water heater. Rear pintle hook.

DIMENSIONS:- Inside body, Length 8' 2", Width 5' 2½". Overall Length, 16' 3"; Width 6' 2", height 7' 10".

WEIGHTS:- Laden, F.A. 1 ton 2 cwt. R.A. 1 ton 17 cwt.

DODGE ½-TON 4 x 4 CARRYALL

S.M.2222, 3.

Similar to British Type Heavy Utility body except only 2 side doors (instead of 4) and fixed roof (no sliding section).

Seating for 8; 3 on the front seat, 2 on centre seat and 3 on rear seat. Bottom half of rear panel lowers to form tailboard. Spare tyre carried in recess on side panel. No rear pintle hook.

OVERALL DIMENSIONS:- Length 15' 5", Width 6' 2", Height 7' 0".

LADEN WEIGHTS:- F.A. 1 ton 2¼ cwt., R.A. 1 ton 12 cwt.

Load approx. 1,300 lbs.

CHEVROLET 1½-TON 4 x 4

Normal Control. Wheelbase 12' 1".

ENGINE:- Make Chevrolet A.Y. 6 cyl. O.H.V. Petrol. Bore 3 9/16". Stroke 3 15/16". Capacity 235.5 c.ins., 3.8 litres. Max. B.H.P. 83.5 @ 3,000 v.p.m. Max. Torque 2034 lbs. ins.

CLUTCH:- Single dry plate. 10¾" dia.

GEARBOX:- 4 speed. Overall Ratios (High) 47.1 to 1, 23.3 to 1, 11.4 to 1, 6.67 to 1, r. 46.5 to 1.

TRANSFER CASE:- 2 speed. Ratios, 1.00 to 1 and 1.94 to 1. Front axle drive can be disengaged.

FRONT AXLE:- Full floating. Hypoid 6.67 to 1.

REAR AXLE:- " " " " "

SUSPENSION:- Semi-elliptic f. and r. with auxiliary rear. Hydraulic Shock absorbers all round.

BRAKES:- Hydraulic vacuum servo assisted on all wheels. Hand transmission brake.

STEERING. Left Hand Drive. Recirculating ball and sector.

WHEELS & TYRES:- 20 x 7 wheels. Tyres 7.50 - 20, 8 ply dual rear.

ELECTRICAL EQUIPMENT:- 6 volt.

CAPACITIES:- Fuel 25 galls. Cooling 3½ galls.

S.M. 2227. G.S.
S.M. 2229. Lipper.

CHEVROLET 1½-TON 4 x 4 G.S. BODY

Standard U.S. Army type Cargo (G.S.) body with detachable canvas tilt and bows. Folding troops seats fitted to sides of body. 2 towing hooks at front. Rear pintle hook.

WINCH:- Garwood 2U512 mounted at front. Capacity 10,000 lbs., 300 ft. of cable.

DIMENSIONS

	OVERALL	INSIDE BODY
Length	19' 2½"	9' 0"
Width	7' 2"	6' 0"
Height	8' 9"	

Shipping Height: 7' 3½" top cab. Cab does not "split".

WEIGHTS

	UNLADEN	LADEN (Load 3,000 lbs.)
F.A.W.	-	1 ton 18 cwt.
R.A.W.	-	3 tons 4 cwt.
GROSS	3 tons 15¼ cwt.	5 tons 2 cwt.

CHEVROLET 1½-TON 4 x 4 END TIPPER

All steel body with folding troop seats, detachable bows and tilt. Forward mounted winch. Hydraulic hoist. DIMENSIONS:- As G.S. body above. WEIGHTS:- Approx. 2 cwt. more front, 4 cwt. more rear than G.S.

STUDEBAKER CHASSIS TYPES

THE BASIC CHASSIS is the standard U.S. Army Studebaker model U.S.6. which is a

1. 2½-TON 6 x 6

Available in two wheelbases, 148" or 162".

Both types are available either with or without a winch mounted at the front.

This basic chassis is also available rated as a

2. 5-TON 6 x 4

This is the same as the 2½ ton 6 x 6 except for the fitting of a conventional non-driving front axle and the omission of the low gear in the transfer case. But it is rated as a 5 tonner for road work only, whereas the 2½ ton 6 x 6 is a cross country rating.

3. 7-TON SEMI-TRAILER

The prime mover used for the semi-trailer is the Studebaker 2½ ton 6 x 4 chassis, short 148" wheelbase. Except for fitment of turn-table, semi-trailer brake connections, non-driving front axle and omission of low transfer gear, this is basically the same vehicle as the standard 2½ ton 6 x 6 148" wheelbase chassis.

STUDEBAKER CHASSIS TYPES ARE COVERED IN THE FOLLOWING PAGES:-

STUDEBAKER CHASSIS AND BODY TYPES AND SUPPLY MECH. NOS.

2½-TON 6 x 6 148" w.b.

Body type: 9' Cargo (G.S.) with winch on S.M.2154, 2216, 2381.

2½-TON 6 x 6 162" w.b.

Body type: 12' Cargo (G.S.) without winch on S.M.2216, 2230, 2422, 2479.

Body type: 12' Cargo (G.S.) with winch on S.M.2216, 2231.

Body type: 750 gall. Tanker (625 Imp. galls.) on S.M.2218.

Body type: Stores on S.M.2296.

Body type: Workshop on S.M.2440.

5 -TON 6 x 4 162" w.b.

Body type: 12' Cargo (G.S.) without winch. On S.M.2217, 2230, 2310, 2314, 2514.

7-TON ARTICULATED SEMI TRAILER

Body type: Platform. On S.M.2304, 2306.

STUDEBAKER U.S.6 2½-TON 6 x 6

Normal control. Wheelbases 148" & 162".
Bogie centres 44".

ENGINE:- Make, Hercules JXD 6-cyl. Petrol. Bore 4". Stroke 4¼". Capacity 320 c.ins. 5.34 litres. Max. B.H.P. 86 @ governed speed of 2600 r.p.m. Max. torque 3024 lbs. ins.

CLUTCH:- Single dry plate. 12" dia.

GEARBOX:- 5 speed and 1 reverse. Ratios:- 6.06 to 1, 3.5 to 1, 1.8 to 1, 1.00 to 1, overdrive 0.8 to 1, R. 6.0 to 1.

TRANSFER BOX:- 2 speed. Ratios 1.15 to 1 and 2.6 to 1. Front axle drive automatically engaged in low ratio. Can be engaged in high ratio.

FRONT AXLE:- Spiral bevel. Ratio 6.6 to 1.

REAR AXLES:- Spiral bevel. Ratio 6.6 to 1.

SUSPENSION:- Front, semi-elliptic with hydraulic shock absorbers. Rear bogie, inverted semi-elliptic, radius rods between axles.

BRAKES:- Foot: hydraulic vacuum servo assisted on all wheels. Hand: transmission.

STEERING:- Left hand drive. Cam and lever.

WHEELS & TYRES:- Disc 20 x 7, dual rear. Tyres 7.50-20 all round.

ELECTRICAL EQUIPMENT:- 6 volt.

CAPACITIES:- Fuel 32 galls. Cooling 4.3 galls.

PERFORMANCE FIGURES:- B.H.P. per ton 12.6. Tractive effort per ton (100%) 2635 lbs./Ton. M.P.G. 8. Radius 250 miles.

STUDEBAKER U.S. 6¼-5 TON 6 x 4

(Also sometimes shown as Studebaker 2½ - 5 ton)

Normal control. Wheelbase 162".
Bogie centres 44"

Exactly same specifications as 2½ ton 6 x 6 model shown on facing page except:-

TRANSFER BOX:- Single speed. Ratio 1.15 to 1.

FRONT AXLE:- Make, Clark. Reverse Elliott type non-driving.

PERFORMANCE FIGURES:- B.H.P. per ton:- 9.9 Tractive effort (100%) 901 lb./ton. M.P.G. 7. Radius 220 miles.

NOTE:- U.S. Army Specification stipulates that this 6 x 4 model is to be capable of conversion to the standard 6 x 6, by the substitution of driving type front axle and propeller shaft and extra low gears for transfer case.

BODY:- Exactly same as 12' cargo body on 6 x 6 model as detailed on page 417.

WEIGHTS

UNLADEN	LADEN (10,000 lbs.)	
	F.A.	1 ton 19½ cwt.
	R.A.	6 tons 15¾ cwt.
4 tons 6 cwt.	GROSS	8 tons 15¼ cwt.

Add 5¾ cwt. to F.A. weight when winch is fitted.

STUDEBAKER 2½-TON 6 x 6
9' CARGO BODY (G.S.)

On short 148" w.b. chassis.
U.S. Army Field Artillery Tractor

BODY:- All steel cargo (G.S.) body with detachable bows and tilt. Troop seats full length either side which can be folded back flat into body sides. Two towing hooks at front. Rear pintle. Two spare wheels carried behind cab.

CAB:- All enclosed steel panel. Seats 2.

WINCH:- (when fitted - see page 413). Make Heil mounted at front of chassis. Capacity 10,000 lbs. 300 ft. cable.

DIMENSIONS

	OVERALL	INSIDE BODY
Length	19'3"	9' 0"
Length with winch	20'6"	- -
Width	7'4"	6' 6"
Height	8'10"	5' 2"

STRIPPED SHIPPING HEIGHT:- Tilt detached 7'3".

WEIGHTS (less winch)*
(Laden wt. same as long chasses)

UNLADEN	LADEN (5,000 lbs.)
F.A.W.	1 ton 18½ cwt.
R.A.W.	4 tons 19¾ cwt.
Gross	6 tons 18¼ cwt.

*Add approx. 5½ cwt. to front axle wt. on models fitted with winch.

STUDEBAKER 2½ TON 6 x 6
12' CARGO BODY

On long 162" w.b. chassis

BODY:- Generally similar to 9' body illustrated on facing page except:-
Length 12' against 9'.
Body extends forward to back of cab.
One spare wheel only fitted; carried under front right hand corner of body
Petrol tank located under front left hand corner of body.
Body, like short chassis, is fitted with detachable bows and tilt, folding troop seats, front towing hooks, rear pintle.

CAB:- All enclosed steel panel. Seats 2.

WINCH:- (when fitted, see page 413). Make Heil, mounted at front of chassis. Capacity 10,000 lbs. 300 ft. cable.

DIMENSIONS

	OVERALL	INSIDE BODY
Length	20' 10½"	12' 0"
Length with winch	22' 1¼"	
Width	7' 4"	7' 0"
Height	8' 10"	5' 2"

STRIPPED SHIPPING HEIGHT:- Tilt detached. 7' 3".

WEIGHTS (Less winch) *

	UNLADEN	LADEN (5,000 lbs.)
F.A.W.		1 ton 18½ cwt.
R.A.W.		4 tons 19¾ cwt.
GROSS	4 tons 13½ cwt.	6 tons 18¼ cwt.

* Add approx. 5½ cwt. to front axle wt. on models fitted with winch.

7-ton SEMI - TRAILER

(To couple with Studebaker 6 x 4 148" w.b. Motive Units)

MADE:- HIGHWAY TRAILER COMPANY. (Similar types built to the same specification by the following makers:- Whitehead & Kales, W.C. Nabors, Kentucky Mfg. Co., Carter Mfg. Co., Edwards Iron Works. Specification stipulates interchangeability of parts and units with other models).

SEMI-TRAILER DETAILS

WHEELBASE:- 12' 1" (King Pin to Axle semi-trailer).

TRACK:- 5' 7".

SPRINGS:- Semi-elliptic with auxiliary.

BRAKES:- Bendix duo-servo internal expanding operated by vacuum servo from motive unit. Vacuum tank on semi-trailer. If semi-trailer accidentally disconnected from motive unit, vacuum brakes are automatically applied. No hand brake is fitted, but chock blocks supplied to hold trailer when disconnected.

TURNTABLE (FIFTH WHEEL):- Semi automatic type; coupled automatically, uncoupled manually. Type, Dayton Universal 33" diameter.

RETRACTABLE UNDER CARRIAGE:- (Landing Gear) Folding type legs, crank operated. In addition semi trailer is fitted with auxiliary support legs at forward end to prevent nose-diving when uncoupled with an uneven load.

TYRES:- Semi Trailer, 7.50 - 20, 10 ply, dual with spare carried forward end of body. Motive Unit; 7.50 - 20, 8 ply, dual rear, spare carried behind cab.

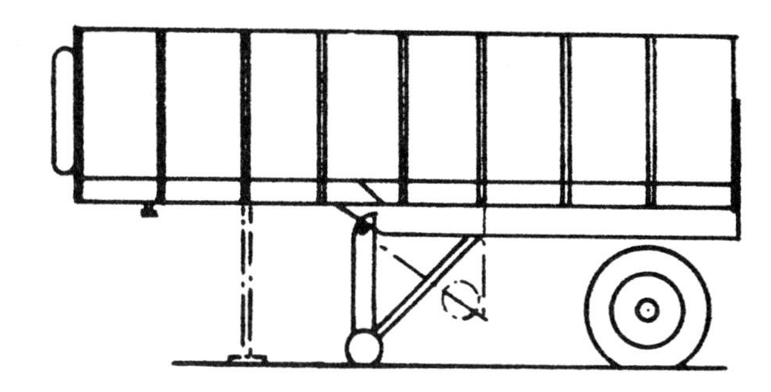

BODY DETAILS
7-ton SEMI - TRAILER

to couple with Studebaker 6 x 4 148" w.b. Motive Unit

BODY:- Steel sides 42" high, flat floor. Tarpaulin cover. Drop tail board 2' high.

DIMENSIONS:- (Semi-Trailer).

	OVERALL	INSIDE BODY
Length	17' 2"	16' 0"
Width	7' 6"	7' 0"
Height	8' 1½"	3' 6"
Loading height from ground	4' 4½"	

OVERALL DIMENSIONS COUPLED TO MOTIVE UNIT

Length 29' 6". Width 7' 6". Height 8' 1½".

ESTIMATED WEIGHTS (coupled to Motive Unit)

	UNLADEN	LADEN (14,000 lbs.)
F.A.	1 ton 16-cwt.	1-ton 19-cwt.
R.A.	3-ton 1-cwt.	5-ton 18-cwt.
S.T.A.	1-ton 12-cwt.	5-ton 0-cwt.
GROSS	6-ton 9-cwt.	12-ton 17-cwt.

WEIGHTS:- Semi Trailer Unladen.
On King Pin 17-cwt. On Axle 1-ton 12-cwt.

PROVISIONAL DETAILS
OTHER BODY TYPES ON STUDEBAKER
2½ TON 6 x 6

PETROL TANKER

Capacity 750 U.S. galls. = 625 Imp. Galls.

Two separate elliptical tanks each 375 U.S. Galls., size 64" long x 54" wide x 34" high. Tanks are identical and interchangeable.

Emergency shut-off valve on each tank. Flame and spark resistor. Gravity petrol dispenser with hose, nozzles and couplings carried in compartment box at the rear. Cans of 5 U.S. gall. capacity, strainer funnels etc.

Note: Tank, gauge rod, etc. all calibrated in U.S. gallons (1 U.S. gallon = .833 Imperial gallon.)

STORES BODY

Basic body is a standard 12' cargo body (as described on page 417) which is fitted up with six spare parts bins, arranged three either side.

WORKSHOP BODIES

Basic body is a standard 12' cargo body (as described on page 417. Full Details of equipments not available, but certain types have generating sets and lathes. Generator driven by P.T.O. not fitted as Studebaker P.T.O. is not suitable.

G.M.C. CCKW 353 2½-ton 6 x 6
and 5-ton 6 x 4

GENERAL NOTES.
The G.M.C. 2½-ton Chassis is available in the same variations as the Studebaker Model US6 on page 412.
i.e.
As a 2½-ton 6 x 6
(1) In 2 wheelbases
164" Model CCKW 353
145" " " 352
(2) With or without front mounted winch (Make, Garwood ZV-512 or Heil JJ-104-B. Capacity 10,000 lbs. 300 ft. cable).

As a 5-ton 6 x 4 Model CCW. 353.
Same basic chassis as 2½-ton 6 x 6, 164" except -
Non-driving front axle.
Low 2.6 to 1 range of transfer gear omitted. Drive is permanently through High 1.15 to 1 range.

BODY TYPES:-
12' Cargo Body with 1 spare tyre on 164" w.b. chassis (CCKW 353 and CCW 353)
9' Cargo Body with 3 spare tyres on 145" w.b. Chassis (CCKW 352)

Bodies are similar to those fitted to Studebaker, see page 416.

G.M.C. CCKW 353 2½-ton 6 x 6

Normal Control Wheelbase 164" Bogie Centres 44".

ENGINE:- Make G.M.C. 270. 6-cyl. Petrol. Bore 3.25/32" Stroke 4" Capacity 269.5 c. ins. 4.41 litres. Max B.H.P. 95 @ 3000 r.p.m. Max torque 2580 lbs. ins. @ 1000 r.p.m.

CLUTCH:- Single dry plate, 10¾" dia.

GEARBOX:- 5 speed forward, 1 reverse. Ratios: 6.06 to 1, 3.5 to 1, 1.8 to 1, 1.00 to 1. Overdrive: 0.79 to 1, r. 6.00 to 1.

TRANSFER BOX:- 2 speed. Ratios 1.16 to 1 and 2.63 to 1. Front axle drive can be disengaged.

*Front Axle:- Make Wisconsin F-30 - B.28 Spiral Bevel Ratio 6.6 to 1.

*REAR AXLES:- Make Wisconsin SBB. 700 Spiral Bevel Ratio 6.6 to 1.

SUSPENSION:- Front, semi-elliptic. Rear inverted semi-elliptic.

BRAKES:- Foot, Hydraulic vacuum servo assisted. Hand, on transmission.

STEERING:- Left Hand Drive. Recirculating ball.

WHEELS AND TYRES:- 20 x 7 Tyres 7.50-20, 8 ply, dual rear.

CAPACITIES:- Fuel 32 gall. Cooling 4 galls.

PERFORMANCE FIGURES:- (2½ (short) ton load)

B.H.P. per ton 14.1
Tractive Effort per ton (100%) 2335 lbs./ton.
Maximum speed: 40 m.p.h.
Petrol consumption 8 m.p.g. Radius 256 miles.

* Certain models are fitted with Chevrolet Hypoid Axles. Same ratio.

DIAMOND T 4-TON 6 x 6
MODEL 968A

Normal Control Wheelbase 151"

Bogie centres 52"

ENGINE:- Make Hercules RXC. 6 cyl. Petrol. Bore 4 5/8". Stroke 5¼". Capacity 8.6 litres.
Max. B.H.P. 119 @ 2,200 R.P.M.
Max. torque 4,740 lbs. ins. R.A.C. Rating 51.34.

CLUTCH:- Lipe Z 42-5 15".

GEARBOX:- 5 speed forward and reverse. Ratios. 7.08 to 1, 3.82 to 1, 1.85 to 1, 1.00 to 1. Overdrive, 0.768 to 1. Reverse 7.08 to 1.

TRANSFER BOX:- Wisconsin T76. Ratios 1.00 to 1, 1.72 to 1.

FRONT AXLE:- Wisconsin F 2090-W. Ratio 8.435 to 1.

REAR AXLES:- Timken SFD.154W. Spiral bevel, Ratio 8.435 to 1.

SUSPENSION:- Front. Semi-elliptic. Rear, Inverted semi-elliptic with upper and lower torque rods to each axle.

BRAKES:- Foot. Bendix Westinghouse. Hand, on transmission. 2-line Trailer brake connections.

STEERING:- Left hand drive. Ross T 71.

WHEELS AND TYRES:- 20 x 8 Spoke, 9.00/20, 10 Ply Dual rear. Turning circle 73'.

CAPACITIES:- Fuel 50 galls. Cooling 10 galls.

PERFORMANCE FIGURES:- B.H.P. per ton. 10.1 Tractive effort per ton (100%) 2,190 lbs. ton.

Maximum speed 40 m.p.h.

M.P.G. 3.3. Radius 165 miles.

BODY DETAILS

CARGO BODY ON DIAMOND T.

MODEL 968 .A. 4-TON 6 x 6

U.S. ARMY MEDIUM ARTILLERY TRACTOR

BODY DETAILS:- Steel cargo (G.S.) body with detachable bows and canvas tilt. Folding troop seats along sides of body. Generally similar to body fitted to Mack 6 ton 6 x 6 as illustrated on page 429.

CAB:- All enclosed, steel panelled. Seats 2.

SPECIAL FITMENTS:- Winch, mounted at front. Make, Garwood (Mead Morrison) 3.U.615. Capacity 15,000 lbs. 300 ft. cable.

DIMENSIONS

	OVERALL	INSIDE BODY
Length	22' 4"	11' 0"
Width	8' 0"	7' 4"
Height	9' 6"	5' 6"

WEIGHTS (Estimated)

	UNLADEN	LADEN (8,000 lbs.)
F.A.		3 tons 10 cwt.
R.A.		8 tons 5 cwt.
GROSS	8 tons 4¼ cwt.	11 tons 15 cwt.

WRECKER (BREAKDOWN)

ON DIAMOND T. 969.A.

4-TON 6 x 6

Chassis details as Diamond T.968A on page 424

CAB:- All enclosed, steel panelled. Seats 2.

WINCH:- Mounted at front. Make, Garwood (Mead Morrison) 3-U-615. Capacity 15,000 lbs. 300 ft. cable.

WRECKER DESCRIPTION:- Make, Ernest Holmes, with two swivel booms and power winches mounted on a structural frame carried on chassis side rails. Two telescopic brace legs are fitted to either side of the structural frame. When lifting from the side the brace leg is lowered under the load boom. The other boom is swung out on the opposite side and is used with the ground anchor as an anchor boom.

For light loads, the booms can be locked together to pull from the rear of the wrecker.

The booms have separately operated power winches, each with 200 ft. cable.

WRECKER (BREAKDOWN) ON DIAMOND T. 969.A. 4-TON 6 x 6 (Cont.)

WRECKER BODY. Garwood steel body. 2 Spare Wheels carried in body.

EQUIPMENT (Main Items)

Air Compressor
Welding set with 2,500 cu. ft. Acetylene and 2,000 cu. ft. oxygen cylinders, etc.
Welders Tool Set.
Recovery Equipment, including ground anchors, towbars, snatch blocks, tow ropes and chains etc.

OVERALL DIMENSIONS:- Length 24' 2". Width 8' 3½". Ht. 9' 8".

WEIGHTS

	UNLADEN	LADEN
F.A.	3 tons 5½ cwt.	3 tons 12¾ cwt.
R.A.	5 tons 4¾ cwt.	6 tons 5¼ cwt.
GROSS	8 tons 1 cwt.	9 tons 18 cwt.

MACK N.M.5. 6-TON 6 x 6

Normal Control. Wheelbase 14' 9"
Bogie Centres 4' 4"

ENGINE:- Make, Mack EY military engine. Petrol. 6 cyl. Bore 5". Stroke 6". Capacity 707 c. ins. (11.6 litres). Max. B.H.P. 170 @ governed speed of 2,100 r.p.m. Max. torque 6,660 lbs. ins. (Net 6,228 lbs. ins.)

CLUTCH:- Lipe 15", single dry plate.

GEARBOX:- 5 speed and 1 reverse. Overall Ratios (High) 50.0 to 1, 33.5 to 1, 19.1 to 1, 10.6 to 1, 7.33 to 1, r. 59.6 to 1.

TRANSFER BOX:- 2 speed. Ratios direct and 2.55 to 1.

FRONT AXLE:- Double reduction, spiral bevel and spur. Ratio 7.33.

REAR AXLES:- Double reduction, spiral bevel and spur. Ratio 7.33.

SUSPENSION:- Front, semi elliptic. Rear inverted semi-elliptic.

BRAKES:- Foot, Westinghouse Compressed air. Hand, transmission. 2 pipe line air trailer brake connections, front and rear. Warner electric trailer brake control. Gun cylinder (As on Scammell) to be fitted as extra on British orders.

STEERING:- Left hand drive. Ross cam and lever.

WHEELS AND TYRES:- Budd disc 22" x 8" rims. Tyres 10.00 - 22, 12 ply. dual rear.

ELECTRICAL:- 66 v. lighting, 12 v. starting.

TRACK:- f. 6' 1¼", r. 6' 0¼".

CAPACITIES:- Fuel 67 galls. Cooling 10½ galls.

PERFORMANCE FIGURES:- (Load 6 short tons) B.H.S. per ton 10.9. Tractive effort per ton (100%) 2,705 lbs./ton. Max. speed @ governed r.p.m. 36 m.p.h. Petrol consumption (estimated) 35 m.p.g. Radius 240 miles.

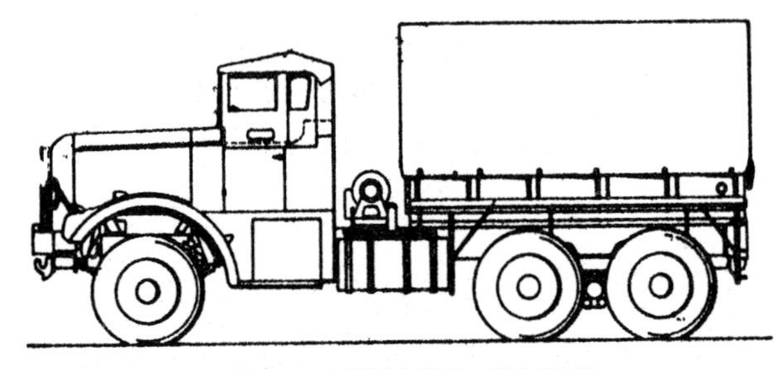

HEAVY ARTILLERY TRACTOR

(or 6 ton load carrier)

ON MACK N.M.5. CHASSIS

NOTE:- The U.S. army use this same vehicle both as a Heavy Artillery Tractor and a load carrier. Models produced for British W.D. are additionally being fitted with gun cylinder (See Brakes opposite.)

BODY:- Cargo body with folding troop carrier seats along side, detachable bows and tilt. Spare wheel carried in body.

CAB:- Open type with folding canvas top and detachable side screens. Windscreens can be folded flat on to bonnet.

SPECIAL FITMENTS:- Garwood winch model U.S.4.S. mounted between rear of cabs and body. Capacity 25,000 lbs. Pintle hook front and rear.

DIMENSIONS

	OVERALL	INSIDE BODY
Length	24' 2½"	11' 0"
Width	8' 0"	7' 4"
Height	9' 7"	5' 0"

WEIGHTS

	UNLADEN	LADEN (12,000 lbs.)
F.A.	4 tons 5 cwt.	4 tons 9 cwt.
R.A.	6 tons 0 cwt.	11 tons 4 cwt.
GROSS	10 tons 5 cwt.	15 tons 13 cwt.

STRIPPED SHIPPING HEIGHT: 8' 0" (approx.)

WHITE 6-TON 6 x 6

Normal Control. Wheelbase 15' 5"
Bogie Centres 4' 4"

ENGINES:- Hercules HXC. 6 cyl. Petrol. Bore 5¼" Stroke 6". Displacement 779 cu. ins. (12.75 litres). Max. B.H.P. 150 @ 2,100 v.p.m. Max. Torque 6,680 lbs. ins.

CLUTCH:- Lipe. Z40SX. 15" dia.

GEARBOX:- 4 speed and reverse. Overall Ratios (High) 47.72 to 1, 23.9 to 1, 12.9 to 1, 7.33 to 1, r. 53.1 to 1.

TRANSFER BOX:- 2 speed ratios, direct and 2.55 to 1.

FRONT AXLE:- Double reduction, spiral bevel and spur. Ratio 7.33 to 1.

REAR AXLES:- Double reduction, spiral bevel and spur. Ratio 7.33 to 1.

SUSPENSION:- Front semi-elliptic. Rear inverted, semi-elliptic (Timken SD - 353 - W Constant Parallelogram).

BRAKES:- Westinghouse compressed air. Hand, Transmission. 2 pipe line air trailer brake connections, front and rear. Warner electric trailer brake control.

STEERING:- Left Hand Drive. Turning Circle 82'.

WHEELS & TYRES:- Budd disc. 22" x 8". Tyres 10.00 - 22, 12 ply dual rear.

ELECTRICAL:- Starter 12 volts. Lighting 6 volts. CAPACITIES:- Fuel 64 gallons. Cooling 13½ galls.

PERFORMANCE FIGURES:- (Load 6 short tons) B.H.P. per ton 9.8. Tractive effort per ton (100%) 2,605 lbs. ton. Max. speed @ governed r.p.m. 36 m.p.h. Petrol consumption (estimated) 3.5 m.p.g. Radius 230 miles.

WHITE 6-TON 6 x 6 G.S.

BODY DETAILS

BODY:- Cargo body with detachable bows and canvas tilt. Folding troop carrier seats at sides. Dimensionally the body is similar to that fitted to the Mack N.M.5. and shown on page 429, the White body however is built to an earlier design.

CAB:- All enclosed, steel panelled type.

SPECIAL FITMENT:- Garwood winch, model U.S. - A - M.B. Mounted amid ship. Capacity 25,000 lbs. 300 ft. cable. Pintle hooks at front and rear. 2 spare wheels and tyres on ½ contract S.M. 2217 one spare only on balance.

DIMENSIONS

	OVERALL	INSIDE BODY
Length	24' 0"	11' 0"
Width	8' 0"	7' 4"
Height	10' 0"	5' 0"

WEIGHTS

	UNLADEN	LADEN (12,000 lbs.)
F.A.		3 tons 18½ cwt.
R.A.		11 tons 7½ cwt.
GROSS	9 tons 19 cwt.	15 tons 6 cwt.

For S.M. Nos. See following pages describing body types.

U.S. MILITARY PATTERN

G.M.C. AFKX - 352 1½ - 3-TON 4 x 4

Forward control Wheelbase 10' 11"

ENGINE:- Make G.M.C. Petrol 6-cyl. Bore 3.23/32". Stroke 3.13/16". Capacity 248.5 cu. ins. (4.04 litres). Max. B.H.P. 89.5 @ 3,000 r.p.m. Max. Torque 2340 lbs. ins. @ 1,000 r.p.m.

CLUTCH:- Single dry plate 10¾" dia.

GEARBOX:- 4 speed and 1 reverse. Overall ratios (High) 46.5 to 1, 22.9 to 1, 11.3 to 1, 6.6 to 1, r. 46.0 to 1.

TRANSFER CASE:- 2 speed. Ratios direct and 1.87 to 1. Front axle drive can be disengaged.

FRONT AXLE:- Spiral bevel. Ratio 6.6 Bendix Weiss joints.

REAR AXLE:- Spiral bevel. Ratio 6.6 to 1.

SUSPENSION:- Semi-elliptic, front and rear, with auxiliary rear. Shock absorber f. & r.

BRAKES:- Foot, Hydraulic vacuum servo assisted. Hand, mechanical on rear wheels.

STEERING:- Left hand drive; Recirculating ball.

WHEELS & TYRES:- 20 x 7. Tyres 7.50 - 20, 8 ply, dual rear.

ELECTRICAL:- 6 volt.

CAPACITIES:- Fuel 25 gallons. Cooling 4¾ galls.

PERFORMANCE FIGURES:- (based on gross laden wt. of 6 tons) B.H.P. per ton 14.9. Tractive effort per ton (100%) 1975 lbs./ton. Maximum speed 42 m.p.h. M.P.G. 8 (estimated) Radius 240 miles.

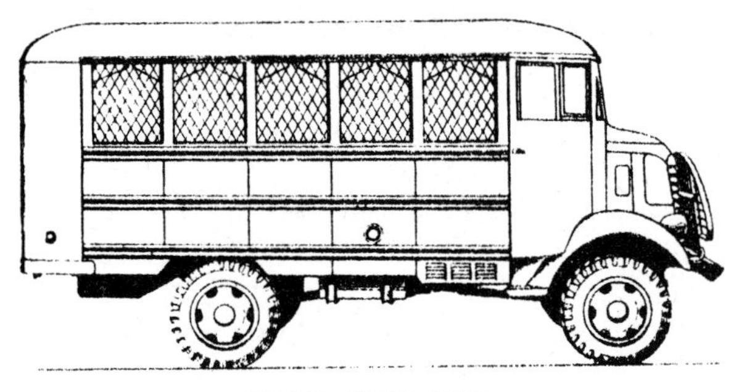

HOUSE TYPE BODY

(U.S. Nomenclature – Bus Type)

On G.M.C. 1½-3 ton AFKX 352. Understood that this chassis is now obsolescent. Later types will be on 2½ ton 6 x 6 chassis.

Basic body for following types:-

- Automotive Repair
- Machine shop
- Small Arms Repair
- Spare Parts
- Tank Maintenance
- Tool and Bench Truck
- Welding

See next pages for equipment details

BASIC BODY:- Forward control design, no partition between main body and driver's seats, 5 drop windows either side protected by removable grills. Built in heater in body, or can be used as fan only. Small fan on dash. Interior lighting wired for 110 volt to be taken from generator on models so fitted or can be tapped from outside source on other models. Pilot lighting from vehicle batteries.

DIMENSIONS

	OVERALL	INSIDE BODY
Length	19' 6"	*12' [illegible]"
Width	8' 0"	7' 6"
Height	1[illegible]' 0"	6' 4"

*Length available behind drivers seat.

WEIGHTS (see next pages)

TOOL & BENCH TRUCK

On GMC AFKX 352
S.M. 2545, 2523

Two benches either side measuring 11' 3" x 2' 1¼". Centre gangway 3' 1¼" wide. 110 v. 60 cycle bench grinder and 2 vises fitted to benches. Wide range of tools carried in locked bins under benches either side.

WEIGHTS

UNLADEN	LADEN
F.A.	2 tons 0¼ cwt.
R.A.	3 tons 7¼ cwt.
GROSS	5 tons 7½ cwt.

SPARE PARTS TRUCK

On GMC AFKX 352
S.M. 2522

Two benches either side 11' 0" long x 1' 6" wide surmounted either end by tiers of trays 2' 2" long. Bins under benches for parts.

WEIGHTS

	*UNLADEN	LADEN (Estimated)
F.A.	1 ton 15¼ cwt.	2 tons 0 cwt.
R.A.	3 tons 8 cwt.	4 tons 0 cwt.
GROSS	5 tons 3¼ cwt.	6 tons 0 cwt.

*With vehicle equipment.

AUTOMOTIVE REPAIR

on GMC 1½-3-ton AFKX 352

S.M. 2518

Body as page 433 main items of equipment are:-
5 k.w. 110 volt, 60 cycle generator single phase AC driven from P.T.O. on transfer case. 3 benches, Battery Charger, Air Compressor, Electric Drills ¼" x ½". Grinders:- cylinder, brake cylinder, valve seat. Piston pin hole.
Axle straightening press, Hydraulic Ram 10 ton, Valve refacer, valve spring tester, vice and a very wide range of small tools.

MACHINE SHOP

on GMC 1½ - 3-ton AFKX 352

S.M. 2520

Body as page 433. Main items of equipment are:-
5 k.w. 110 volt, 60 cycle generator, single phase AC, driven from P.T.O. on transfer case. 3 Benches, Electric Drill 1 - ¼", Grinder, twist drill 7" bench type.
Lathe, gap, 14" - 21" swing, 6' bed, complete with accessories.

SMALL ARMS REPAIR TRUCK

on GMC 1½ - 3-ton AFKX 352
S.M. 2521

Body as page 433, main items of equipment are:-

Portable generating set, 1½ k.w., 110 volt AC, 60 cycle. Two benches (as on Tool and Bench Truck) with boxes of spare parts stored beneath.

Portable electric drill, 3/8"; Bench grinder 6"; vise; wide range of gauges and tools for servicing U.S. small arms.

TANK MAINTENANCE TRUCK

on GMC 1½ - 3-ton AFKX 352
S.M. 2524

Body as page 433, main items of equipment are:-

5 k.w., 110 volt, 60 cycle generator.

Portable electric drill ½"; bench grinder 7", portable grinder 6", Armature Testing Growler.

Portable Testers as follows:- 2 containing ammeter, voltmeter, rheostat, vacuum and fuel pump gauge, compression gauge, variable spark gap; 1 exhaust gas combustion analyzer; 1 spark plug & 1 valve spring tester. Vice + very wide range of small tools.

TRUCK, WELDING, M3

On G.M.C. 1½ - 3 ton AFKX 352

S.M.2526

Body as page 433, main items of equipment are:-

Generator, welding, 200 ampere driven from P.T.O. on transfer case. One Bench; one welding Table, portable electric drill, portable forge, portable electric grinder 6", power hacksaw, anvil, vise.

OXY - acetylene welding and cutting outfit.

Wide range of tools and equipment.

S.M.2342, 2525, 2660.

WARD LA FRANCE 4-TON 6 x 6

Heavy Breakdown (Wrecker M.1)

ENGINE:- Make, Continental 22 R. 6-cyl. Petrol. Bore 4½" Stroke 5¼" Displacement of 501 cu. ins. (8.2 litres) Max. B.H.P. 138 @ governed speed of 2,400 r.p.m. Max. torque 4368 lbs. ins. @ 1200.

CLUTCH:- Single dry plate 14".

GEARBOX:- 5 speeds forward, 1 reverse. Overall. Ratios (High) 58.5 to 1, 28.9 to 1, 14.2 to 1, 8.27 to 1, overdrive 6.42 to 1, and 58.8 to 1.

TRANSFER BOX:- 2 speed Direct and 2.55 to 1.

FRONT AXLE:- Timken F-3200 W.-X-3 Ratio 8.27.

REAR AXLES:- Intermediate, Timken 75350. Rear, Timken 75351. Ratio 8.27.

BRAKES:- Westinghouse Compressed Air.

STEERING:- Ross T-74

WHEELS & TYRES:- Budd 20 x 9-10 Tyres 11.00-20 dual rear.

ELECTRICAL:- 12 volt. Dual ignition magneto and coil.

TURNING CIRCLE:- 83' Track f. 6'1" r. 6'10".

GROUND CLEARANCE:- 12".

CAPACITIES:- Fuel 83 galls. Cooling 6½ galls.

PERFORMANCE FIGURES
B.H.P. per Ton 10.6
Tractive effort per ton (100%) 2560 lbs. Ton
Max. Speed 45 m.p.h. (overdrive).
M.P.G. 3 Radius 250 miles.

HEAVY BREAKDOWN (WRECKER M. 1)
ON WARD LA FRANCE 4-TON 6 x 6

(All details as to recovery capacity are as per manufacturers statement)

DESCRIPTION:- Equipped with 3 power winches.

1. Front winch, capacity 29,000 lbs., 300 ft. cable.
2. Rear Winch, capacity 47,500 lbs., 350 ft. cable.
3. Crane Winch, lifting capacity as detailed below.

CRANE:- Has three basic controls.

1. 2 speed and reverse power winch (as 3 above).
2. Hand operated "Boom raising " or derricking gear.
3. Hand operated swinging or slewing gear.

The jib can be slewed through an arc of 180° on level ground.

The jib is equipped with telescopic support legs which may be attached to the rear of the body, or, for maximum lift, extended to the ground. In addition telescopic body ground jacks can be extended to ground for certain operations.

Crane Capacities

Lifting and travelling: 10,000 lbs. at the rear. Lifting Only (Side or rear): 20,000 lbs. For this operation telescopic support legs are extended to ground.

MAIN ITEMS OF EQUIPMENT

Oxy-Acetylene Welding Outfit.
V towbar with Whiffle tree.
Snatch blocks, trail spades, ground anchors. tow cables and wide range of special recovery tools and equipment.

OVERALL DIMENSIONS

Length 26'6" Width 8'3" Height 10'2"

WEIGHT

Fully equipped, full tanks, driver and mate, but no load on jib. 13 tons 0 cwt.

DATA BOOK OF WHEELED VEHICLES

PART 5

TANK TRANSPORTERS

BRITISH AND IMPORTED TYPES

Published by D.G.F.V., Ministry of Supply.

TANK TRANSPORTERS

GENERAL NOTES

This section contains data covering all British and Imported tank transporters. Certain low loading trailers, designed primarily for the carriage of heavy road making machinery, are included, as they can be used as emergency tank transporters although not fitted with special tank lashing tackle.

TRANSPORTER AND RECOVERY TYPES

All transporters of from 20 tons upwards (except low loading trailers mentioned above) are now being equipped with snatch blocks for multiplying winch pull in loading work.

TYPES OF TANKS CARRIED

The chart on page 504 shows the types of tanks capable of being carried by different transporters. It should be read in conjunction with the notes on page 505.

WEIGHT DATA

All unladen weights in this Tank Transporter Section are for the vehicle with full screw, fuel tanks and fully equipped with all tools and tackle ready for the road.

Laden weights are as unladen plus the weight of the tank specified.

INDEX SEE NEXT PAGE

INDEX

* Designed primarily for carrying heavy road making equipment.

NOTES TO CHART OF TANKS THAT CAN BE CARRIED

1. This list is based on details supplied by D.T.D. and to some extent on actual trials with tanks at W.V.E.E. These tanks are usually in the experimental stage of development and consequently there may be discrepancies between these and production types.

2. 'O.K.' indicates that the transporter has been specially designed to carry the tank specified and that the gear is suitable.

3. 'Yes' indicates that the transporter has not been specifically designed to carry the tank, but as far as is known, it can carry it if occasion demands. In these cases it may be necessary to improvise with lashing tackle, as the tackle provided has not been designed specifically for the tank indicated.

4. (E) after 'Yes' indicates that such tanks can only be carried in an emergency. e.g., Rogers and Mk.II 40 ton trailers will need careful routing with T.3 tank owing to high axle loads (26 tons).

5. Figures given against Scammell 30-ton indicate overall heights for low bridge clearances. Scammell 30 ton has the highest loading height. Heights for other transporters can be calculated by deducting the difference between their platform heights given under "Loading Dimensions" and the Scammell.

6. Scammell 20 and 30 ton types and all Macks and Whites are unsuitable for operation on hairpin bends owing to large turning circles.

7. Where Grants, Lees or Shermans are to be carried on routes with low bridges, the Rogers trailer should be used owing to its low platform height.

PROVISIONAL ONLY

UTILITY OF TANK TRANSPORTERS

TANKS CARRIED

Tank Transporter	T.3.	Churchill	Grant	Lee	Sherman	Matilda	Cromwell	Valentine	Crusader	Coven- anter	Stuart	Dragons	Bren Carriers
40 Ton Mk. I	No	O.K.	Yes	Yes	Yes	Yes, with tracks. No, without tracks	No	Barely, with tracks. No, without tracks	Yes	Yes	Yes	No	No
40 Ton Mk. II	Yes (E)	O.K.	Yes	Yes	Yes	Yes	Yes	Yes	Yes	Yes	Yes	Yes	No
Rogers 40 T.	Yes (E)	O.K.	Yes	Yes	Yes	Yes	Yes	Yes	Yes	Yes	Yes	Yes	No
Scammell 30 T.	No	No	Yes 15'4"	Yes 15'4"	Yes 14'0"	O.K. 13'4"	Yes 13'2"	O.K. 12'9"	Yes 12'8"	Yes 12'8"	Yes 14'0"	Yes	No
Albion 20 T.	No	No	No	No	No	No	No	O.K.	O.K.	O.K.	Yes	Yes	Yes (2)
Scammell 20 T.	No	No	No	No	No	No	No	O.K.	O.K.	O.K.	Yes	Yes	No
Federal 20 T.	No	No	No	No	No	No	No	O.K.	O.K.	O.K.	Yes	Yes	?
Mack EXBX	No	No	No	No	No	No	No	Yes	Yes	Yes	Yes	Yes	Yes
White 922	No	No	No	No	No	No	No	Yes	Yes	Yes	Yes	Yes	Yes
White 920	No	No	No	No	No	No	No	Yes (E)	No	No	Yes	Yes	Yes
Mack NR	No	No	No	No	No	No	No	No	No	No	O.K.	Yes	Yes
Crane Light Recovery	No	No	No	No	No	No	No	No	No	No	No	Yes	Yes

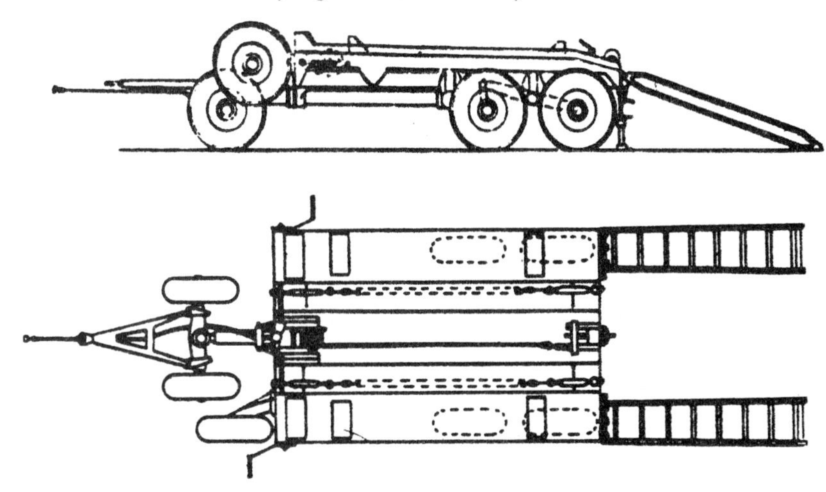

WEIGHTS (laden 7 ton tank)

	Front Axle	Rear Bogie	Gross
Unladen	19 cwt.	2 tons 2 cwt.	3 tons 1 cwt.
Laden	2 tons 14 cwt.	7 tons $8\frac{1}{2}$ cwt.	10 tons $2\frac{1}{2}$ cwt.

LOADING DIMENSIONS

In the plan view above three rectangular tank locating chocks are shown on the tank runways. The front ones (near winch handles) are fixed stops. The middle and rear chocks are capable of vernier adjustment to accommodate various lengths of tank tracks. Maximum length of track that can be accommodated (fixed front to rearmost position of rear chock) = 9' $11\frac{1}{2}$".

Overall length of runway = 12' 4"
Inside width of runway = 1' 9"
Loading height of runway (rear end) = 3' 4" unladen; 3' $2\frac{1}{2}$" laden. Runway inclined 3" in 12' 0".

Overall width across runways 7' 6", inside 4' 0".

LOADING RAMPS: Length 8' 1". Incline 1 in 3.2 (31 degrees). Inside width, front 14"; rear splayed end, $15\frac{3}{4}$".

BRIDGING DIMENSIONS

Centre towing eye to F. axle 4' $5\frac{1}{4}$".
Wheelbase (F.A. to C.L.R. bogie) 11' 6".
Rear bogie centres 3' 4".
Track, front 3' 7"; rear 6' 3".

SHIPPING DIMENSIONS

Overall Length 20' 0"
" width 7' $8\frac{1}{2}$"
" height 4' 5"

TRAILER, 7½ TON 6-WHEELED, LIGHT RECOVERY

Make: Cranes

TOWING TRACTORS

(a) Early type trailers with cable operated brakes. Can be towed by any tractor fitted with gun cylinder attachment.

(b) Later type trailers fitted with vacuum servo-operated brakes (see "Brakes" below) Leyland Retriever and Dodge V.K.60 3-ton 6 x 4 Breakdown. Certain of these tractors have been specially modified by the fitment of single pipe line vacuum servo trailer brake connection.

ABRIDGED SPECIFICATION OF TRAILER

GENERAL CONSTRUCTION:- Central tubular under-frame supporting two channels which form runways for tank tracks.

SUSPENSION:- Front, Axle rocks about centre line of main tube. Rear wheels are mounted on stubs at the ends of inter-connected swinging arms.

BRAKES:- Early Types. On rear bogie wheel operated by Bowden cable from tractor. Latest Types. Single line vacuum servo operated from tractor. Early types are also being converted. Separate hand parking wheel operating on rear bogie wheels.

WHEELS AND TYRES:- 6.00 - 13 split type. 8 stud fitting. Tyres 10.50 - 13. Spare wheel and tyre carried at nearside front of trailer.

WINCH:- Hand winch is fitted with 50 ft. of 5/8" diameter steel cable for use when power winch is not available. 2 winch sheave rollers at front, large roller at rear.

HOLDING DOWN TACKLE and adjustable tank locating chocks are provided.

GROUND CLEARANCE: 19".

LOADING RAMPS are carried between the two main channels. Reversible ramps, one side slotted, other smooth. Adjustable legs fitted at rear to support rear cross member when loading.

MACK MODEL NR4. 13-TON TRANSPORTER

6 x 4. Wheelbase 16' 8½". Bogie Centres 4'7".

ENGINE:- Make, Mack E.D. Diesel, 6 cyls. Bore, 4 3/8" Stroke 5½". Capacity, 519 c. ins. B.H.P. 131 @ governed speed 2000 R.P.M. Max torque lbs. ins. 4572 @ 1300 R.P.M.

CLUTCH:- Mack 15" Single Dry Plate.

GEARBOX:- 5 Speeds and reverse.

AUXILIARY BOX:- 2 speeds.

AXLES R. Mack Double reduction. Ratio 9.02.

TRANSMISSION:-

Auxiliary Ratio / Gear Box Ratio	Direct	1.38	Overall Ratio	Direct	Auxiliary
1.	6.74	9.30	1.	60.89	83.88
2.	3.82	5.27	2.	34.45	47.53
3.	1.92	2.65	3.	17.31	23.90
4.	1.00	1.38	4.	9.02	12.35
5.	0.78	1.08	5.	7.03	9.74
R.	6.80	9.39	R.	61.33	84.79

SUSPENSION:- F. Semi Elliptic with Houdaille shock absorbers.
R. Inverted Semi Elliptic.

BRAKES:- Foot:- Compressed air servo on all wheels.
Hand:- Mechanical on transmission.

STEERING:- Left Hand Drive. Worm and Sector.

TYRES:- 11.00-24 12 ply front. 14.00-20 18 ply Single Rear, M & S tread. (Some are 16 ply).

ELECTRICAL:- 12 v. lighting. 24 v. starting.

FUEL CAPACITY:- 2 Tanks total 125 galls.

COOLING SYSTEM CAPACITY:- 10½ galls.

This chassis differs from Mack NR Load Carrier in following main items:-

	TANK CARRIER	LOAD CARRIER
Longer Frame:-	208" behind cab	202" behind cab
Rear Tyre Equipment:-	14.00/20, 18 ply M & S tread	14.00/20 16 ply
Suspension:- (Heavier but fewer leaves)	Springs 8 leaves Front " 15 leaves Rear	Springs 11 leaves Front " 17 leaves Rear

PERFORMANCE FIGURES

(Based on gross wt. of 22.25 tons laden with American M.3. tank)

B.H.P. per Ton = 5.8

TRACTIVE EFFORT

Max. @ 100/% efficiency = 16,900 lbs.

Max. available on good roads when fully laden = 14,350 lbs. (Limiting factor engine torque)

T.E., per Ton (100%) = 756 lbs./ton
" " " (85%) = 642 " "

Adhesive wt. (Ratio of wt. on driving wheels to gross wt.) = .78

MAX. SPEEDS (@ governed engine r.p.m.)

GEAR	AUXILIARY Direct	1.38 to 1
1	4.5 m.p.h.	3.3 m.p.h.
2	7.9 "	5.8 "
3	15.7 "	11.4 "
4	30.3 "	22.0 "
5	39.0 "	28.2 "

Average Speed (estimated) = 20 m.p.h.

MAX GRADEABILITY

(Based on Tractive Effort. 50 lbs./Ton Rolling Resistance)
26.5% or 1 in 3.8

MANOEUVRABILITY

Turning Circle (estimated) 82 ft.

FUEL CONSUMPTION

(Estimated) 5 m.p.g.

Radius of Action = 625 miles.

CAB:- Enclosed steel panelled type, seats 3. Insulated roof. Full opening windscreen.

BODY:- Specially reinforced tank transporter body, quite different to Mack NR cargo body on Page 281. Steel guide rails along outer edge of body. Adjustable tank locating chocks. Tank holding down gear.

RAMPS:- Two one-piece ramps which are carried on floor of body and placed in position manually. Tie rods for holding ramps in alignment.

12 ton hydraulic jack for supporting rear of frame when loading. Rear pintle hook. No power winch is fitted to this vehicle.

SHIPPING DIMENSIONS

Overall length	27' 4"
" width	8' 6"
" height	8' 6"

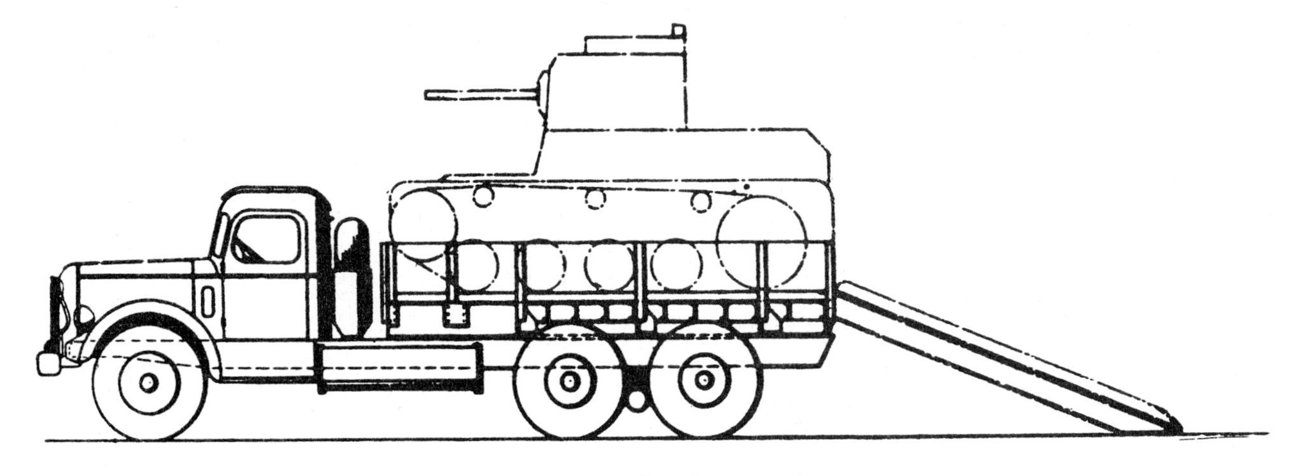

MACK NR 13 TON TRANSPORTER

WEIGHTS

	F. Axle	R. bogie	Gross
Unladen	4 tons 2½ cwts.	6 tons 18 cwts.	11 tons 0½ cwt.
Laden (American M.3 tank)	4 tons 15 cwts.	17 tons 10 cwts.	22 tons 5 cwts.

LOADING DIMENSIONS

Length of tank carrier platform	15' 3"
Length from foot of front-locating chock to foot of rear adjustable chock. (Note: Chocks not shown in drawing)	11' 1"
Width across body inside guide rails.	7' 3"
Height of platform (unladen)	5' 0"
" " (laden)	4'10½"

LOADING RAMPS

Width across ramps inside guide rails to ramps	7' 3"

No inner guide rails to ramps. Each ramp 12" wide.
Angle of ramp 22 degrees.

BRIDGING DIMENSIONS

Wheelbase to C.L. rear bogie	16' 8½"
Rear bogie centres	4' 7"
Track front	6' 5"
" rear	6' 1"
Width over rear tyres	7' 4½"

GENERAL NOTES

WHITE 920 AND MACK EXBX 18 TON TRANSPORTERS AND WHITE RUXTALL (922)

18 TON RECOVERY

These vehicles are standard U.S. commercial type chassis (mostly originally ordered by the French Government) which have been fitted with British built transporter bodies.

They differ in chassis components, dimensions and weights, but the basic design is similar.

RAMP DESIGN

The White 920 and Mack EXBX were originally fitted with detachable, folding ramps ("Bees Knees") operated by 2 hand winches as illustrated on page 521.

These are now being replaced by built-up type loading ramps, which are carried in sections on the floor of the Lorry. This later type ramp has been fitted to all White Ruxtall models and is illustrated on page 522.

POWER WINCHES

Gar Wood Winches are fitted to all White Ruxtall models. A few Mack EXBX types have also a power winch.

Mack EXBX details page 514
White 920 " " 516
White Ruxtall " " 518

MACK EXBX 18-TON TRANSPORTER

6 x 4 Wheelbase 17' 5¼" Bogie Centres 4' 4½"

ENGINE:- Make, Mack EO. Petrol, 6 cyls. Bore, 4 3/8" Stroke, 5¾" Capacity, 8.54 litres. B.H.P. 131 @ governed speed 2000 R.P.M. Max. torque lbs/ins. 4584 @ 1000 R.P.M. R.A.C. 45.9 H.P.

CLUTCH:- Single Dry Plate.

GEAR BOX: 5 speeds and reverse.

Overall			Gear
7.33	8.17	9.02	
59.0	65.80	72.60	8.05
33.5	37.35	41.25	4.57
19.1	21.32	23.55	2.61
10.6	11.83	13.07	1.45
7.33	8.17	9.02	1.00
59.6	66.40	73.33	8.13

REAR AXLES:- Double reduction. Full floating. Ratios, 7.33-1; 8.17-1; 9.02-1, Mack axles 8.17 ratio with power divider. Timken axles 7.33 and 9.02 ratios, no power divider.

SUSPENSION:- Front, Semi-elliptic with Houdaille shock absorbers. Rear, inverted semi-elliptic.

BRAKES:- Foot, Compressed air serve on all wheels. Hand, Transmission.

STEERING:- Left Hand Drive. Worm and Sector.

TYRES:- 10. 10.50/22 Dual rear (7.33 axle). 10. 11.00/24 Duel rear (8.17 and 9.02 axles).

ELECTRICAL:- 12 V Lighting and Starting.

FUEL CAPACITY:- 2 Tanks total of 57 gals.

COOLING SYSTEM CAPACITY:- 10 gals.

MACK EXBX
PERFORMANCE FIGURES

(Based on gross vehicle wt. of 29.5 tons laden with 18 ton Tank)

B.H.P. per Ton = 4.64

	Tractive Effort	
Axle Ratio	= 7.33 & 8.17*	9.02
Max @ 100% efficiency	= 12,360 lbs.	14,450 lbs.
Max available on good roads when fully laden	= 10,500 lbs.	12,280 lbs.
T.E. per Ton (85%)	= 358 lbs.	420 lbs.

Adhesive Wt. (Ratio of wt. on driving wheels to gross vehicle wt.): '.76.

MAX SPEEDS @ governed engine r.p.m.

Gear	Axle Ratio 7.33	8.11	9.02
1	4.55	4.28	3.88
2	8.0	7.57	6.85
3	13.9	13.2	12.0
4	25.2	23.9	21.5
5	36.6	34.5	31.2

MAX GRADEABILITY. (Based on Tractive Effort)

	14.6%	17.4%
or	1 in 6.8	1 in 5.7

MANOEUVRABILITY:
Turning Circle, right 82' 3"; Left 87' 9"

FUEL CONSUMPTION:

(Estimated) 4 m.p.g.

* Approx. same as 7.33 ratio owing to larger tyres.

WHITE MODEL 920
18-TON TANK TRANSPORTER

6 x 4. Wheelbase.17'8", Bogie Centres 4'5"

ENGINE:- Make, White 529. Petrol, 6 cyls. Bore, 4 5/8" Stroke, 5¼" Capacity, 8.67 litres. B.H.P. 134 @ governed speed 2400 r.p.m. Max. torque lbs/ins. 4680 @ 1000 r.p.m.

CLUTCH:- White, 15" Single Dry Plate.

GEAR BOX:- 5 Speeds and reserve.

Ratios	Overall	Gear
1.	65.49	6.37
2.	32.40	3.15
3.	17.88	1.74
4.	10.27	1.00
5.	8.10	0.78
R.	65.7	6.40

REAR AXLES:- Timken spiral bevel. Double reduction. Ratio 10.27 - 1.

SUSPENSION:- Front, semi-elliptic. Rear, inverted semi-elliptic.

BRAKES:- Foot, Compressed air servo on all wheels. Hand, Transmission.

STEERING:- Left Hand Drive. Worm and Lever.

TYRES:- 10. 10.50/22 Dual rear.

ELECTRICAL:- 12 V. Lighting and Starting.

FUEL CAPACITY: 2 Tanks total of 84 gals.

COOLING SYSTEM CAPACITY:- 6½ gals.

WHITE 920
PERFORMANCE FIGURES

(Based on gross vehicle wt. of 29.7 tons laden with 18 ton Tank)

B.H.P. per Ton = 4.5

Tractive Effort

Max. @ 100% efficiency = 14,500 lbs.

Max. available on good roads = 12,350 lbs. when fully laden (Limiting factor engine torque)

T.E. per Ton (100%) = 488 lbs./ton.

" " " (85%) = 415 " "

Adhesive Wt. (Ratio of wt. on driving wheels to gross wt.) = .77

MAX SPEEDS @ governed r.p.m.

GEAR	
1	4.29 m.p.h.
2	8.67 "
3	15.7 "
4 Direct	27.4 "
5 Overdrive	34.7 "

MAX GRADEABILITY

(Based on Tractive Effort. 50 lbs./ton Rolling Resistance)

16.6% or 1 in 6

MANOEUVRABILITY

Turning Circle 85' 9" left and right.

FUEL CONSUMPTION (Estimated) 4 m.p.g.

WHITE-RUXTALL MODEL 922.
18-TON TANK TRANSPORTER

ENGINE:- Make, White 529. Petrol, 6 cyls. Bore, 4 5/8" Stroke, 5¼" Capacity, . 8.67 litres. B.H.P. 134 @ governed speed 2400 R.P.M. Max. torque lbs/ins. 4680 @ 1000 R.P.M.

CLUTCH:- White, Single Dry Plate.

GEAR BOX:- 5 Speeds and reverse White 36B.

AUXILIARY BOX:- 3 Speed.

REAR AXLES:- Full floating. Helical bevel drive. Ratio 11.67-1.

TRANSMISSION:-

Auxiliary Ratios		0.747	Direct	2.62
Gear Ratios	(1.	4.75	6.37	16.68
	(2.	2.35	3.15	8.25
	(3.	1.29	1.74	4.55
	(4.	0.74	1.00	2.62
	(5.	0.58	0.78	2.04
	(R.	4.78	6.40	16.76
Overall Ratios	(1.	55.50	74.30	194.60
	(2.	27.40	36.60	96.20
	(3.	15.20	20.30	53.20
	(4.	8.72	11.67	30.30
	(5.	6.86	9.20	22.80
	(R.	55.80	74.60	195.60

SUSPENSION:- Front and Rear, Semi-elliptic. Rear springs have now been replaced by solid beam suspension.

BRAKES:- Foot, Compressed air servo on all wheels. Hand, Transmission.

STEERING:- Left Hand Drive.

TYRES:- 10. 11.25/24 Duel Rear.

ELECTRICAL:- 12 V Lighting and Starting.

FUEL CAPACITY:- 2 Tanks total of 73 gals.

COOLING SYSTEM CAPACITY:- 6¼ gals.

WHITE RUXTALL MODEL 922
PERFORMANCE FIGURES

(Based on gross vehicle wt. of 31.05 tons laden with 18 ton Tank)

B.H.P. per Ton = 4.3

Tractive Effort

Max. @ 100% efficiency =. 39,900 lbs.

Max. available on good roads when fully laden = 34,000 lbs. (Limiting factor engine torque)

T.E. per ton (100%) = 1,286 lbs./ton

" " " (85%) = 1,094 " "

Adhesive Weight (Ratio of wt. on driving = .806 wheels to gross vehicle wt.)

MAXIMUM SPEEDS @ governed r.p.m.

Aux.Ratios:-	.747	1.00	2.68	
	5.5	4.1	1.5	m.p.h.
	11.1	8.2	3.5	"
	19.9	14.9	5.7	"
	34.7	26.0	10.0	"
	44.2	33.0	13.3	"

MAXIMUM GRADEABILITY

(Based on Tractive Effort. 50 lbs./ton Rolling Resistance) = 1 in 1.95 or 51%.

MANOEUVRABILITY

Turning Circle 88 feet.

FUEL CONSUMPTION. (estimated) 4 m.p.g.

WHITE 920 & MACK EXBX

Bridging Dimensions

White 920

Wheelbase to C.L. R. Bogie	17' 8"
R. Bogie centres	4' 5"
Track, front	6' 4"
Track, rear	6' 2"
Width over rear tyres	8' 2"

Mack EXBX

Wheelbase to C.L. R. Bogie	17' 5¼"
R. Bogie centres	4' 4½"
Track, front	6' 7½"
Track, rear	6' 0½"
Width over rear tyres	8' 0"

Cabs (White 920 and Mack EXBX)

Both types are fitted with standard commercial type steel panelled cabs. Seating 3 men.

LOADING RAMPS:- Folding type ramps, lowered by 2 hand operated winches. Braced at hinge by triangular support. See also new type loading ramp on page 522.

ELEPHANTS FEET:- Hinged legs with elephants feet at rear directly under point of attachment of ramp.

TANK LOCATING GEAR:- Adjustable chocks provided to fit into platform.

TRANSPORTER BODY ON MACK EXBX (Illustrated) AND WHITE 920

WITH EARLY TYPE FOLDING RAMPS

(All vehicles are being fitted with new design ramps as shown overleaf.)

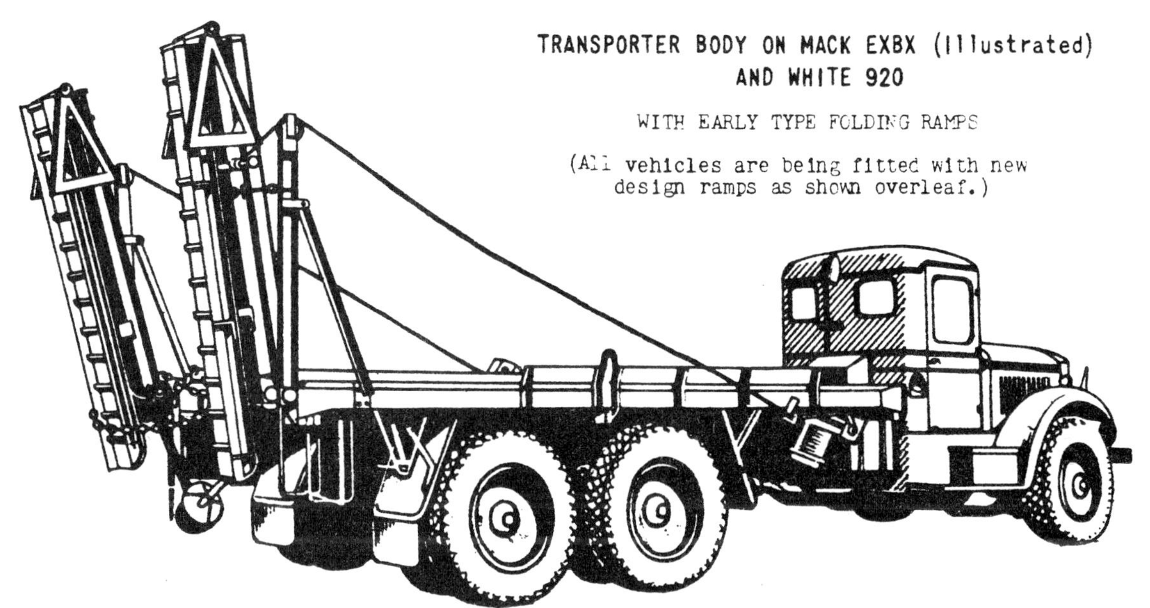

WHITE 920. WEIGHTS

	Unladen.	Laden (18-ton Tank)
F. Axle	3 tons 15 3/4 cwts.	7 tons 2 1/4 cwts.
R. Bogie	7 tons 8 1/2 cwts.	22 tons 11 3/4 cwts.
Gross	11 tons 4 cwts.	29 tons 14 cwts.

MACK EXBX. WEIGHTS

	Unladen	Laden (18-ton Tank)
F. Axle	2 tons 19 1/4 cwts.	6 tons 13 cwts.
R. Bogie	8 tons 1 3/8 cwts.	22 tons 17 cwts.
Gross	11 tons 1 cwt.	29 tons 10 cwts.

WHITE 920. DIMENSIONS

	Overall	Inside Body
Length	31' 11"	17' 0"
Width	10' 3"	8' 8"
Height	10' 11 1/2" ramps folded.	
Loading Height	4' 6"	

MACK EXBX. DIMENSIONS

	Overall	Inside Body
Length	30' 10 1/2"	17' 0"
Width	10' 1"	8' 8"
Height	11' 4" ramps folded.	
Loading Height	4' 4"	

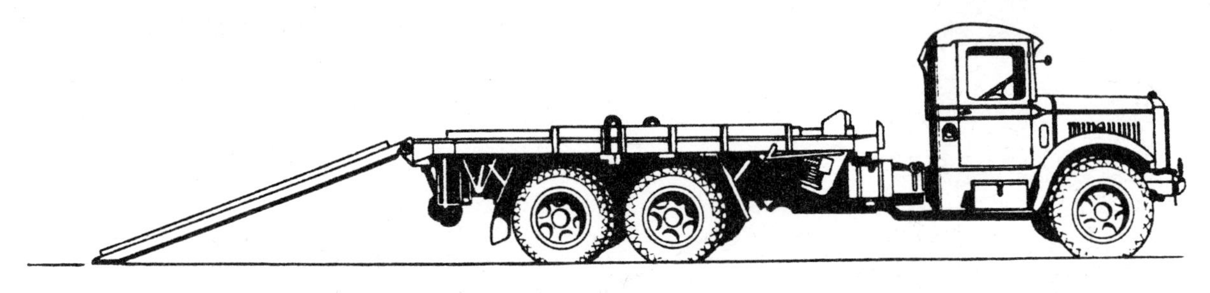

RAMPS ON WHITE RUXTALL 922 AND NEW TYPE RAMPS ON
WHITE 920 AND MACK EXBX

The illustration above shows the Mack EXBX Transporter fitted with the new built-up, non-folding type loading ramps.

Each ramp consists of two side girders with a reversible top; top has smooth surface one side, slotted the other. The ramps are carried in sections under the floor of the lorry and placed in position manually.

This type of ramp is being fitted to all White 920 and Mack EXBX Transporters in place of the folding type ramps shown overleaf.

All White Ruxtall Recovery models have built-up type loading ramps as above.

FOR WEIGHTS AND DIMENSIONS (MACK EXBX AND WHITE 920)

See Data on page 521, except for the following items which are changed by reason of new ramps.

WHITE 920	
Overall Length	29' 10"
Overall Height (Cab)	7' 10½"

MACK EXBX	
Overall Length	29' 5½"
Overall Height (Cab)	8' 3"

WHITE RUXTALL 922

CAB:- All enclosed, steel panelled cab, seating three.

LOADING RAMPS:- See description opposite.

ELEPHANTS FEET:- Hinged legs with elephants feet at rear directly under point of attachment of ramp.

WINCH:- Make, Garwood 5 M.B. 500 ft. Cable.

TANK LOCATING GEAR:- Adjustable chocks provided to fit into platform.

BRIDGING DIMENSIONS:-

Wheelbase	18' 10"
Bogie Centres	4' 4"
Track, front	6' 2½"
Track, rear	6' 1½"
Overall width over rear tyres	8' 1½"

WEIGHTS

	Unladen	Laden
Front Axle	4 tons 4¼ cwts.	6 tons 0 cwt.
Rear Bogie	8 tons 11½ cwts.	25 tons 1 cwt.
Gross	12 tons 15¾ cwts.	31 tons 1 cwt.

DIMENSIONS

	Overall	Inside Body
Length	30' 9½"	16' 11"
Width	9' 9½"	9' 4"
Height	8' 2½"	-

TRAILER, 20 TON 16 WHEELED LOW LOADING

MAKE: Multiwheeler also made by S.M.T. to the Multiwheeler Design.

IMPORTANT:- Designed primarily for the carriage of heavy road-making machinery. Can be used for transport of tanks, but no special holding-down tackle is provided other than "D" type side shackles.

TRAILER - ABRIDGED SPECIFICATION

GENERAL:- 16 wheels arranged as 8 twins. Two oscillating half axles front and rear. Welded frame. Hardwood flooring.

SUSPENSION:- Underslung semi-elliptic.

STEERING:- Front wheels mounted on twin turn-tables. Drawbar pivoted at centre and connected by steering rods to effect Ackerman type steering up to 45 degree angle in unison with drawbar articulation.

BRAKES:- Early types, cable operated from tractor on all wheels. Later types, 2 line air pressure operated. Eventually all types will be converted to 2 line system. Hand brake operated by wheel from brakeman's seat at front offside and inter-connected with main brake linkage.

WHEELS AND TYRES:- Total of sixteen, fitted in twin formation. Tyres 29 x 8 R.H.S. 2 spares carried.

LOADING GEAR:- Four 2-piece ramps carried on platform. Ramp support brackets with ground plates. Two 12 ton hydraulic jacks. Wheel chocks. Sheave - roller, for winch cable guide, mounted at front and slightly offset.

EQUIPMENT COMPARTMENTS:- Located under main body at front.

BODY:- Main section, flat platform except for metal chock rails or kerbs which protrude approx. 3½" along outside edges of body and are splayed at the rear (see plan)

TURNING CIRCLE:- 62 feet.

GROUND CLEARANCE:- (Min.) 6½" under spring U bolts.

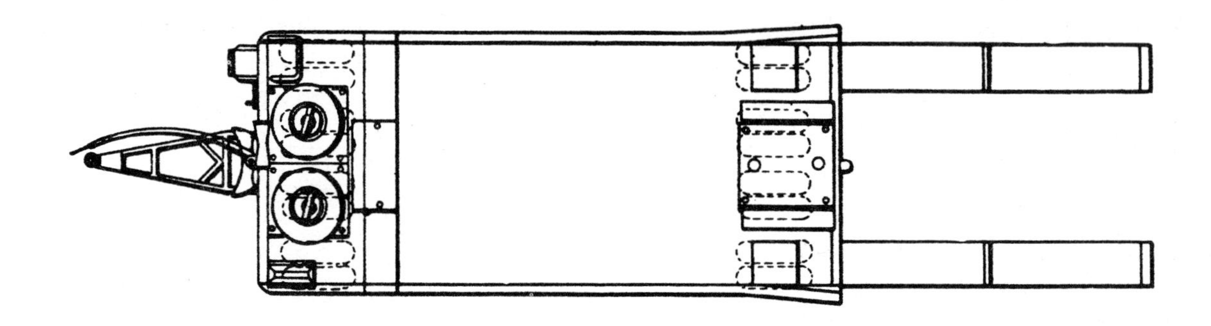

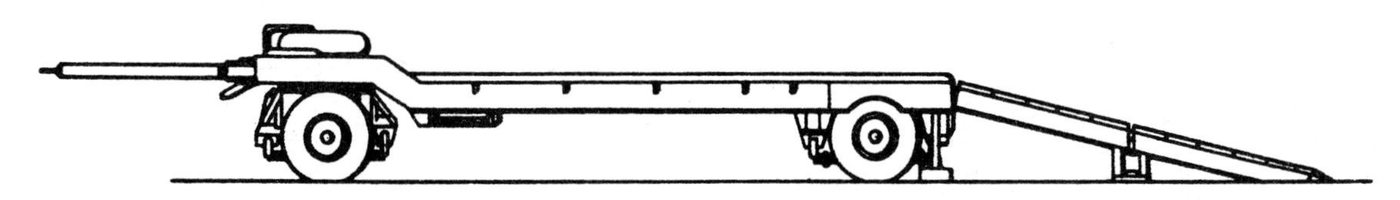

TRAILER, 20 TON 16 WHEELED - MULTIWHEELER, S.M.T.

WEIGHTS

	F. AXLE	R. AXLE	GROSS
Unladen	4 tons 1 cwt.	3 tons 3 cwts.	7 tons 4 cwts.
Laden (20 tons)	13 tons 1 cwt.	14 tons 3 cwts.	27 tons 4 cwts.

LOADING DIMENSIONS

Overall length of body, including raised front section = 21' 0".
Length of flat platform section = 16' 0½"
Width, inside chock rails. 9' 0"
Loading Ht. unladen = 3' 0"
Length, rear of trailer to end of ramps measured on ground = 11' 6"
Width of single ramp = 13"
(4 complete 2 piece ramps to each trailer)

BRIDGING DIMENSIONS

Length, centre of drawbar eye to front of trailer = 6' 1".
From front of trailer to F. axle = 2' 1½".
Wheelbase = 16' 6 "
R. axle to rear of trailer = 2' 3½"
Track, measured across outside pair of duals = 7' 0⅛"
Track measured across inside pair of duals = 2' 2¼"
Overall width across outside edges of outside tyres = 8' 7½"
Dual Tyre spacing = 10¼"

SHIPPING DIMENSIONS:- Overall length (inc. drawbar, less ramps) = 27' 5"
Overall width (across splayed rear section) = 9' 5½"
Overall height (to top of brakesman's seat) = 4' 5"

TRAILER, 20 TON 8 WHEELED LOW LOADING

Make: Crane

TRAILER, 18 TON 8 WHEELED LOW LOADING

Make: Carrimore

These two trailers have been designed primarily for the carriage of heavy road-making machinery. They can both be used for the transport of tanks, but no special tank holding down tackle is provided, other than "D" type shackles along the sides.

GENERAL DESIGN:- (both Trailers)

Two oscillating half axles front and rear.
Steering by single turntable.
Each half axle fitted with 2 (single) wheels.

LOADING FROM REAR:- (on both types) is effected by jacking up rear of frame, retracting rear half axles, and lowering frame to ground. Short ramps provided to fit up to rear of frame.

In addition (on Crane model only) the ramps can be fitted up to effect side loading without retracting rear axles.

TYRES:-

Crane:- 13.50 - 20 tyres (singles)
Carrimore:- 11.25 - 20 tyres (singles)

BRAKES:-

CRANE:- Single pipe line air pressure operated on all wheels. (Can be operated by tractor fitted with 2 line system). Hand brake, mechanical operated by wheel from brakesman's seat.

CARRIMORE:- Two pipe line air pressure operated on all wheels. Hand, mechanical on front wheels.

SHEAVE-ROLLER:- for winch cable fitted to Crane model only.

TRAILER 18 TON 8 WHEELED LOW LOADING (MAKE: CARRIMORE)

(Similar but smaller design to Crane model shown overleaf)

WEIGHTS

	UNLADEN	LADEN
F.A.	3 tons 16 cwts.	12 tons 6 cwts.
R.A.	3 tons 1 cwt.	12 tons 1 cwt.
Gross	6 tons 17 cwts.	24 tons 17 cwts.

DIMENSIONS

	OVERALL	LOAD SPACE
Length	28' 8"	12' 0"
Width	8' 6"	8' 6"
Height	4' 10"	

Ht. of platform from ground unladen 1' 11".

BRIDGING DIMENSIONS:- Wheelbase 18' 8".
Length Centre towing eye to f. axle 7' 7".
Track over outer wheels, f. 6' 6"; R. 7' 1".
Track over inner wheels, f. 1' 10"; R. 2' 4½".

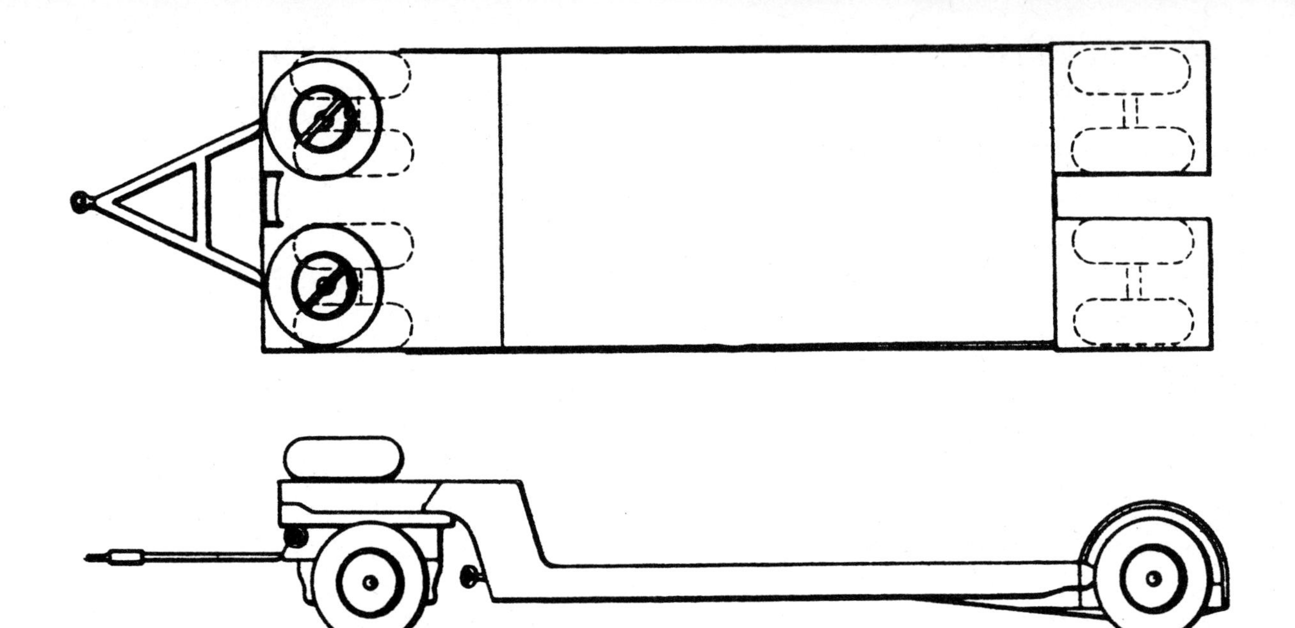

TRAILER, 20 TON 8 WHEELED LOW LOADING (MAKE CRANE)

20 TON CRANE DETAILS
as illustrated above
(for Carrimore see 527A over)

WEIGHTS

	UNLADEN	LADEN
F.A.	6 tons 9 cwts	15 tons 5½ cwts.
R.A.	4 tons 18 cwts.	16 tons 5½ cwts.
Gross	11 tons 7 cwts.	31 tons 11 cwts.

DIMENSIONS

	OVERALL	LOAD SPACE
Length	36' 5"	17' 0"
Width	9' 1"	9' 1"
Height	5' 2¼"	

Ht. of platform from ground 2' 6¼" unladen.

BRIDGING DIMENSIONS:- Wheelbase 25' 4".
Length centre towing eye to f. axle 8' 5".
Track over outer wheels, f. 7' 9"; R. 7' 10".
Track over inner wheels, f. 2' 9"; R. 2' 10".

ALBION, CX24. 20-TON SEMI-TRAILER TANK TRANSPORTER

6 x 4 - 4

(For Wheelbase of motive unit and semi-trailer see plan on page 531)

ENGINE:- Make, Albion Petrol EN248.B. 6 cyls. Bore 4½" Stroke 6". Capacity 10487 c.c. B.H.P. 140 @ governed speed 2100 R.P.M. Max torque lbs. ins. 5500 @ 900 R.P.M. R.A.C. 54.15 H.P.

CLUTCH 1: Single Dry Plate.

GEAR BOX:- 4 speed and reverse.

AUXILIARY BOX:- 2 Speed. Early models, ratios direct and 1.84 to 1. Later models, direct and 1.36 to 1.

AXLES:- Full floating. Overhead worm drive. Ratio 9.33 - 1.

TRANSMISSION:-

		Direct	1.84	1.36
Auxiliary Ratios		Direct	1.84	1.36
Gear Box Ratios	1.	5.79	10.65	7.87
	2.	2.90	5.35	3.94
	3.	1.63	3.00	2.22
	4.	1.00	1.84	1.36
	R.	7.63	14.00	10.37
Axle Ratios		9.33 - 1		
Overall Ratios		Direct	1.84	1.36
	1.	54.00	99.20	73.43
	2.	27.05	49.80	36.80
	3.	15.22	28.00	20.70
	4.	9.33	17.20	12.70
	R.	71.2	131.00	96.82

SUSPENSION:- F. Semi Elliptic. R.2 Inverted Semi Elliptic.

BRAKES:- Foot. Compressed air servo on all wheels. (Motive Unit and semi-trailer). Hand. Mechanical on transmission.

STEERING:- Right Hand Drive. Worm and Sector.

TYRES:- 36 x 8 Singles on front axles, 36 x 8 twins on all other axles tractor and semi trailer, 2 spares carried under semi-trailer.

ELECTRICAL:- 24V Starter. 12V Lighting.

FUEL CAPACITY:- 2 Tanks total of 100 gals.

COOLING SYSTEM CAPACITY:- 8¼ gals.

ALBION 20-TON TRANSPORTER PERFORMANCE FIGURES

(Based on gross weight of 34.2 tons laden with Crusader Tank)

		(1.84 Aux.)	(1.36 Aux.)
B.H.P. per ton	= 4.09		
Tractive Effort per Ton	(100%) =	868 lbs.	644 lbs.
Tractive Effort per Ton	(85%) =	738 lbs.	548 lbs.

Adhesive Weight (Ratio driving wheel to gross vehicle weight) = .41

Tractive Effort maximum available on good roads when fully laden:-
Early models (1.84 Aux.) 25,200 lbs.
Later models (1.36 Aux.) 18,750 lbs.
(Limiting factor engine torque)

MAX. SPEEDS

Gear	Aux. High (1.00-1)	Aux. Low * 1.84 to 1	1.36 to 1
1	4.35 m.p.h.	2.63 m.p.h.	3.12 m.p.h.
2	8.7 "	4.7 "	6.4 "
3	15.5 "	8.4 "	11.6 "
4	25.2 "	13.5 "	18.5 "

(* Early models, ratio 1.84 to 1. Later models, ratio 1.36 to 1).

Average speed (100 miles circuit) = 19.0 m.p.h.

MAX. GRADEABILITY
1 in 3.1 (Early models 1.84 Aux.)
1 in 4.25 (Later models 1.36 Aux.)

MANOEUVRABILITY

MINIMUM TURNING CIRCLE (OUTER)

Left Hand	Right Hand
69'1"	69'2"

MINIMUM TURNING CIRCLES (INNER)

Left Hand	Right Hand
36'4"	37'3"

FUEL CONSUMPTION
100 miles circuit = 2.89 m.p.g. (At average speed of 19.0 m.p.h.).
Radius = 290 miles.

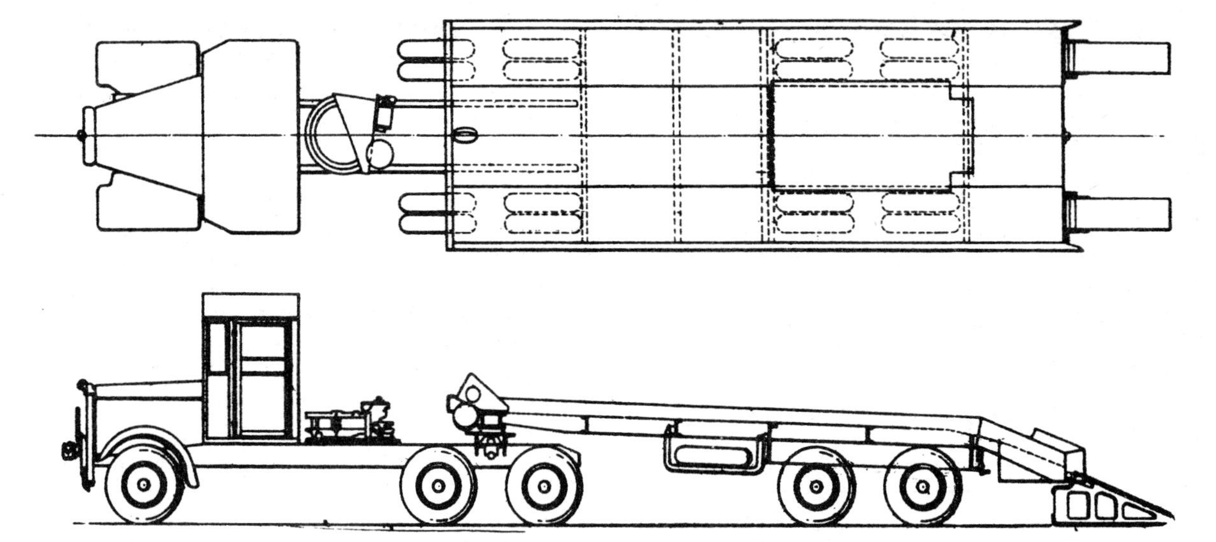

ALBION 20 TON TRANSPORTER

WEIGHTS (Laden Crusader Tank)

	F. Axle	R. Bogie	S. Trailer Bogie	Gross
Unladen	3 tons $10\frac{1}{4}$ cwt.	5 tons 19 cwt.	5 tons $13\frac{1}{4}$ cwt.	15 tons $2\frac{3}{4}$ cwt.
Laden	3 tons $12\frac{1}{4}$ cwt.	14 tons $1\frac{3}{4}$ cwt.	16 tons $11\frac{1}{2}$ cwt.	34 tons $5\frac{1}{2}$ cwt.

LOADING DIMENSIONS

Length of tank carrier runways (from rear of fixed front stop block to rear of horizontal section of runway).	19' 4"
Overall width of runways inside the outer guide rails.	8' 9"
Width of each runway.	2' 2"
Width inside runways.	4' 1"
Loading Ht. of runway at rear	4' 0"
at front	4' 7"
Loading ramps, width.	1' 3"
Width over ramps.	7' 9"
" inside "	5' 3"

Angle of ramp 30 degrees.

BRIDGING DIMENSIONS

Wheelbase, tractor, F.A. to C.L.r. bogie.	14' 9"
Rear bogie centres.	4' 6"
Wheelbase, semi trailer, from point of attachment to C.L. semi trailer bogie.	16' 0"
Rear bogie centres.	4' 6"

Semi trailer is mounted directly over C.L. of r. bogie on tractor.

TRACK

F. axle tractor.	6'7 7/8"
R. bogie tractor.	6'3 1/16"
R. bogie semi trailer.	6'3 1/16"
Width over tyres, r. bogies.	$7'10\frac{1}{2}"$

ALBION 20 TON SEMI-TRAILER TRANSPORTER

CAB:- The experimental "greenhouse type" cab shown on the drawing opposite, has been replaced by a more modern cab with sloping windscreen, seating 3 men.

WINCH:- This Albion model is fitted with the Scammell 8 ton vertical spindle winch. 600 ft. of 13/16" dia.rope. Torque limiting device is set for a pull of 8 tons.

RAMPS:- Hinged triangulated type.

TANK LOCATING GEAR:- Adjustable chocks on tank runways.
W.D. Drawbar gear fitted to front of motive unit.

SHIPPING DIMENSIONS

Overall length (Tractor and S.trailer with ramps in raised position).		43'9¾"
Overall length tractor.		22'0"
" " semi-trailer.		27'6"
Overall width tractor		7'10"
" " semi trailer.		9'5"
Overall Height.	(top of cab).	8'6"
" "	(S. trailer, ramp raised in position)	4'6"
	(S. trailer, front stop block)	6'2"

FEDERAL 604 20-TON SEMI-TRAILER TRANSPORTER 6 x 4 - 4

(Motive Unit Also made by REO to Federal design)
(REO model No. is 28 x S)
(For Wheelbase of motive unit and semi-trailer, see plan on page 535)

ENGINE:- Make. Cummins H.B. 600. C.I. 6 cyls. Bore. 4 7/8" Stroke 6" Capacity 11.03 litres. B.H.P. 150 @ governed speed 1800 R.P.M. Max torque lbs. ins. 6000 @ 1300 R.P.M.

CLUTCH:- W.C. Lipe. 15". Single Dry Plate.

GEARBOX:- 4 speeds and reverse.

AUXILIARY BOX:- 2 Speed.

AXLES:- Double reduction. Full Floating. Ratio 10.27 - 1.

TRANSMISSION:-

		Direct	
Auxiliary Ratio			2.28
Gear Box Ratio	1.	3.72	8.48
	2.	2.05	4.69
	3.	1.00	2.28
	4.	0.77	1.75
	R.	4.84	11.03
Axle Ratio		10.27	
Overall Ratio			
	1.	38.10	86.80
	2.	20.90	47.70
	3.	10.20	23.40
	4.	7.90	18.00
	R.	49.70	113.20

SUSPENSION:- F and R Semi-Elliptic. R. Inverted.

BRAKES:- FOOT. Compressed air servo on all wheels.
HAND. Mechanical on transmission.

STEERING:- Left Hand Drive. Ross T7A. Cam and Lever.

TYRES:- Motive Unit 10. 10.00/20. 12 Ply on Tractor. Dual rear.
Semi Trailer. 8. 10.00/15. 14 Ply Special low platform trailer tyres.

ELECTRICAL:- Starter 24 V. 12 V Lighting.

FUEL CAPACITY:- 2 Tanks total of 64 gals.

COOLING SYSTEM CAPACITY:- 8 1/5th gals.

SEMI-TRAILER:- Make Trailmobile.

FEDERAL 20 TON TRANSPORTER
PERFORMANCE FIGURES

(Based on gross weight of 34.2 tons)

B.P.H. per ton = 4.3

Tractive Effort (100%) = 1010 lbs./ton.
" " (85%) = 858 lbs./ton.

Tractive Effort, Maximum available on good roads when fully laden, 23,300 lbs. (Limiting factor adhesion)

Adhesive weight (Ratio of driving wheel weight to gross weight) = .38

MAX. SPEEDS

GEAR	AUXILIARY 1.00 - 1	2.28 - 1
1	5.2 m.p.h.	2.3 m.p.h.
2	9.5 m.p.h.	4.2 m.p.h.
3	19.5 m.p.h.	8.6 m.p.h.
4	25.2 m.p.h.	11.1 m.p.h.

Average Speed (estimated) = 19.0 m.p.h.

MAX. GRADEABILITY (CALCULATED)

Based on adhesive weight and .8 adhesion = 1 in 3.6
" " " " " .7 adhesion = 1 in 4.1

MANOEUVRABILITY

Minimum Turning Circles (outer)
Left 63' Right 65'

FUEL CONSUMPTION

Estimated 4 m.p.g,
Radius of Action 256 miles.

CAB:- Enclosed steel panelled type. Seats 2.

WINCH:- Make: Mead Morrison-Garwood. Type 4 MB. Nominal capacity 25,000 lbs. 300 ft. of cable. Winch roller and sheave guides on semi trailer. Single snatch block.

RAMPS:- Hand operated of hinged triangulated construction.

TANK LOCATING GEAR:- In addition to adjustable inside guides on the tank runways there are adjustable chock blocks on the runways. Equipment box fitted to front of semi trailer. Pintle hook fitted to front and rear of tractor.

BRIDGING DIMENSIONS

Wheelbase (F. axle to C.L. rear bogie) 14' 0". Bogie Centres 4' 4". C.L. tractor bogie to semi trailer axle 22' 9". Semi trailer is mounted 3" forward of tractor bogie centres.

TRACK:-		
	F. Axle Tractor	6' 7½"
	R. Bogie Tractor	6' 2"
	Dual tyre spacing (tractor)	12½"
	Width over rear tyres (Tractor)	8' 0"
Track across semi trailer outer wheels		7' 7½"
Track across semi trailer inner wheels		2' 1½"
Dual tyre spacing, semi-trailer		11¼"
Overall width across tyres semi trailer		9' 6"

Angle of Approach: Tractor 45 degrees.
Angle of Departure: Tractor 45 degrees: Semi Trailer 30 degrees.

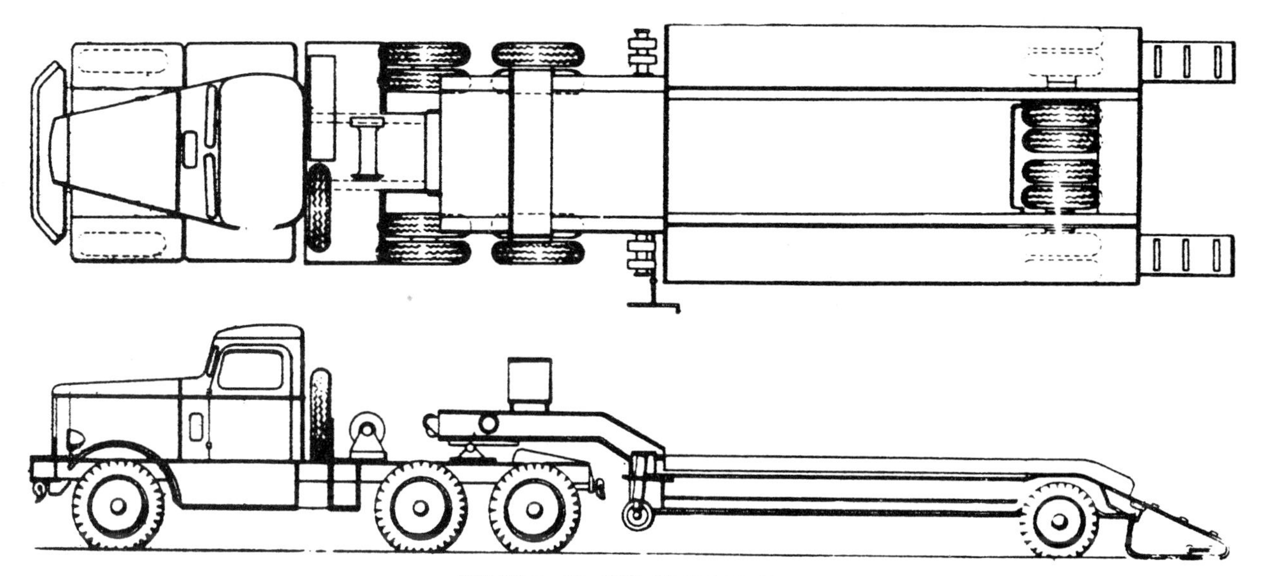

FEDERAL 20-TON TRANSPORTER

WEIGHTS (laden 20 tons)

	F. Axle	R. Bogie	S. Trailer Axle	Gross
Unladen	3 tons 11½ cwt.	7 tons 18½ cwt.	2 tons 12 cwt.	14 tons 2 cwt.
Laden	3 tons 15 cwt.	13 tons 11½ cwt.	16 tons 17 cwt.	34 tons 3½ cwt.

LOADING DIMENSIONS

Length of tank carrier runway (from rear of swan neck to front of cutaway portion over semi trailer axle). 14' 0"
Overall width across runways 9' 6"
" " inside is adjustable from 5' 1" to 5' 7"
The inside guide angles to runways are reversible, each in section measuring 6" x 4½ x ½", either leg can be bolted to floor. No outer guide angles.
Ht. of runway from ground laden 37½" at front; 35½" at rear.

RAMPS:- Length 42", Width 18". Width over ramps 8' 8". Width inside ramps 5' 4" Ramp angle 25 degrees.

SHIPPING DIMENSIONS

Overall Length,	tractor and semi trailer	43' 2"
" "	tractor only	21' 11"
" "	semi trailer	27' 9"
Overall Width,	tractor	8' 0"
" "	semi trailer	9' 6"
Overall Height,	tractor	8' 0"
" "	semi trailer	7' 3"

(detached from tractor on landing wheels)

6 x 4 - 8 Normal Control

(For Wheelbase of motive Unit and semi-trailer see plan on page 539)
Motive Unit is similar to 30 ton model, except for tyre equipment.

ENGINE:- Make Gardner. 6LW 6 cyls. C.I. Bore 4½" Stroke 6" Capacity 8369 c.c. 8.4 litres. B.H.P. 102 @ governed speed 1700 R.P.M. Max torque lbs. ins. 4164 @ 1100 R.P.M.

CLUTCH:- Borg and Beck 16" Single Dry Plate.

GEAR BOX:- 6 speeds and reverse.

Ratios:	Overall	Gear
1.	181.00	7.12
2.	112.00	4.43
3.	70.20	2.75
4.	40.80	1.61
5.	25.50	1.00
6.	15.90	0.62
R.	232.00	9.10

REAR AXLE:- Double Reduction. Spiral Bevels and Double helical spur gears. Ratio 13.86 - 1. Final drive by Scammell gear train bogie by pinions on differential shafts meshing via idlers with gear wheels on road wheel shafts. Overall axle ratio 25.5 to 1.

SUSPENSION:- F. and R. Semi-elliptic.

BRAKES:- FOOT: Compressed air servo on all wheel .
HAND: (1) Neate brake on driving wheels.
(2) Transmission brake.

STEERING:- Right Hand Drive. Cam and Roller (For turning circles see performance figures).

TYRES:- Front 2. 10.50/20. Rear 4. 15.00/20. Semi Trailer 8. 10/50/20. Singles all round, motive unit and semi-trailer.

FUEL CAPACITY:- 1 Tank total of 54 gals.

COOLING SYSTEM:- 7½ gals.

SCAMMELL 20-TON PERFORMANCE FIGURES

(Based on gross weight of 35.14 tons laden with 20 ton tank)

B.H.P. per ton 2.87.

Tractive effort (100%) = 960 lbs/tons.
" " (85%) = 816 lbs/tons.

Adhesive Weight (Ratio driving wheel to gross vehicle weight) = .395.

Tractive Effort, maximum available on good roads when fully laden, 25,250 lbs. (Limiting factor, adhesion).

MAX. SPEEDS

Gear		
1	-	1.35 m.p.h.
2	-	2.2 "
3	-	3.5 "
4	-	6.0 "
5	-	9.7 "
6	-	15.5 "

MAX. GRADEABILITY (Estimated)

(Based on co-efficient of adhesion of .8, dry concrete road) 1 in 3.5

MANOEUVRABILITY

Minimum Turning circle (outer)

	Left hand	Right hand
	71' 8"	73 ft.
(Inner)	25' 7"	27' 6"

FUEL CONSUMPTION

Estimated - 4 m.p.g.
Radius of action - 220 miles

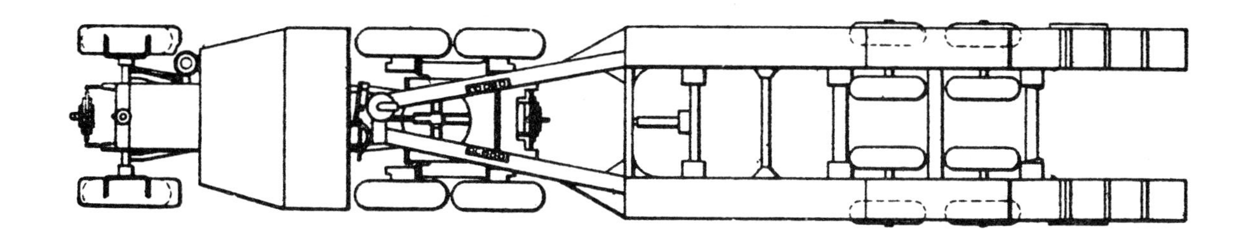

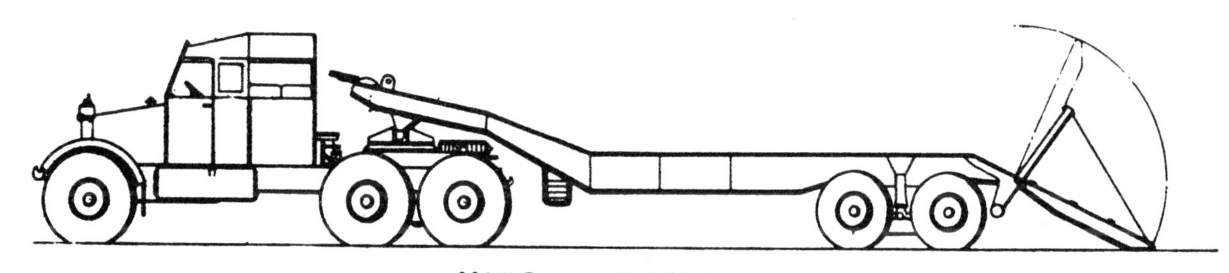

SCAMMELL 20 TON RECOVERY

Note: Front tyres are 10.50 - 20, same size as semi-trailer tyres.

WEIGHTS

	F. AXLE	R. BOGIE	SEMI-TRAILER BOGIE	GROSS
Unladen	3 tons 0 cwt.	7 tons 8¾ cwt.	5 tons 1 cwt.	15 tons 10 cwt.
Laden	3 tons 5 cwt.	14 tons 1¾ cwt.	18 tons 1 cwt.	35 tons 8 cwt.

LOADING DIMENSIONS

Length of main tank carrier runway	17' 7½"
Overall width across runways	8' 8"
" " inside "	5' 0"
Width of each runway	1' 10"

C.G. of 20 ton tank is carried 7' 10" forward of semi-trailer bogie centres.

Height of runway from ground (laden) 4' 2 5/8"

LOADING RAMPS:- Length 6' 6". Overall width across ramps 8' 8". Width of each ramp 1' 10". Ramp angle 25 degrees.

BRIDGING DIMENSIONS

Wheelbase (F. axle to C.L. rear bogie)	15' 0"
Bogie centres (Tractor)	4' 6½"
C.L. tractor bogie to C.L. Trailer Bogie	22' 6"
Trailer bogie centres	4' 5"

Semi-Trailer mounting is 9" forward of tractor bogie centres.

TRACK:- F. Axle Tractor 6' 9½"
R. Bogie " 6' 7½"
Semi Trailer (Outer wheels) 8' 1½"
" " (Inner ") 3' 1½"
Width over tyres r. bogie tractor 7' 10½"
" " " " " trailer 9' 2"

SCAMMELL 20 TON SEMI-TRAILER

CAB:- Enclosed steel panelled type seating 3 in front compartment, with an extension at the back seating a further 4 men.

WINCH:- Scammell 8 ton vertical spindle type winch fitted with 600 ft. of 13/16" dia. rope. Torque limiting device is set for a pull of 8 tons.

RAMPS:- Ramps are carried in position shown in side elevation. They are lowered by 2 small hand winches, the cables running over detachable stanchions fitted into the rear sloping section of the runway.

TANK LOCATING GEAR:- Adjustable bollards (or tank locating chocks) are provided to fit the inner sides of the tank runways.

Equipment boxes carried under floor of semi-trailer.

Motive Unit is fitted with W.D. type drawbar gear front and rear.

SHIPPING DIMENSIONS

Overall Length (Tractor & Trailer with Ramps in raised position)	49' 3"
Overall Length (Tractor)	22' 0"
" " (Semi Trailer)	35' 7"
Overall width (Tractor)	8' 7"
" " (Semi-Trailer)	9' 2"
Overall Height (top of cab)	9' 5"
" " (top of ramps in raised position)	9' 4"

SCAMMELL 30-TON SEMI-TRAILER RECOVERY

6 x 4 - 8 Normal Control

(For Wheelbase of motive Unit and semi-trailer see plan on page 543)

ENGINE:- Make, Gardner, C.1. 6 cyls. Bore, 4¼" Stroke, 6" Capacity, 8369 c.c. 8.4 litres. B.H.P. 102 @ governed speed 1700 r.p.m. Max, torque lbs./ins. 4164 @ 1100 R.P.M.

CLUTCH:- Borg and Beck 16" Single Dry Plate.

GEAR BOX:- 6 speeds and reverse.

Ratios:	Overall	Gear
1.	181.00	7.12
2.	112.00	4.43
3.	70.20	2.75
4.	40.80	1.61
5.	25.50	1.00
6.	15.90	0.62
R.	232.00	9.10

REAR AXLE:- Double Reduction. Spiral Bevel and Double helical spur gears. Ratio 13.86 - 1. Final drive by Scammell gear train bogie by pinions on differential shafts meshing via idlers with gear wheels on road wheel shafts. Overall axle ratio 25.5 to 1.

SUSPENSION:- Front and Rear, Semi-elliptic.

BRAKES:- Foot, Compressed air servo on all wheels. Hand, (1) Neate brake on driving wheels. (2) Transmission brake.

STEERING:- Right Hand Drive. Cam and Roller (For turning circles see performance figures).

TYRES:- Front 2. 13.50/20. Rear 4. 15.00/20. Semi Trailer 8. 13.50/20. Singles all round, motive unit and semi-trailer.

FUEL CAPACITY:- 1 Tank total of 54 gals.

COOLING SYSTEM:- 7¼ gals.

SCAMMEL 30-TON
PERFORMANCE FIGURES

(Based on gross weight of 49.75 tons, laden with Matilda Tank)

B.H.P. per ton = 2.05

Tractive effort (100%) = 664 lbs/ton.

" " Maximum available good roads fully laden, 27,770 lbs. (Limiting factor adhesion).

Adhesive Weight (Ratio of weight on driving wheels to gross weight) = .31

MAX SPEEDS

Gear		
1	-	1.35 m.p.h.
2	-	2.2 "
3	-	3.5 "
4	-	6.0 "
5	-	9.7 "
6	-	15.5 "

Average speed (100 miles circuit) = 9 m.p.h.

MAX. GRADEABILITY:- On the Scammel 30-ton this is governed by steering ability. Weight is transferred from front Axle (due to draft pull on ball joint when ascending hill) causing front wheels to lose steering ability. Therefore maximum gradient climbable on a hill involving curves is 14% or 1 in 7.

On a dead straight hill the vehicle can successfully negotiate a gradient of 22% or 1 in 4.5 (Dry concrete surface).

MANOEUVRABILITY

Minimum Turning Circle (outer) left hand - 79 ft. right hand - 81 ft.

Minimum Turning circle (inner). 38½ ft.

FUEL CONSUMPTION

100 miles circuit - 3.6 m.p.g.

Radius of Action - 180 miles

CAB:- Enclosed steel panelled type seating 3 in front compartment, with an extension at the back seating a further 4 men.

WINCH:- Scammell 8 ton vertical spindle type winch fitted with 600 ft. of 13/16" dia. rope. Torque limiting device is set for pull of 8 tons. Snatch block with four-part tackle at front end of semi-trailer.

RAMPS:- Are carried in position shown in side elevation. They are lowered by 2 small hand winches, the cables running over detachable stanchions which can be placed in the semi-trailer floor.

JACKS:- 2 hydraulic type are provided to support rear of semi-trailer when loading. They can be repositioned to jack up semi-trailer frame for changing a wheel, or for detaching the semi-trailer from the motive unit.

TANK LOCATING GEAR:- Adjustable bollards (or tank locating chocks) are provided to fit on the inner sides of the tank runways, and the ramps when loading.

Equipment boxes carried under semi-trailer. Motive Unit is fitted with W.D. type drawbar gear front and rear.

SHIPPING DIMENSIONS

*Overall Length (Tractor & Trailer with ramps in raised position)	49' 8"
Overall width tractor	8' 7"
" " S. Trailer	9' 5½"
Overall Height (top of cab)	9' 5"
" " (top ramps in raised position)	10' 11"
*Overall Length Tractor	22' 0"
" " Semi Trailer	36' 6"

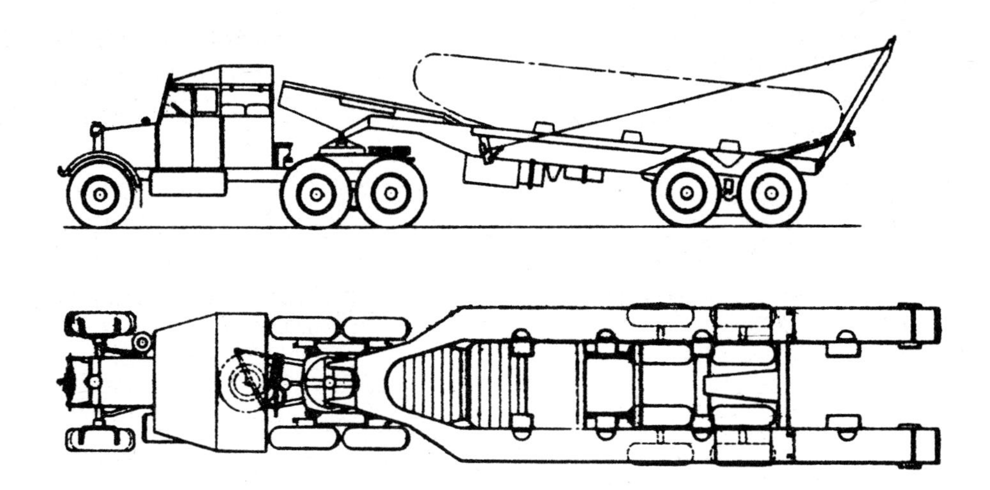

SCAMMELL 30 TON RECOVERY

WEIGHTS

	F. Axle	R. Bogie	Semi-Trailer Bogie	Gross
Unladen	3 tons 12 cwt.	8 tons 0 cwt.	8 tons 6 cwt.	19 tons 18 cwt.
Laden (Matilda)	3 tons 17 cwt.	15 tons 10 cwt.	30 tons 9 cwt.	49 tons 16 cwt.

LOADING DIMENSIONS

Length of main tank carrier runways (from front to adjustable section over rear wheels)	15' 7½"
Overall width across runways	9' 2"
Width of each runway	2' 0"

C.G. of 30 ton tank is carried 6' 4½" forward of semi-trailer bogie centres. Ht. of runway from ground at this point is 5' 1".

C.G. of 20 ton tank carried 7' 6" forward of bogie centres. Ht. of runway from ground at this point is 5' 3".

LOADING RAMPS:- Length 11' 0". Width across 9' 2". Width inside 5' 2". Angle of ramp 18 degrees.

BRIDGING DIMENSIONS

Wheelbase (F. axle to C.L. rear bogie)	15' 0"
Bogie Centres (Tractor)	4' 6½"
C.L. Tractor Bogie to C.L. Trailer Bogie	22' 4"
Trailer Bogie centres	5' 1"

Semi-trailer ball mounting is 9" forward of tractor bogie centres.

TRACK:-	F. Axle Tractor	6' 10"
	R. Bogie Tractor	6' 7½"
	Semi Trailer (Outer wheels)	8' 3"
	" " (Inner wheels)	3' 9"
	Width over tyres R. bogie tractor	7' 10½"
	Width over tyres R. bogie trailer	9' 5½"

DIAMOND T 40 TON
TRANSPORTER TRACTOR MODEL 980
RECOVERY TRACTOR MODEL 981 *(see below)

6 x 4 Normal Control. Wheelbase 14' 11½"
Bogie centres 4' 4"

ENGINE:- Make, Hercules DFXE. C. 1. 6 cyls. Bore, 5 5/8" Stroke, 6" Capacity 14.5 litres, 893 cu. ins. B.H.P. 201 @ governed speed 1600 R.P.M. Max. torque, 8220 lbs./ins. @ 1150 R.P.M.

CLUTCH:- W.C. Lipe 14½" Two dry plate disc.

GEAR BOX:- 4 Speeds and reverse. Fuller 4B86

AUXILIARY BOX:- 3 Speed Fuller 3A86.

REAR AXLES:- Double reduction spiral bevel and spur. Ratio 11.66 - 1.

TRANSMISSION:-

	0.77	Direct	1.99
Auxiliary Ratio	0.77	Direct	1.99
Gear Ratios	4.27	5.55	11.04
	2.51	3.27	6.40
	1.35	1.76	3.50
	0.77	1.00	1.99
Axle Ratio 11.66-1	49.80	64.70	128.70
	29.30	38.10	75.80
	15.70	20.50	40.80
	8.98	11.66	23.20

SUSPENSION:- Front, Semi-elliptic. Rear, Inverted semi-elliptic.

BRAKES:- Foot, Westinghouse compressed air servo on all wheels. Hand, Disc on transmission. Two pipe line trailer brake connections front and rear.

STEERING:- Left hand drive. Ross cam and lever.

TYRES:- 10. 12.00/20 Mud and Snow. Dual rear.

ELECTRICAL:- Starter 24V. Lighting 6 V.

FUEL CAPACITY:- 2 Tanks total of 125 gals.

COOLING SYSTEM CAPACITY:- 13 gals.

*MODEL 981. Recovery Tractor is identical with 980 Transporter, except for longer winch rope and the fact that winch cable guide fittings are provided to pull from the front of the tractor as well as the rear (rear Model 980).

PERFORMANCE FIGURES (SOLO)

(Based on gross vehicle weight of 19 tons when laden with 6.65 tons ballast).

B.H.P. per Ton 10.6
Tractive Effort (100%) - 50,600 lbs. - 2,665 lbs/Ton
" " (85%) - 43,200 lbs. - 2,265 lbs/Ton
Tractive Effort, maximum available on good roads when laden 6.65 tons ballast (.8 adhesion):-
25,000 lbs. - 1,327 lbs/Ton

Adhesive weight (Ratio driving wheel to gross vehicle weight) = .74

MAX. SPEEDS @ GOVERNED R.P.M.

AUX. RATIO	0.77	1.00	1.99
1st Gear	4.00 m.p.h.	3.08 m.p.h.	1.05 m.p.h.
2nd "	6.82 "	5.23 "	2.67 "
3rd "	12.66 "	9.71 "	4.88 "
4th "	22.20 "	17.10 "	8.59 "

Average Speed (100 mile circuit)
Estimated = 15.1 m.p.h.

PERFORMANCE FIGURES (TOWING 40 TON TRAILER)

(Based on gross train weight of 69.4 tons, made up of Diamond T. Tractor laden with 6.65 tons of ballast, and Rogers 40-ton Trailer laden with A.22 Mk. IV tank).

B.H.P. per ton 2.89
Tractive effort (100%) 730 lbs/ton
" " (85%) 620 lbs/ton
Adhesive weight (Ratio driving wheel to gross vehicle weight) .202

GRADEABILITY:-
With adhesive factor of .8 = 1 in 6.8 or 14.75%

MANOEUVRABILITY:-
Minimum Turning circles:-
Outer = 71 ft.
Inner = 36 ft.

FUEL CONSUMPTION:-
Tractor (Solo) = 5.25 m.p.g.
Tractor and Trailer laden = 2.2 m.p.g. Radius 275 miles

DIAMOND T 40 TON TRACTOR

Model 980 - Transporter
" 981 - Recovery

CAB:- All enclosed, steel panelled cab, seats 2. Ventilator in roof. Non-detachable top to cab.

BODY:- All steel, fixed sides, drop tailboard. Overall body dimensions 9'0" long x 8'4" wide. Front section of body measuring 17½" long x 8'4" wide is partitioned off to provide compartment for spare wheel and 2 open bins. Two enclosed bins each 54" long x 18" wide in both rear corners of body.

WINCH:- Make Gar Wood. Model 5M723B. Nominal capacity 40,000 lbs. Fitted with torque control which is set to cut out at pull of 22,400 lbs. on the first layer of the drum.

WINCH CABLE:-
Length (Model 980 Transporter) .... 300 ft.
" (Model 981 Recovery) .... 500 ft.

Diameter of cable 7/8" Breaking strain of cable 53,000 lbs.

Model 980 has winch cable line fittings to pull from rear only.
Model 981 has fittings to pull from front and rear.

SPECIAL FITMENTS
Drawbar gear front and rear.
2 line air trailer brake connections front and rear.

SHIPPING DIMENSIONS

Overall Length 22' 7". Overall width 8' 4"
Overall Ht. top of cab 8' 3"

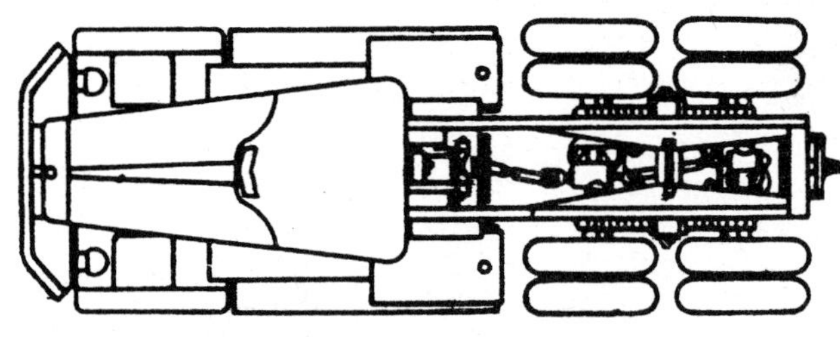

DIAMOND T 40 TON TRACTOR

WEIGHTS

	UNLADEN	LADEN (6.65 tons ballast)
F. Axle	4 tons 17¼ cwt.	4 tons 18 cwt.
R. Bogie	7 tons 1¼ cwt.	14 tons 0½ cwt.
Gross	11 tons 18½ cwt.	18 tons 18½ cwt.

BRIDGING DIMENSIONS

F. Axle to Middle Axle.	153¼"
Middle Axle to R. Axle.	52"
R. Axle to centre Towing Hook.	31¼"
Track, front.	75 7/8"
Track, rear.	74"
Dual tyre spacing.	13½"

Angle of approach 40½ degrees. Angle of Departure 51 degrees.

GROUND CLEARANCES:- F. Axle. 13 1/16"
Belly 17 7/8" R. Axle. 11 1/8"

TRAILER 40 TON 24 WHEELED TRANSPORTER

Makes: Rogers D45 LF1 and Winter Weiss (to Rogers design) S.M. 2114, 2060, 2148, 2309, 2434.

TRAILER 40 TON 24 WHEELED RECOVERY

Make: Winter Weiss (see note below) S.M.2061, 2113.

GENERAL CONSTRUCTION

Transporters, whether made by Rogers or Winter Weiss, are built to the same design. Recovery trailers are similar to transporters except for the addition of snatch blocks for multiplying winch pull.

Frame is of all-welded construction.

FRONT END:- Two oscillating ½ Axles (spring mounted) in line each with 2 dual wheels. Total 8 tyres. Front dolly (or turntable) can assume an angle of 90° to the trailer.

REAR BOGIE:- 2 bogie assemblies each with 2 oscillating axles carrying 2 dual wheels. (Total 16 tyres).

BRAKES:- Westinghouse 2 pipe line air pressure operated on all wheels. Wheel type hand parking brake operating mechanically on all rear bogie wheels.

WHEELS & TYRES:- Wheels, make both types Rogers, size 15 x 7. Tyres 8.25-15 14 ply. 24 running tyres fitted plus 2 spares.

LOADING RAMPS:- Hinged triangular construction.

LOADING TACKLE:- Four adjustable tank locating chock blocks. Two lashing eyes at front 5 lashing "D"s along each side, 4 at front and 2 at rear. Winch cable shieve and roller at front.

ON RECOVERY TRAILERS ONLY:- One anchored snatch block and one movable snatch block carried on central cable trough (or cake-walk). All types are now to be fitted out with snatch blocks.

BRIDGING DIMENSIONS

Length centre of tow bar eye to front axle =95"
Wheelbase (to c.l.r. bogie) =187"
Rear bogie centres =40"

TRACK:- measured across outer pair of duals=86"
measured across inner pair of duals=25"
Dual tyre spacing =10"
Overall width across outer edge of outer tyres =108"

REAR CLEARANCE:- Angle of Departure 24 degrees. Ht. extreme rear of frame from ground (laden) 18"

SHIPPING DIMENSIONS

Overall length (towbar to rear of ramp when ramp is folded up)-30' 2"
Overall width 114" Overall ht. 57"

ACTUAL PACKS AS SHIPPED (SUP)

Gross wt. 25,500 lbs. Net wt. 22,800 lbs.
Three cases:-

1. Frame, Drawbar, etc=287" x 114" x 40", total 757 cubic ft.
2. Front wheel assembly and miscellaneous 107½" x 63" x 42½", 170 cu. ft.
3. Rear bogie assemblies, etc. 112" x 79" x 36", 184 cu ft.

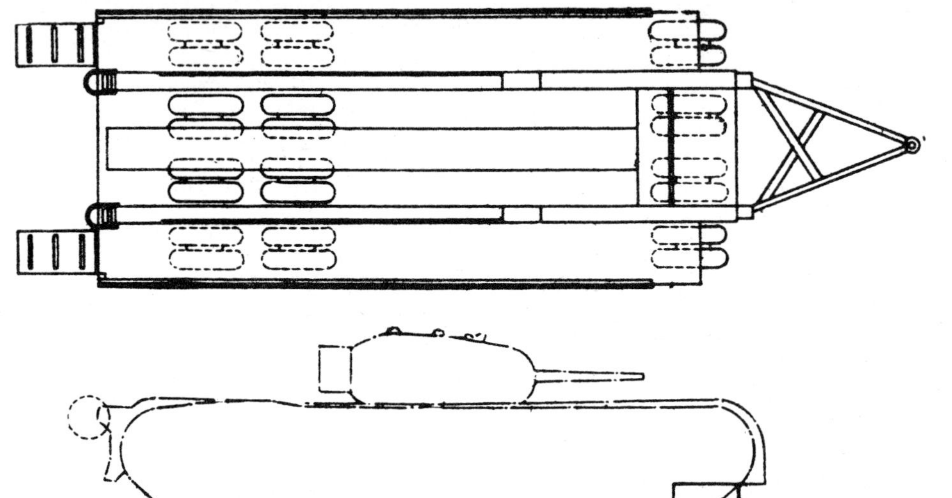

TRAILER 40 TON 24 WHEELED TRANSPORTER AND RECOVERY (ROGERS, WINTER WEISS)

IMPORTANT:- There are no outer guide rails to the runways. Therefore wider tanks than Churchill which may project beyond tank runways can be carried.

WEIGHTS

	UNLADEN	LADEN (Churchill)
F. Axle.	4 tons 7½ cwt.	16 tons 15½ cwt.
R. Bogie	5 tons 15½ cwt.	33 tons 11½ cwt.
Gross	10 tons 2 cwt.	50 tons 7 cwt.

LOADING DIMENSIONS

LENGTH of tank runways from foot of fixed front stop to C.L. rear bogie=169½"

From C.L. rear bogie runway is continued horizontally for approx. 30" and then slopes gradually downwards to meet ramp.

C.L. of heaviest load is designed to be carried 58" forward of rear bogie centres.

OVERALL WIDTH across tank runways=114"

Width over foot of inside guide angles to runway is adjustable from 67" to 61"

Height of guide angles can be altered from 4¾" to 6"

Guide angles are splayed inwards at rear.

LOADING RAMPS:- Each 18" wide. Width over ramps 105". Width between ramp 69" Ramp angle 30°

HT. OF RUNWAYS (laden)=39"

OVERALL HEIGHT Laden Churchill Tank: 11' 6"

SEE OVERLEAF FOR BRIDGING AND SHIPPING DIMENSIONS.

TRAILER 40 TON
24 WHEELED
TRANSPORTER
(BRITISH MARK I)

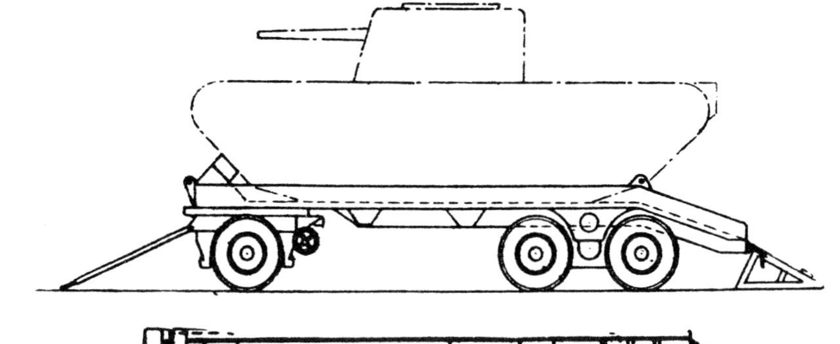

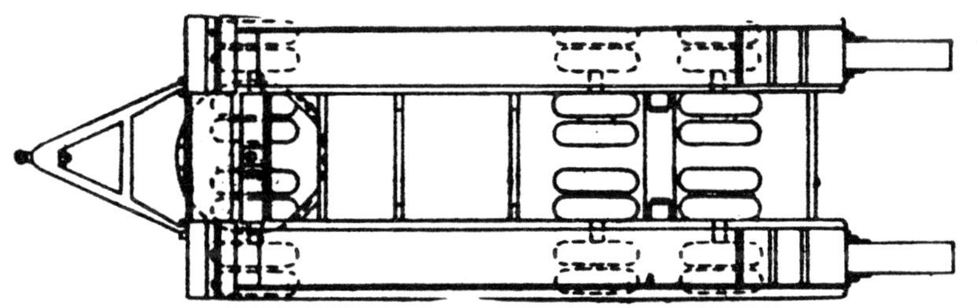

WEIGHTS

	FRONT AXLE	REAR BOGIE	GROSS
UNLADEN	4 tons 16 cwt.	8 tons 16 cwt.	13 tons 12 cwt.
LADEN (Churchill Tank)	15 tons 14 cwt.	37 tons 0 cwt.	52 tons 14 cwt.

LOADING DIMENSIONS

Length of tank runways from foot of fixed front stop block to C.L. rear bogie=14' 6"

From C.L. of rear bogie, runway extends horizontally for 30" and from there slopes downwards to point of ramp attachment.

C.L. of heaviest load is designed to be carried 60" forward of rear bogie centres.

Overall width measured inside the outer guide rails to the runway = 9' 3"

Width measured over the inner guide rails = 5' 0"

Height of inner guide rails 8½"
(Guide rails are not adjustable)

LOADING RAMPS Each 12" wide x 3' 9" long.
Width over ramps 8' 2"
Width between ramps 6' 2"

HEIGHT OF RUNWAYS (laden) 3' 9½" Overall Ht. laden Churchill Tank, 12' 0½"

HEIGHT OF TOWING EYE FROM GROUND=2' 7"

SHIPPING DIMENSIONS

Overall length (tow bar to rear of ramp when ramp is folded up)	31' 11"
Overall length less tow bar	26' 4"
Overall width	10' 0"
Overall height	5' 10"

TRAILER, 40 TON, 24 WHEELED, TRANSPORTER

BRITISH MK.I TYPE

(Originally known as the Crane design, but has also been made by Dysons and other manufacturers. Now superseded by Mk.II. See following pages).

FRONT AXLE:- Eight wheels spring mounted as two pairs of duals. Total 8 tyres. Restricted lock turnable steering limited to 45° each way.

CENTRE AND REAR AXLES:- Each eight wheels, mounted as two pairs of duals, on oscillating ½ axles.

BRAKES:- Westinghouse 2 pipe line air pressure system operating on all wheels. Hand wheels on trailer actuate all front and rear brakes mechanically.

TYRES:- 36 x 8 all round, 2 spares carried on spare hub. Total 24 running tyres, plus two spares.

LOADING RAMPS:- Hinged type of triangulated construction fitted to end of fixed ramp.

LOADING TACKLE:- Adjustable tank, locating chock blocks. Fixed front stop block. Winch cable guide fittings and roller.

BRIDGING DIMENSIONS

Length centre of two bar eye to front axle = 8'5"
Wheelbase (to C.L. rear bogie) =15'0"
Rear bogie centres = 4'6"

TRACK:- Measured across outer pair of duals =7'10½"
Measured across inner pair of duals =2' 7½"
Dual tyre spacing = 10¾"

Overall width across outer edge of outer tyres =9'6½"

REAR CLEARANCE:- Angle of Departure = 20°
Height extreme rear of frame from ground, unladen =18"

TRAILER, 40 TON, 24 WHEELED, TRANSPORTER

BRITISH MK.II TYPE

(Originally known as the Dyson design, but is also being made by Cranes and other manufacturers).

THE MAIN DIFFERENCES OF THIS IMPROVED DESIGN ARE:-
Lower Unladen Wt., 11 tons 16 cwt. as against 13 tons 12 cwt. for MK. I
Welded frame. Ht. of runways 4'0½"
No outer guide rails to the tank carrier runways, so enabling larger tanks than the Churchill (which may project beyond the runways) to be carried.
Inside guide rails to the runways are adjustable (from 5'0" to 5'8" measured across).
The Mark II utilises the same axle, hub and wheel components as the Mk.I Type (see page 550) and has similar:-
Westinghouse 2 pipe line system air pressure operated brakes on all wheels. Hand wheels operate brakes mechanically on all wheels.
36 x 8 tyres. 24 running tyres plus two spare.
Hinged triangular loading ramps.
Front Drawbar

BRIDGING DIMENSIONS

Length centre of tow bar eye to front axle = 8'5"
Wheelbase (to C.L. & Bogie) =15'0"
Rear bogie centres = 4'6"

TRACK:- measured across outer pair of duals = 7'10½"
measured across inner pair of duals = 2'7½"
Dual tyre spacing 10¾"

Overall width across outer edge of outer tyres 9'6½"

REAR CLEARANCE:- Angle of Departure 18 degrees.
Ht. extreme rear of frame from ground 16¾" unladen.

SHIPPING DIMENSIONS

Overall length (tow bar to rear of ramp when ramp is folded up) 32'4½"
Overall length less tow bar 26'6½"
Overall width 10'0"
Overall Ht. 5'8½" unladen.

40-TON TRANSPORTER & RECOVERY TRAILERS

There are four main types of 40 ton Trailers as follows:-

1. The British Mk.I type (originally known as the Crane design) which is now superseded by
2. The British Mk.II type (originally known as the Dyson design)

These two types are now to be known only as Mk.I and Mk.II as Dysons and other manufacturers have built Mk.I trailers to the Crane design and vice versa.

3. The U.S.A. Rogers Transporter Trailer (also made by Winter-Weiss to Rogers design)

4. The U.S.A. Recovery Trailer, which is basically the Roger's Transporter with snatch blocks, although it has been made only by Winter-Weiss.

HOW TO IDENTIFY THE FOUR 40 TON TRAILERS

1. The British Mk.I is the only type with guide rails extending along the outer side of the tank carrier runways.

2. The British Mk.II type has no outer guide rails, but like the British Mk.I the tank carrier runways extend to the front of the trailer terminating in triangular stop blocks.

3. Whereas on the U.S. Rogers trailer the carrier runways extend only as far as the rear of the front wheel and terminates in a sloping stop plate in front of which is the hand brake wheel.

4. The U.S. Recovery trailer is the same as the U.S. transporter type except for the fitment of snatch blocks carried on the central cable trough. All transporter trailers are to be equipped with snatch blocks and hence this distinction between transporter and recovery type trailers will disappear.

TRANSPORTER 30 TON 6 x 4-8 SEMI-TRAILER RECOVERY

TRACTOR: DIAMOND T. 6 x 4 MODEL 980
SEMI TRAILER: SHELVOKE & DRURY

The Tractor is a standard Diamond T Model 980 (as described page 544), modified by the fitment of semi-trailer turntable in place of ballast body.

The semi-trailer rear bogie is unsprung and is similar to that fitted to the British Mk.I and Mk.II 40 ton transporter Trailers. (Pages 550-553).

Semi-trailer tyre equipment is eight 13.50-20 single tyres.

Brakes are 2 pipe line air pressure operated.

Ramps are spring balanced, triangulated type, can be operated by two men.

Semi-trailer can carry any tank up to 30 tons weight with inside track widths varying from 5' 0" to 7' 0".

Winch is standard Diamond T with suitable guides and pulleys to permit loading when trailer is out of line with tractor.

PERFORMANCE FIGURES (Laden Sherman tank)

B.H.P. per ton 3.76.
Tractive effort per Ton (100%) 944 lbs./Ton.
Adhesive Weight (Ratio of driving wheel to gross vehicle weight) .34.

MANOEUVRABILITY
Minimum Turning Circles.
Right, Outer, 76' 3" Inner 36' 6"
Left, Outer, 75' 0": Inner 33' 8".

FUEL CONSUMPTION

Approx. 2¼ to 2½ m.p.g.

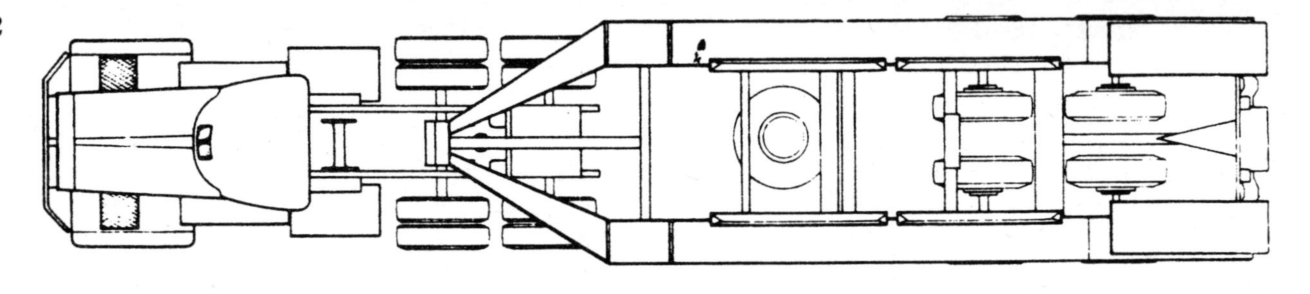

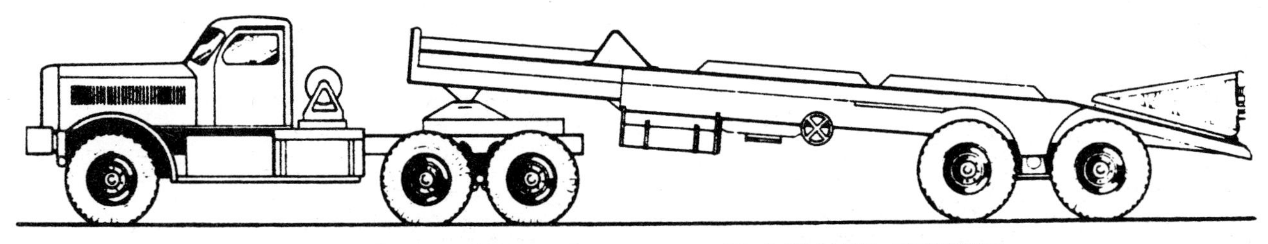

TRANSPORTER 30 TON 6 x 4-8 SEMI TRAILER RECOVERY

DIAMOND T.980 TRACTOR COUPLEL SHELVOKE & DRURY SEMI-TRAILER

WEIGHTS

	UNLADEN	LADEN (SHERMAN)
F. Axle	4 tons 11½ cwt.	5 tons 0 cwt.
Tractor Bogie	8 tons 4½ cwt.	18 tons 2 cwt.
Trailer Bogie	9 tons 13 cwt.	30 tons 6 cwt.
Gross	22 tons 9 cwt.	53 tons 8 cwt.

DIMENSIONS

Overall complete unit
Length 49' 10" (ramp up) Width 9' 8"
Ht. (Tractor) 8' 3"

SHIPPING DIMENSIONS

Tractor (as Diamond T on page 546)
Semi Trailer, uncoupled, Length 35' 5" Width 9' 8"
Ht. 8' 0"

TRAILER, 45 TON, TRACKED RECOVERY

MAKE: BOULTON & PAUL

GENERAL: To be towed by a Churchill A.R.V. and designed to give maximum cross-country performance. Speed is restricted to 5 m.p.h. and trailer has limited life on hard roads, hence it will usually be transported over roads on a wheeled recovery trailer.

Mk. I design has solid suspension.

Mk. II design has rubber blocks interposed between frame and axles carrying track units.

TRACKS: Girder type - Orolo design. Ground pressure 22 lbs. per sq. in. laden.

STEERING: Turntable.

BRAKES: None.

Tank Platform is flat except for chock blocks to locate tank in position.

LOADING RAMPS: Folding type, spring loaded for easy handling. Normally they are raised by trailer winch.

POWER WINCH: mounted at front of trailer. Type Scammell 8 ton vertical spindle type. Driven by Dagenham Ford V.8 engine. Armoured compartment for winch operator.

TRAILER, 45 TON, TRACKED RECOVERY

MAKE: BOULTON & PAUL

LOADING DIMENSIONS

Length of tank platform	20' 0"
Width over " "	10' 3"
Ht. of platform at fixed front chock	6' 0"
Ht. of rear end of platform	4' 3"
Length of ramp extended	8' 6" (measured on ground)
Width of each ramp	2' 0"
Width across ramps (outer positions)	9' 9"
(inner position)	7' 5"

SHIPPING DIMENSIONS

Overall length, inc. drawbar, ramps vertical	34' 8"
Overall length, less drawbar	30' 3"
Overall Width	10' 3"
Overall Height (top of winch comp)	11' 6½"

BRIDGING DIMENSIONS

Length toweye to centre of front track	10' 0"
Length centre front to centre rear track	18' 0"
Width over tracks	9' 3"
" inside "	4' 6½"

WEIGHTS

UNLADEN	LADEN (45 tons)
31 tons 1 cwt.	76 tons 1 cwt.

DATA BOOK OF WHEELED VEHICLES

PART 6

NEW VEHICLES AND MOTOR CYCLES

This section covers new type vehicles and motor cycles produced since the publication of the Third Edition.

Vehicles are not separated according to county of origin in this new section. British, Canadian and U.S. types are all grouped together starting with motor cycles and working up by weight capacity to Tractors and Amphibians.

New Trailers are included in Part 7, pages 701 onwards.

Published by D. G. F. V. Ministry of Supply.

ENFIELD LIGHTWEIGHT MODEL WD/RE 125 C.C.

GENERAL: Special lightweight model with folding handle bars and footrests.

ENGINE: Make, Enfield. Single cylinder, 2 stroke.

Bore 54mm. Stroke 55 mm. Capacity 125 C.C.

Max. B.H.P. 2.6 @ 4500 r.p.m.

CLUTCH: Single dry plate, cork insets.

GEAR BOX: 3 speed. Overall ratios 22.7 to 1, 12.7 to 1, 7.8 to 1.

TRANSMISSION: Primary 3/8" x .225" x 58.
Secondary ½" x .312 x 103.

BRAKES: Internal Expanding. Front drum 3¾" hand operated. Rear drum 5", foot operated.

TYRES: 2.75 x 19 (Early type 2.50 x 19).

WHEELBASE: 4' 0".

GROUND CLEARANCE: 4½".

TANK CAPACITY: 1½ gals. Radius of action 140 miles.

OVERALL DIMENSIONS: Length 6' 2". Width, folded, 1' 4". Height 2' 10½".

UNLADEN WEIGHT: 125 lbs. dry. 137 lbs. full fuel.

JAMES LIGHTWEIGHT
MODEL M.L. 125 C.C.

GENERAL: Special lightweight model with folding handlebars and footrests.

ENGINE: Make, Villiers. Single cylinder, 2 stroke.

Bore 50 mm. Stroke 62 mm. Capacity 122 C.C.

Max. B.H.P. 3.2 @ 3,800 r.p.m.

CLUTCH: Make, Villiers, Single plate with cork inserts.

GEARBOX: Villiers, 3 speed. Overall Ratios. 22.9 to 1, 12.9 to 1, 8.1 to 1.

TRANSMISSION: Primary, 3/8" x .25 x 58.
Secondary ½" x .335 x 105.

TYRES: 2.75-19.

WHEELBASE: 4' 0".

GROUND CLEARANCE: 4 3/8".

TANK CAPACITY: 2¼ gals. Radius of action 200 miles.

OVERALL DIMENSIONS: Length 6' 1". Width (folded) 1' 4". Height 2' 10¾".

UNLADEN WEIGHT: Dry 139 lbs. Full fuel 157 lbs.

TRIUMPH 350 C.C. TYPE 3 H.W.

(Note: This overhead valve engined typs has replaced in production the 3 S.W., side valve type, shown on page 344).

ENGINE: Make Triumph. Single cylinder overhead valve. Bore 70 mm. Stroke 89 mm. capacity 349 c.c. Max. B.H.P. 15.8 @ 5200 r.p.m.

CLUTCH: Multi-plate. Ferodo inserts.

GEAR BOX: 4 speed foot operated. Overall ratios. 17.8 to 1, 13.3 to 1, 8.4 to 1, 5.8 to 1.

TRANSMISSION: Primary ½" x .305" x 74.
Secondary 5/8" x .375" x 90.

BRAKES: Internal expanding 7" drums. Hand operating on front wheel. Foot operating on rear wheel.

TYRES: 3.25 x 19.

WHEELBASE: 4' 5½".

GROUND CLEARANCE: 4".

TANK CAPACITY: Petrol 3¼ galls. Oil 7 pints.

OVERALL DIMENSIONS: Length 7' 0"
Width 2' 7" Height 3' 4".

UNLADEN WEIGHT: with full tanks, 378 lbs.

S.M. 2193, 2220, 2384, 2404

MOTOR CYCLES SOLO

INDIAN 500 C.C. TYPE 741-B

ENGINE: Make: Indian. Twin cylinder V. type. Bore 2½". Stroke 3 1/16" Capacity 30.07 cu. ins. (493 C.C.) Max. B.H.P. 15 @ 4,800 r.p.m.

CLUTCH: Multi-disc Raybestos operating in oil.

GEARBOX: 3 speed. Overall Ratios 12.8 to 1, 7.3 to 1, 5.2 to 1.

TRANSMISSION: Primary, 3/8" Pitch three row chain.
Secondary, 5/8" x 3/8" x 94.

BRAKES: Internal expanding, 7" drums. Hand operating front wheel. Foot-operating on rear wheel.

TYRES: 3.5 x 18.

WHEELBASE: 4' 9½".

GROUND CLEARANCE: 4".

TANK CAPACITY: 2 5/8 Imp. galls. 2 1/8 Imp. qts.

M.P.G. 55 @ average speed of 30 m.p.h. Radius 145 miles.

OVERALL DIMENSIONS: Length 7' 4¼". Width 2' 9". Ht. 3' 3¼".

UNLADEN WEIGHT, full tanks, 475 lbs.

STANDARD 12 H.P. 4 x 2
LIGHT UTILITY

ENGINE: Make, Standard. Petrol 4 cyl. Bore 69.5 mm. Stroke 106 mm. Capacity 1.6 litres. R.A.C. rating. 11.98 h.p. B.H.P. 44 @ 4000 r.p.m. Max torque 820 lbs. ins. @ 2300 r.p.m.

GEARBOX; Four forward 1 reverse. Ratios 3.9 to 1, 2.4 to 1, 1.45 to 1, 1.00 .1; R, 3.9 to 1.

REAR AXLE: Semi floating. Spiral bevel. Ratio 6.29 to 1.

SUSPENSION: Semi-elliptic with hydraulic shock absorbers front and rear.

BRAKES: Hand and foot, Mechanical on all wheels.

WHEELS AND TYRES: Disc wheels 16 x 4.00. Tyres 6.00-16.

WHEELBASE: 9' 0". Track f.4' 1", r. 4' 3½".

TURNING CIRCLE: 40 feet.

ELECTRICAL EQUIPMENT: 12 volt.

CAPACITIES: Fuel 11½ gals. Cooling 2¼ gals.

M.P.G.: 22 Radius 250 miles.

MAX SPEED: 50 m.p.h.

CAB AND BODY
Generally similar to Hillman shown on page 311 except cab fitted with detachable canvas canopy and spare wheel carried on tailboard.

OVERALL DIMENSIONS: Length 13' 10" Width 5' 4½" Ht. 6' 8". Cut down Ht. 4' 6".

WEIGHTS

	UNLADEN	LADEN
F. Axle	12¼ cwt.	15¾ cwt.
R. Axle	11 cwt.	19¼ cwt.
Gross	1 ton 3¼ cwt.	1 ton 15 cwt.

CHEVROLET HEAVY UTILITY
4 x 4 C.8A

WHEELBASE 8' 5"

CHASSIS MODEL 8448.
COMPLETE VEHICLE MODEL 8445

ENGINE: Make, G.M. valve in head. Petrol 6 cylinder. Bore 3½". Stroke 3¾. Capacity 216 cu. ins. R.A.C. 29.4 h.p. Max. B.H.P. 85 @ 3,400 r.p.m. Max. torque 2,040 lbs./ins. @ 1,200 r.p.m. comp. ratio 6.25 - 1.

CLUTCH:- Single dry plate, 10¾ dia.

GEARBOX:- 4 speeds and 1 reverse. Ratios 7.058 to 1, 3.48 to 1, 1.71 to 1, 1.00 to 1, Reverse 6.982 to 1.

TRANSFER CASE:- Single speed with front axle declutch. Ratio 1.00 to 1.

FRONT AXLE:- Spiral bevel Ratio 6.16 to 1.

REAR AXLE:- Spiral bevel Ratio 6.16 to 1.

SUSPENSION:- Front and rear, semi-elliptic, double acting shock absorbers on all wheels.

BRAKES:- Foot, hydraulic on all wheels. Hand, mechanical on rear wheels.

STEERING:- Right hand drive. Recirculating ball. Ratio 23.6 to 1.

TYRES:- 9.25/16 Pneumatic. (Note Run flat fitted to early Ambulance types).

ELECTRICAL EQUIPMENT:- 6 volts.

TURNING CIRCLE:- 39 ft.

GROUND CLEARANCE:- 9 15/16" under axles.

CAPACITIES:- Fuel 25 gals. Cooling 3¼ gals.

M.P.G.:- 10.5. Radius of Action 260 miles.

TRUCK 15 CWT. 4 x 4 G.S.

ON CANADIAN CHEVROLET C. 15A or 8444 CHASSIS

BODY: G.S. type body similar to Canadian 15 cwt. 4 x 2 shown on page 216. Fitted with detachable canvas tilt and hoops. Spare wheel carried between cab and body.

CAB: Latest Canadian Military Pattern cab with reverse slope windscreen. All enclosed. Seats 2. Split top to cab. Roof hatch fitted above passenger seat on early types. A.A. ring on later types.

DIMENSIONS

	OVERALL	INSIDE BODY
Length	14' 3"	6' 5½"
Width	7' 2"	6' 9½"
Height	8' 1"	4' 6"
Stripped		
Shipping Ht.	6' 3"	

WEIGHTS

	UNLADEN	LADEN
F. Axle	1 ton 10¼ cwt.	1 ton 17¾ cwt.
R. Axle	1 ton 10½ cwt.	2 tons 9¼ cwt.
Gross	3 tons 0¾ cwt.	4 tons 7 cwt.

TRUCK 15-CWT. 4 x 3 G.S.
GUY QUAD ANT

CHASSIS: Generally similar to Guy Quad Ant F.A.T. shown on page 83, except:-

(1) No winch fitted.
(2) Springs modified for load carrier role.
(3) Fuel tank capacity 15½ gals.

TYRES: 10.50/20.

BODY: Similar to standard 15 cwt. 4 x 2 G.S. type shown on page 18A except wider. Space between cab and body for spare wheel.

CAB: All enclosed. Seats 2. Folding windscreen, folding canvas top and detachable side curtains. A.A. ring in cab roof.

W.D. type drawbar gear at rear. Ht. of hook 2' 4½".

DIMENSIONS

	OVERALL	INSIDE BODY
Length	14' 9"	6' 5½"
Width	7' 5"	6' 7"
Height	7' 11"	4' 6"

WEIGHTS

	UNLADEN	LADEN
F. Axle	1 ton 11¼ cwt.	1 ton 18 cwt.
R. Axle	1 ton 8¾ cwt.	2 tons 7¼ cwt.
Gross	3 tons 0 cwt.	4 tons 5¼ cwt.

TRUCK 15-CWT. 4 x 4 G.S.

MAKE CANADIAN FORD. F.15A.
CHASSIS MODEL C.291W.Q.

SEMI-FORWARD CONTROL WHEELBASE 8' 5"

ENGINE:- Make, Ford. Mercury. Petrol V.8. Bore 3.19". Stroke 3.75". Capacity 3.91 litres. Max. B.H.P. 95 @ 3600 r.p.m. Max. torque 2112 lbs. ins. @ 1800 r.p.m.

CLUTCH:- Single dry plate 11" dia.

GEARBOX:- Four speed. Ratios 6.4 to 1, 3.09 to 1, 1.69 to 1, 1.00 to 1, rev. 7.82 to 1.

TRANSFER BOX:- Single speed. Ratio direct.

AXLES:- Ratio 6.5 to 1.

SUSPENSION:- Semi-elliptic front and rear with hydraulic shock absorbers.

BRAKES:- Foot, Hydraulic. Hand, mechanical on rear wheels.

STEERING:- Right hand drive.

WHEELS AND TYRES:- Wheels divided 6.00 x 16. Tyres 9.00-16.

GROUND CLEARANCE:- 8½" @ axles. 17" @ belly.

TURNING CIRCLE:- 52 ft.

TRACK:- 5' 10½".

CAPACITIES:- Fuel 25 gals. Cooling 5 gals.

M.P.G.:- 10. Radius of action 250 miles.

G.S. BODY FITTED TO CANADIAN FORD
Model F.15A. 15 cwt. 4 X 4

The body is identical to that fitted to Chevrolet 15-cwt. 4 x 4 described on page 613 Cab is Canadian Military Pattern with reverse slope windscreen.

WEIGHTS AND DIMENSIONS are for all practical purposes identical to the Chevrolet 15 cwt. 4 x 4 shown on page 613.

TRUCK 15-CWT. 4 x 2 A.A.

Illustration shows Mk.II body on Canadian Chassis.

Similar bodies and equipment have been mounted on the following 15 cwt. 4 x 2 chassis.

Bedford M.W.
British Ford W.O.T. 2.
Canadian Ford F.15 (or C291WF).
Canadian Chevrolet C.15 (or 8421).

BASIC FEATURES applicable to all types.
Platform body equipped with octagonal base plate and clamping brackets to take either latest universal mounting or Hazard Baird mounting for 20 mm. gun.
Locker at front of body for 4 x 60 round circular magazines or 8 x 20 round Polsten magazines. Gun is loaded on vehicle by small hand winch up tubular ramps.
Stowage for tools, gun ramps and spare gun wheel on Mk.IA type (Bedford and British Ford). Mk.II type (Canadian Ford and Chevrolet) additionally provides stowage for a spare vehicle wheel.
WEIGHTS:- In every case within comparable 15 cwt. G.S. loading.

OVERALL DIMENSIONS	Mk.IA (Bedford)	Mk.II Can. Ford and Chevrolet
Length	14' 4"	14' 2"
Width	6' 10"	7' 0"
Height	7' 0"	7' 2"

British Ford as above but 6" longer.

TRUCK 15-CWT. 4 x 2 OFFICE

ON FORD W.O.T. 2.H. CHASSIS

This later type Office Body has superseded in production the earlier Morris type shown on page 23.

Externally the new Office body looks like a standard 15 cwt. G.S. truck with tilt.

The superstructure and the canvas tilt are however of special design and can be erected to form an all-enclosed tent extending on either side of the body.

The body proper seats 4 clerks with a full width desk at the front and two additional fixtures to carry typewriters. The rear of the body houses kit lockers at the sides and the space in the centre is used to carry folding chairs, tables and stools.

The tent extensions when erected at the sides of the body each measure 6' 10" long x 6' 6" wide x 6' 3" ht. The main body measures 6' 5½" long x 5' 10½" wide x 4' 7" ht. Interior light fixtures are provided and the spare wheel carrier is mounted on the tailboard.

OVERALL DIMENSIONS

Length 15' 9¼". Width 6' 5½". Ht. 7' 4½".

WEIGHTS

	UNLADEN	LADEN
F. Axle		1 ton 2¼ cwt.
R. Axle		1 ton 7½ cwt.
Gross	2 tons 9 cwt.	3 tons 0 cwt.

TRUCK 15-CWT. 4 x 2 WIRELESS

The Early Type on the Morris C.S.8 chassis is shown on page 21.

The latest Type (No.1 Mk.III body) is mounted both on the Morris C.4 Mk.II chassis (details as page 620) and on the Guy Ant chassis (details as page 16).

It is equipped with the No.19 Wireless set.

It differs in the following respects from the early type:-

(1) All steel (instead of timber) construction with radiused corners (rounded edges to roof).
(2) Improved weight distribution. Wheelbases of Morris C.4 and Guy chassis are longer and hence there is less overhang.
(3) Map table fitted offside front of body. Exhaust pipe from generating set compartment arranged to exhaust above roof.

As before, it has a generating set compartment at offside rear, auxiliary dynamo driven off P.T.O. and vehicle is fully screened.

LADEN WEIGHTS

	Morris C.4 (Est.)	Guy Ant.
F. Axle	1 ton 6½ cwt.	1 ton 6 cwt.
R. Axle	2 tons 6¼ cwt.	2 tons 5¾ cwt.
Gross	3 tons 12¾ cwt.	3 tons 11¾ cwt.

OVERALL DIMENSIONS

	Morris C.4	Guy Ant.
Length	14' 9"	14' 3"
Width	6' 9"	6' 5"
Height	8' 9"	8' 5"

N.B. Early Guy Ant models (in this role only) had coachbuilt cab with hatch in roof. Later models standard cab.

TRUCK 15 CWT. 4 x 2 LIGHT WARNING

ON GUY AND FORD W.O.T.2 CHASSIS

Similar to latest Truck 15 cwt. 4 x 2 Wireless described opposite, except:-

(1) Generating set compartment is omitted.
(2) Auxiliary dynamos not fitted.
(3) Chore Horse battery charger is fitted.
(4) Vehicle has standard suppression but is not fully screened.

LADEN WEIGHTS

	Guy Ant	Ford
F. Axle	1 ton 5½ cwt.	1 ton 5½ cwt.
R. Axle	2 tons 5 cwt.	2 tons 2¼ cwt.
Gross	3 tons 10½ cwt.	3 tons 7¾ cwt.

OVERALL DIMENSIONS

	Guy	Ford
Length	14' 3"	15' 4"
Width	6' 5"	6' 9"
Height	8' 5"	8' 5"

TRUCK 15-CWT. 4 x 2 P.C.C. EQUIPMENT

ON GUY ANT CHASSIS

Similar to the latest Truck 15 cwt. 4 x 2 Wireless described opposite. Includes generating set compartment. Excludes auxiliary dynamo. In this role Guy chassis is fitted with folding canvas cab.

Dimensions:- Length 14' 3". Width 6' 5". Ht. 8' 8".

Laden Weight: F. Axle 1 ton 6 cwt. R. Axle 2 tons 0¾ cwt. Gross 3 tons 6¾ cwt.

MORRIS C.4 15 CWT.
4 x 2 CHASSIS

The Morris C.4 15 cwt. 4 x 2 (four cylinder engine) replaced the Morris C.S.8 (six cylinder engine) in production.

There have been two Marks.

(1) The C.4 Mk.I Chassis 8' 2" Wheelbase which is similar to the C.S.8 chassis shown on page 17, except for following:-

ENGINE:- Make, Morris. Petrol 4-cyl. Bore 100 mm. Stroke 112 mm. Capacity 3.5 litres.
R.A.C. rating 24.8 h.p. Max. B.H.P. 60 @ 3000 r.p.m.
Max. torque 1750 lbs. ins. @ 1250 r.p.m.

TYRES:- 9.00-16 pneumatic (Early C.S.8 chassis were fitted with Runflat tyres).

COOLING SYSTEM CAPACITY:- 4½ gallons.

(2) The C.4 Mk.II chassis, 8' 11" Wheelbase which in turn, superseded the C.4 Mk.I chassis.

It differs from the Mk.I chassis only in that the rear axle has been moved back to give a longer wheelbase and reduced overhang at the rear.

Note:- Chassis modified for Air Compressor and Wireless roles are distinguished as follows:-

C.4 A.C. = Air Compressor Chassis. Addition of power take off to main gearbox to drive compressor.

C.4 W.T. = Wireless Chassis. Addition of auxiliary generator driven by P.T.O.

LORRY 30-CWT. 4 x 2
SLAVE BATTERY

ON BEDFORD OX CHASSIS

GENERAL:- Externally this vehicle resembles 30 cwt. G.S. Lorry shown on page 28. It has two main functions:-

(a) To act as a "slave" or help in starting service vehicles.
(b) To act as a spare battery carrier and charging plant, and as a minor repair depot for wiring etc.

For function (a) the vehicle carries a bank of slave batteries, together with suitable selector switches tapping 30, 20 or 10 volts to suit vehicles with 24, 12 or 6 volt electrical systems.
For function (b) suitable racks are provided for carrying an assortment of spare batteries. Two 12-volt dynamos are provided for charging spare batteries. These dynamos, together with the slave battery charging dynamos, are driven by a power take-off from the gearbox. A bench and set of tools are fitted.

BODY:- G.S. type with detachable cover and hoops.

CAB:- Enclosed steel panel cab seating two.

OVERALL DIMENSIONS:- Length 16' 3". Width 7' 1½". Ht. 9' 9".

LADEN WEIGHT:- 4 tons 15 cwt. (Estimated)

LORRY 3 TON
4 x 2 SLAVE BATTERY

Similar equipment to Lorry 30-cwt. 4 x 2 Slave Battery, but mounted on the Bedford OY 3 ton 4 x 2 chassis. Larger body can accommodate more batteries than 30-cwt. type.

OVERALL DIMENSIONS:- Length 20' 5". Width 7' 1½". Ht. 10' 2¼".

LADEN WEIGHTS:- F. Axle 1 ton 19 cwt.
R. Axle 3 tons 1½ cwt.
Gross 5 tons 0½ cwt.

FORD F.602L(C.298WFS)
3 TON 4 x 2

SEMI FORWARD CONTROL WHEELBASE 13' 2"

GENERAL:- This is the latest type Canadian Ford 3 ton 4 x 2 chassis which has replaced in production the earlier EC098T model shown on page 221. Externally this chassis resembles the Canadian Ford 3 ton 4 x 4, it has the same semi-forward control military pattern cab and in the G.S. role mounts a similar wheelarch body.

ENGINE:- Ford, Mercury. Petrol V.8 Bore 3.19". Stroke 3.75". Capacity 3.91 litres. R.A.C. 32.5 h.p. Max. B.H.P. 95 @ 3600 r.p.m. Max. Torque 2112 lbs./ins. @ 1800 r.p.m.

CLUTCH:- Single dry plate, 11" dia.

GEARBOX:- 4 speed. Ratios 6.4 to 1, 3.09 to 1, 1.69 to 1, 1.00 to 1, rev. 7.82 to 1.

TRANSFER BOX:- Not fitted.

FRONT AXLE:- Tubular type. Non-driving.

REAR AXLE:- 2 speed. Ratios 6.33 and 8.81.

SUSPENSION:- Semi-elliptic with hydraulic shock absorbers front and rear.

BRAKES:- Foot, Hydraulic vacuum servo assisted. Hand, mechanical on transmission.

STEERING:- Right hand drive.

WHEELS AND TYRES:- Wheels, divided 6.00 x 16. Tyres 10.50-16 singles.

ELECTRICAL EQUIPMENT:- 6 volt.

TURNING CIRCLE:- 65' 0".

CAPACITIES:- Fuel 24½ gals. Cooling 5 1/8 gals.

M.P.G.:- 9. Radius of Action 220 miles.

S.M.2671

LORRY 3 TON 4 x 2 G.S.

ON CANADIAN FORD F.602L (C.298WFS) CHASSIS

Externally similar to latest Canadian 3 ton 4 x 4 model shown on page 627. Has similar body and cab.

OVERALL DIMENSIONS:- 20' 3" x 7' 6" x 9' 10". Stripped shipping Ht. 6' 3". (Cab split).

LADEN WEIGHTS:- F. Axle 2 tons 4½ cwt.
R. Axle 4 tons 10½ cwt.
Gross 6 tons 15 cwt.

S.M.2972

AMBULANCE 4 STRETCHER 4 x 2

On Canadian Ford F.602L (C.298WFS) Chassis modified for this role by fitment of special ambulance springing.

Body Lindsey House Type. Has accommodation for driver and mate in cab with attendant and 4 stretcher cases or 10 sitting cases in main body. Has similar stretcher raising gear to Austin Ambulance shown on page 81. Equipment includes medical lockers, interior heater and fan.

OVERALL DIMENSIONS:- Length 19' 1". Width 7' 6". Ht. 9' 2".

LADEN WEIGHTS:-

	(4 stretcher cases)	(10 sitting cases)
F. Axle	2 tons 4 cwt.	2 tons 6 cwt.
R. Axle	3 tons 12 cwt.	3 tons 18 cwt.
Gross	5 tons 16 cwt.	6 tons 2 cwt.

S.M.2871

LORRY 3 TON 4 x 2
PETROL 800 GALS.

ON CANADIAN FORD F.602L (C.298 WFS CHASSIS)
(Chassis details page 622)

For petrol tanker role chassis is modified by fitment of 10.50-20 tyres.

TANKER:- 2 equal compartments. Capacity 800 Imperial gallons. Equipped with

Two lengths of 2" x 112" hose
" " " 1¼" x 112" "

One length of 1¼" x 48" hose
Four lengths of ½" "

Complete with necessary couplings and two 2", one 1¼" and four ½ nozzles.

One semi-rotary hand pump is fitted and there is gravity can filling apparatus (two taps) at the rear. Superstructure is mounted on tank to take camouflage cover.

OVERALL DIMENSIONS

Length 19' 9". Width 7' 2". Ht. 7' 7".
No cut down height.

WEIGHTS

	UNLADEN	LADEN
F.A.	1 ton 16 cwt.	2 tons 7½ cwt.
R.A.	2 tons 6 cwt.	4 tons 18 cwt.
Gross	4 tons 2 cwt.	7 tons 5½ cwt.

LORRY 3 TON 4 x 2 TIPPER

ON CANADIAN DODGE T.110L-6 CHASSIS

THE CHASSIS is a short wheelbase version of the Dodge T.110-L5 3 ton 4 x 2 chassis shown on page 266.

That is, it has
A 2 speed rear axle, ratios 6.33 and 8.81 to 1.
10.50-16 tyres, singles rear.

And differs in
Wheelbase:- 11' 4".
Turning Circle:- 62'.
M.P.G.10. Radius of action 150 miles.

THE BODY is of all steel welded construction. Capacity 3 cubic yards. Fitted with hydraulic operated tipping gear.

DIMENSIONS

	OVERALL	INSIDE BODY
Length	17' 7"	7' 0"
Width	7' 1"	6' 0"
Height	7' 0"	2' 0"

WEIGHTS

	UNLADEN	LADEN
F. Axle	1 ton 7 cwt.	1 ton 12 cwt.
R. Axle	1 ton 16 cwt.	4 tons 14 cwt.
Gross	3 tons 3 cwt.	6 tons 6 cwt.

STRIPPED SHIPPING HT. 6' 4".

BODY TYPES FITTED TO CANADIAN FORD AND CHEVROLET LORRY 3 TON 4 x 4 G.S.

BODIES:- The illustration on page 226 shows an English body of wooden construction which was fitted to early Canadian Ford 3 ton 4 x 4 models only.

The majority of Ford and all Chevrolet 3 ton 4 x 4 lorries have Canadian built bodies of the following types.

First Canadian bodies were of welded steel construction with 20" deep sides. Then followed welded steel with 30" sides. These were followed by knock-down steel or knock-down wooden bodies again with 30" sides. The illustration on the facing page is typical of the latest Canadian design. Dimensions are applicable to all Canadian built bodies.

CABS:- Early types were as shown on page 226. Later types have the new Canadian Military Pattern cab with reverse slope windscreen. Both seat 2 men.

WEIGHTS of the latest model are roughly 10 cwt. heavier than the first types produced. This is partly due to increase in chassis wt. and partly in body wt.

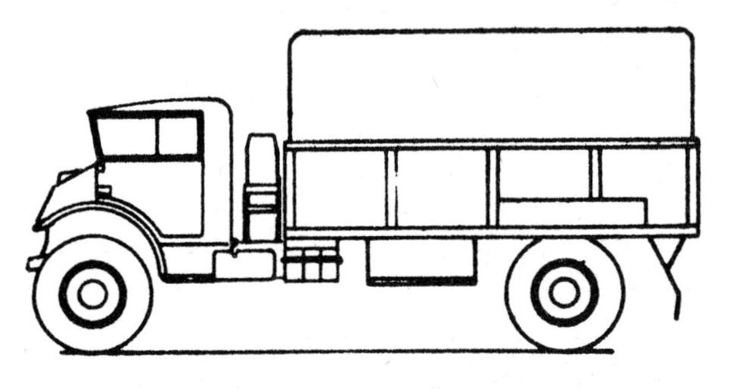

LORRY, 3-TON 4 x 4 G.S.

ON CANADIAN CHEVROLET C.60L CHASSIS AND CANADIAN FORD F.60L CHASSIS

BODY:- General Service body. Floor is flat except for low wheel arches. Fixed body sides 30" high. Detachable Superstructure and canvas tilt.

CAB:- Canadian Military Pattern model 13 cab. Seats 2. Detachable top to reduce shipping height. Reverse slope windscreen.

SPECIAL FITMENTS:- Spring drawbar gear.

DIMENSIONS (Both types)

	OVERALL	INSIDE BODY
Length	20' 3"	12' 0"
Width	7' 6"	6' 8"
Height	10' 3"	6' 0"

STRIPPED SHIPPING HEIGHT:- 6' 6" top of cab and superstructure removed.

WEIGHTS

(Chevrolet. Ford is approx. 2 cwt. lighter)

	UNLADEN	LADEN
F.A.		2 tons $17\frac{1}{2}$ cwt.
R.A.		4 tons 15 cwt.
Gross	4 tons $0\frac{1}{2}$ cwt.	7 tons $12\frac{1}{2}$ cwt.

LORRY 3 TON 4 x 4 WINCH

MAKE: CANADIAN CHEVROLET C.60S
(CHASSIS MODEL 8442)

CHASSIS:- Similar to standard Chevrolet C.60L 3 ton 4 x 4 shown on page 224, except wheelbase is 11' 2".

BODY:- Externally is similar to standard Canadian 30-cwt. 4 x 4 body shown on page 219. Body is welled type. Winch is mounted in forward end of welled section. Make Gar Wood 3 M. Capacity 5 tons. Length of rope 200 ft. Spare tyre is carried in front of body at offside.

W.D. type drawbar at rear.

DIMENSIONS

	OVERALL	INSIDE BODY
Length	17' 3"	9' 8"
Width	7' 2"	6' 8"
Height	9' 10"	6' 0"

* UNLADEN WEIGHT

F. Axle	2 tons 4 cwt.
R. Axle	2 tons 3 cwt.
Gross	4 tons 7 cwt.

*Note: Although rated as a 3 tonner, actual payload that can be carried is restricted by short body length and by fact that rear welled section of body must be kept clear if winch is to be used.

ON KARRIER K.6 CHASSIS

BODY:- G.S. body with detachable cover and hoops. Flat floor.

CAB:- Enclosed, seating 1 man and driver. Split top to cab.

WINCH:- Mounted under body and driven by P.T.O. from transfer box. 120 ft. Plough steel cable 2" circumference. Max. pull about 4½ tons.

DIMENSIONS

	OVERALL	INSIDE BODY
Length	18' 11½"	10' 9"
Width	7' 2½"	6' 6"
Height	10' 4 1/8"	6' 0"

LADEN WEIGHTS:- F. Axle 2 tons 18 cwt.
R. Axle 4 tons 10 cwt.
Gross 7 tons 8 cwt.

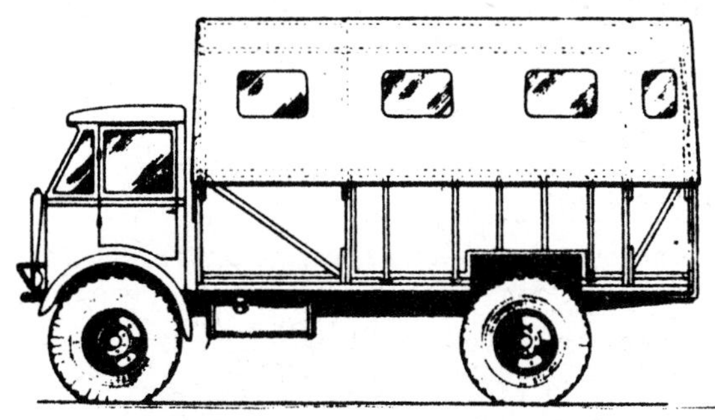

LORRY 3 TON 4 x 4 MACHINERY

FLAT FLOOR TYPE

for "B", "H", "I 30", "J", "M" and "X" roles

ON ALBION F.T11 CHASSIS

BODY. known as No.4 Mk.V is commonly called the flat floor type, though on the Albion it incorporates low wheel arches. The sides of the body are hinged to fold down (1) horizontally to give extra floor space or (2) double fold to form benches for working on from ground level.

Weather protection is by canvas tilt on detachable superstructure. With cab split, superstructure and machinery removed height can be cut down to 6' 10".

W.D. type drawbar gear at rear.

DIMENSIONS

	OVERALL	INSIDE BODY
Length	20' 8"	14' 4"
Width	7' 9½"	6' 9"
Height	10' 2"	6' 0"

WEIGHTS

	UNLADEN	LADEN (X Role)
F. Axle	2 tons 11½ cwt.	3 tons 17¼ cwt.
R. Axle	2 tons 5 cwt.	5 tons 2½ cwt.
Gross	4 tons 16½ cwt.	8 tons 19¾ cwt.

* The X Role is the heaviest. Other roles roughly 10 cwt. lighter.

LORRY 3 TON 4 x 4 MACHINERY

FLAT FLOOR TYPE

for "B", "H", "130", "J", "M" and "X" roles

on Ford W.O.T. 6 Chassis

CHASSIS. Standard Ford W.O.T. 6 modified by fitment of two side petrol tanks (capacity of each 15½ gals.), repositioned drawbar gear and rearranged exhaust.

BODY. known as "Machinery, No.4, Mk.VI", is basically similar to that mounted on the Albion 3 ton 4 x 4 and described opposite.

DIMENSIONS

	OVERALL	INSIDE BODY
Length	20' 3"	14' 4¼"
Width	7' 4½"	6' 9⅜"
Height	10' 1½"	6' 0"

WEIGHTS

	UNLADEN less machinery	LADEN (X role, the heaviest)
F. Axle	2 tons 3 cwt.	3 tons 9 cwt.
R. Axle	2 tons 2 cwt.	4 tons 19 cwt.
Gross	4 tons 5 cwt.	8 tons 8 cwt.

Other roles are approx. 10 cwt. lighter.

LORRY 3 TON 4 x 4 MACHINERY HOUSE TYPE

for D1, F and Z roles

MOUNTED ON FORD W.O.T.6. 3 TON 4 x 4 CHASSIS

THE CHASSIS is standard Ford W.O.T.6 except for 2 side petrol tanks each 15½ gals. capacity, repositioned exhaust and repositioned drawbar gear.

THE BODY, known as "Body Machinery House Type No.8", differs from the earlier 3 ton 6 wheel Machinery House Body shown on page 69 in the following respects.

(1) It is a flat floor body except for low wheel arches (in place of well type body).

(2) Opening flaps either side at front of body, but no front door. Rear entrance door only.

(3) Generator not fitted. (This was not carried in later type 3 ton 6 x 4 models.)

DIMENSIONS

	OVERALL	INSIDE BODY
Length	19' 11"	14' 0"
Width	7' 4½"	6' 8"
Height	10' 6"	6' 9"

WEIGHTS

	UNLADEN (less all equipment)	LADEN *(Estimated heaviest role)
F.A.	2 tons 5 cwt.	2 tons 18 cwt.
R.A.	2 tons 1 cwt.	4 tons 8 cwt.
Gross	4 tons 6 cwt.	7 tons 6 cwt.

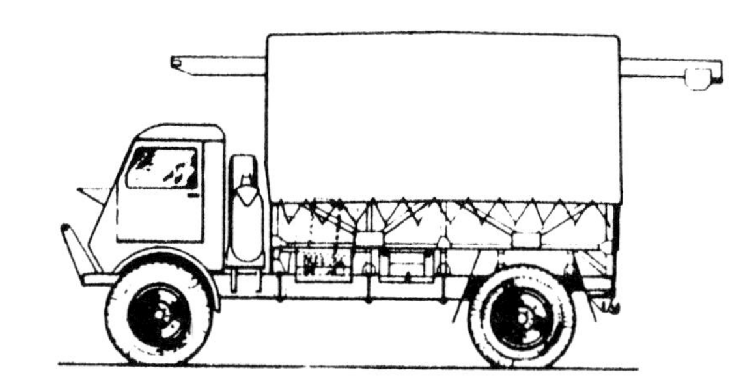

LATEST TYPE

LORRY 3 TON 4 x 4 MACHINERY 24 K.W. R.E.

Mounted on Ford W.O.T. 6. 4 x 4 chassis

BODY: is fitted with overhead gantry with travelling hand-operated block capable of a max. lift of 2½ tons. (Similar to type fitted to 3 ton 6 x 4 Breakdown Lorries.)

At front end of body is a 24 k.w. generator driven by P.T.O.

DIMENSIONS

	OVERALL	INSIDE BODY
Length	19' 11"	12' 6"
Width	7' 6"	6' 10"
Height	10' 6"	6' 0"

LADEN WEIGHT: 7 tons 10¾ cwt.

EARLY TYPE

LORRY 3 TON 6 x 4 MACHINERY 24 k.w. R.E.

ON LEYLAND RETRIEVER CHASSIS

Similar type body and similar equipment to 3 ton 4 x 4 type above

OVERALL DIMENSIONS: Length 20' 11", Width 7' 6", Height 11" 4".

LADEN WEIGHTS: F. Axle 3 tons 7½ cwt.
R. Bogie 6 tons 8½ cwt.
Gross 9 tons 16 cwt.

LORRY 3 TON 4 x 4 KITCHEN

On Bedford 3 ton 4 x 4 chassis

This is a conversion of a standard Bedford Q.L. G.S. body to carry the following equipment.

(1) Cabinet cross front of body, containing water tank, sink, drain pipe, shelf, 5 lockers and 2 drawers.

(2) Cooker along offside of body, comprising standard petrol burner arranged to heat a set of 5 standard food containers.

(3) Storage lockers along nearside with 6 insulated compartments for keeping food warm (Actually 2 of these 6 insulated compartments are located under sink at front end.)

The tailboard can be let down to form a platform and a set of steps is provided.

OVERALL DIMENSIONS	
Length	19' 8"
Width	7' 5"
Height	9' 9"

Weights: Not available, but within G.S. Lorry loading.

LORRY 3 TON 4 x 2 X-RAY

On Bedford OY 3 ton 4 x 2 chassis

This is a revised edition of the early X-Ray lorry produced on Albion 3 ton 4 x 2 chassis.

The House Type body is divided into two compartments. The front generator compartment has an entrance door on the nearside and a flap on the offside which can be let down to form a horizontal platform so that the generating set may be serviced outside the vehicle.

There is a rear entrance door to the rear compartment.

Body is 6' 0" wide inside x 6' 3", high. Front compartment is 4' 0" and rear compartment is 8' 7" in length.

Overall Dimensions: 20' 3" x 7' 2" x 9' 0" high.

LORRY 3 TON 6 x 4 CRANE
TURNTABLE OR COLES CRANE

On Austin 3 ton 6 x 4 K.6 chassis

This vehicle mounts the Coles E.M.A. Mk.VI Series II Petrol Electric Crane and is similar to the early type Leyland & Crossley Coles Crane vehicles shown on page 78, except power winch is not fitted and crane capacity is increased (see below).
Chassis is modified by fitment of solid beam rear suspension.

Overall Dimensions: Length 23' 4", Width 7' 2", Height 12' 0".

Fully Equipped WTS. (No load on jib)

F. Axle	1 ton 19¾ cwt.
R. Axle	7 tons 15 cwt.
Gross	9 tons 14 cwt.

Crane Duty:-

1 ton	@	17'0"	radius
1½ "	"	13'6"	"
2 "	"	11'0"	"
2½ "	"	9'3"	"
3 "	"	7'9"	"

Maximum height of lift 20'6"

LORRY 3 TON 4 x 4
RECORDER A/A Mk. I

(Formerly known as "Westex Recorder")

A House type body mounted on the Austin 3 ton 4 x 4 chassis.

Overall Dimensions: Length 19' 3", Width 7' 6", Height 9' 10".

Laden Weights: Within comparable G.S. Lorry loadings.

LORRY 3 TON 4 x 4
MOBILE OPS. ROOM

On Bedford Q.L. chassis

A standard R.A.F. Signals type body (R.A.F. type "J").

Overall Dimensions: Length 19' 8", Width 7' 6", Height 9' 10".

Laden Weights (W.D. Role)

F. Axle.	2 tons 14 cwt.
R. Axle.	3 tons 11 cwt.
Gross.	6 tons 5 cwt.

LORRY 3 TON 6 x 4

BREAKDOWN GANTRY

ON AUSTIN 3 TON 6 x 4 CHASSIS

BODY: Similar Breakdown body and gantry to that shown on Leyland Retriever chassis on page 68, except that there is no space between back of cab and front of body.

CHASSIS: Similar to standard Austin 3 ton 6 x 4 (on page 64) except that:-

(1) Heavier springs are fitted.
(2) Rear of frame shortened.
(3) 5 ton winch is fitted.
(4) Single line vacuum servo trailer brake connection is fitted (as one of the roles of this Gantry Breakdown is to tow Trailer 7½ ton 6 whld. Light Recovery).
(5) 10½ cwt. of ballast fitted front end of chassis. Half of this ballast is detachable and is carried in body when Breakdown is not lifting and towing.

WEIGHTS

	UNLADEN	LADEN
	(including 20 cwt. of equipment and 10½ cwt. of ballast)	(50 cwt. load suspended on gantry)
F. Axle	2 tons 14½ cwt.	1 ton 7½ cwt.
R. Bogie	4 tons 3½ cwt.	8 tons 0¼ cwt.
Gross	6 tons 18 cwt.	9 tons 8 cwt.

DIMENSIONS

	OVERALL	INSIDE BODY
Length	20' 3"	12' 3"
Width	7' 6"	6' 10½"
Height	11' 0¼"	6'6"

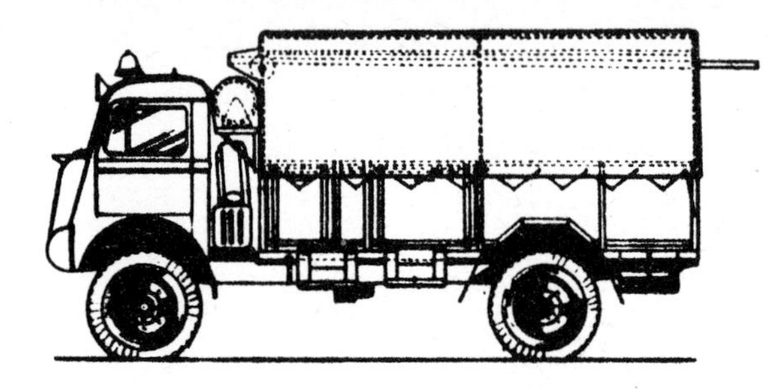

LORRY 3 TON 4 x 4 FIRE TENDER

ON BEDFORD QL 3 TON 4 x 4 CHASSIS

This vehicle has two functions. (1) It has a water tank, pump and first aid hose reel which can be got into action instantly on arriving at a fire. (2) It tows a trailer pump and has lockers for stowing hose, nozzles and other fire fighting equipment required for trailer pump, with fittings to carry ladders etc.

The basic body of this vehicle is externally similar to a standard G.S. Lorry with tilt, except that it is slightly longer and has four lockers cut in the front of the body.

The main equipment of the vehicle consists of a 200 gallon water tank mounted centrally in the forward end of the body, a first aid hose reel mounted between the body and cab and a pump driven from P.T.O.

Fixtures are provided to carry ladders under the tilt and there are fittings to carry suction hose and nozzles for the trailer pump.

For shipping purposes, the trailer pump can be stowed in the body of vehicle.

OVERALL DIMENSIONS: Length 20' 2". Width 7' 6". Height 10' 1".

STRIPPED SHIPPING HT.: 7' 0"

LADEN WEIGHT: F. Axle 2 tons 15 cwt.
R. Axle 3 tons 19¾ cwt.

LORRY 3 TON 4 x 4 CIPHER OFFICE
" " " COMMAND H.P.
" " " COMMAND L.P.
" " " MOBILE CARRIER TERMINAL
" " " T.E.V.
" " " WIRELESS (I)
" " " WIRELESS (R)
" " " WIRELESS H.P.

CHASSIS. All the above types have been mounted on the Bedford QLR 3 ton 4 x 4 chassis (except latest T.E.V. types, see note at foot).

Bedford Q.L.R. chassis is similar to standard QL chassis shown on page 48, except for special electrical equipment and suppression and fitment of 660 watt aux. dynamo driven off P.T.O.

GENERAL. From the general vehicular and dimensional aspect, all the above can be simplified down to three basic vehicles. Thus:-

(1) The Original Type shown opposite.
(2) The New Body type for Command H.P. & L.P. & Wireless. H.P. roles shown on page 642.
(3) The new T.E.V. types described below.

NEW T.E.V. TYPES: have a 16' body and are mounted on a specially modified Bedford QL chassis embodying a chassis frame extension ff similar to that used on the Bedford QLT Troop Carrier Chassis. There are 3 new T.E.V. types, known respectively as Type A, Type B and Type C.

OVERALL DIMENSIONS: Length 22' 3". Width 7' 6". Ht. 10' 0". Weight (Type C the heaviest), 6 tons 15 cwt.

A STILL LATER T.E.V. (Div) type, known as Body 15 unfitted, has a similar body shell to the Body 2 unfitted shown on page 642.

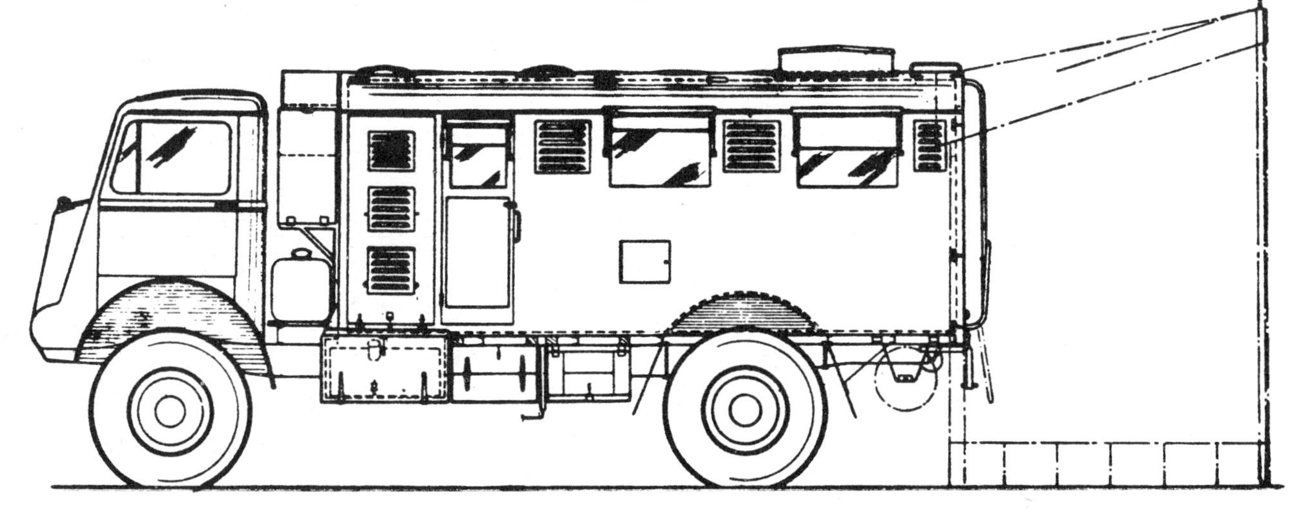

THE ORIGINAL TYPE BEDFORD 3 TON 4 x 4 WIRELESS HOUSE TYPE

USED FOR THE FOLLOWING ROLES

CIPHER OFFICE
*COMMAND H.P.
*COMMAND L.P.
MOB. TERMINAL CARRIER
**T.E.V. (Div. and Corps.)
WIRELESS (I)
WIRELESS (R)
*WIRELESS H.P.

*Superseded in production by later body type shown on page 642.

**Superseded in production by new T.E.V. type described at foot of facing page

This illustration is typical of the House Type body used for the roles listed at the left. The interior furniture, partitioning and signals equipment vary with the different roles.

The tent shown erected at the rear of vehicle was installed for the Wireless (R) role and has since been added to later types of the other roles.

OVERALL DIMENSIONS:

Length 20' 6". Width 7' 7". Height 9' 7".

WEIGHTS

	UNLADEN less all signals and Vehicle Equipment	LADEN Wireless H.P. the heaviest role
F. Axle	1 ton 19½ cwt.	3 tons 1 cwt.
R. Axle	1 ton 19½ cwt.	3 tons 15¼ cwt.
Gross	3 tons 19¼ cwt.	6 tons 16¼ cwt.

Less Signals Equipment is known as :-

"LORRY 3 TON 4 x 4 COMMAND H.P. BODY 2 UNFITTED"

This body with standard furniture and appropriate signals equipment covers both Command H.P., L.P. & Wireless H.P. roles. In addition to features listed opposite, this new body has an improved "L shaped" tent which can be erected along the nearside and rear of vehicle.

OVERALL DIMENSIONS: Length 20' 0". Width 7' 6". Height 9' 4".

INSIDE BODY DIMENSIONS: Length front comp. 6' 3" rear comp. 6' 3½" Width 6' 8". Headroom 5' 3".

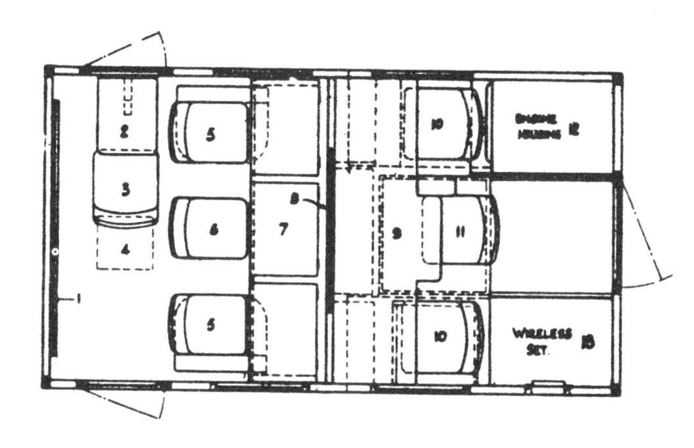

WEIGHTS

	UNLADEN	LADEN Command H.P. E.53 the heaviest role
F. Axle	2 tons 5½ cwt.	2 tons 18¼ cwt.
R. Axle	2 tons 13¼ cwt.	4 tons 11¾ cwt.
Gross	4 tons 19¾ cwt.	7 tons 10 cwt.

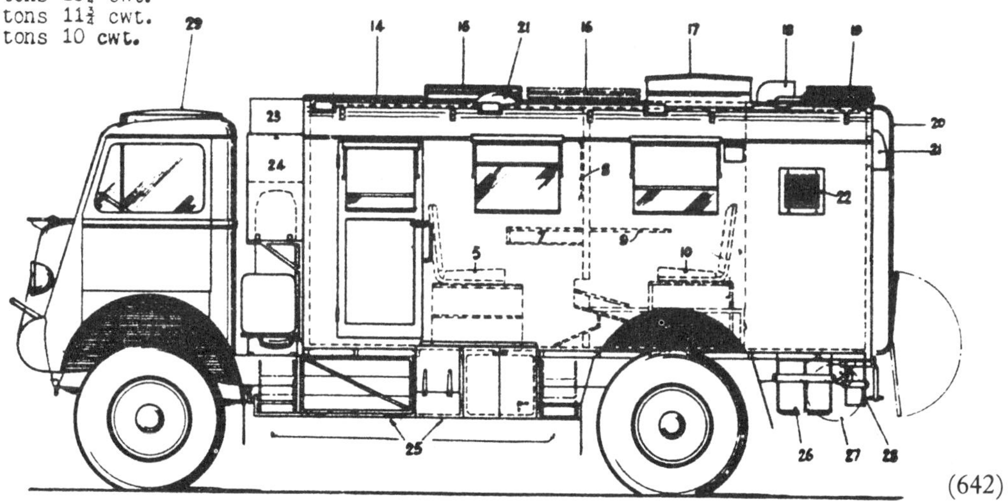

LATEST TYPE
LORRY 3 TON 4 x 4 COMMAND H.P.

AND WIRELESS H.P.

(KNOWN AS "COMMAND H.P. BODY 2 UNFITTED)

STAFF COMPARTMENT (See Plan view)
(1) Full width double sided map board.
(2) Hinged table on offside door.
(3) Revolving seat with
(4) Folding stand for Commander.
(5) Two sliding seats with lockers.
(6) Revolving seat.
(7) Full width desk for 3 men.
(8) Sliding glass window in partition.

OPERATING COMPARTMENT (See Plan view)
(9) Full width operators desk.
(10) Two operators seats with stowage underneath.
(11) Central operators seat with drawer underneath.
(12) Compartment for Auxiliary Engine.
(13) Wireless set.

EXTERIOR (See side elevation)
(14) Sliding glass window in roof.
(15) Luggage Grid.
(16) Two folding tables.
(17) Locker housing the tent.
(18) Cowling for vehicle exhaust.
(19) Silencer for auxiliary engine exhaust.
(20) Roof ladder.
(21) Cowlings for 12 volt extractor fans.
(22) Louvred grille for fan fitted in Wireless set.
(23) Stowage for engine ramps and aerial masts.
(24) Full width kit locker.
(25) Locker units for: 3 cable drums, 8 batteries non-skid chains and 20 gallon drinking water tank, rectifier box, tools and fuel tank for auxiliary engine.
(26) Jerrican (on nearside).
(27) The 4th Cable drum (on offside).
(28) Oil can.
(29) A.A. Ring.

LORRY 3 TON 4 x 2 WATER
500 GALLONS

ON BEDFORD OYC. 3 TON 4 x 2 CHASSIS

Similar to 350 gallon type shown on page 118 except:-

Fitted with a power operated pump (capacity 1000 g.p.h.) in addition to hand pumps and filtering equipment as on 350 gallon Tanker.

NOTE: First 500 gallon types produced had hand pumps only.

OVERALL DIMENSIONS: Length 20' 1". Width 6' 7". Ht. 7' 10". Height superstructure removed 6' 7".

WEIGHTS

	UNLADEN	LADEN
F.A.	1 ton 7½ cwt.	1 ton 16¾ cwt.
R.A.	1 ton 14¾ cwt.	4 tons 2¼ cwt.
Gross	3 tons 2¼ cwt.	5 tons 19 cwt.

LORRY 3 TON 4 x 2 WATER
450 GALLONS

ON CANADIAN DODGE T.110 L.5 CHASSIS

Chassis details see page 266.

Tank Capacity 450 imperial gallons. Equipped with a 1000 g.p.h. power operated pump, two hand operated pumps and complete filtering equipment.

OVERALL DIMENSIONS: 21' 0" x 7' 6" x 7' 9".

LADEN WEIGHT: 6 tons (Estimated)

GENERAL NOTES

CAMERA LORRY
DARK ROOM LORRY
PRINTING LORRY
PHOTO-MECHANICAL LORRY
ENLARGING AND RECTIFYING LORRY
AUTO-PROCESSING LORRY

THE CAMERA LORRY C.V.2 is mounted on a 3 ton 6 x 4 chassis. It is designed to work in conjunction with the Dark Room and as such has a bellows in the side to couple with its working mate. See page 648.

THE DARK ROOM LORRY P.V.2 is also mounted on a 3 ton 6 x 4 chassis. (See page 648). It was previously known as Processing Lorry, but now renamed Dark Room to distinguish it from Auto Processing below.

THE PRINTING LORRY has expandable sides to the body. Early types were mounted on 3 ton 6 x 4 chassis, latest types are on the Foden 10 ton 6 x 4 chassis. (See page 646).

THE PHOTO MECHANICAL LORRY like the Printing Lorry has expandable sides. Early types were mounted on 3 ton 6 x 4 and latest types are on 10 ton 6 x 4 chassis. (See page 647).

THE ENLARGING AND RECTIFYING LORRY
and
THE AUTO-PROCESSING LORRY are two new types now being developed. They will be generally similar to the 10 ton 6 x 4 Printing Lorry. First requirements have been met by modifying 3 ton 6 x 4 Camera Lorries.

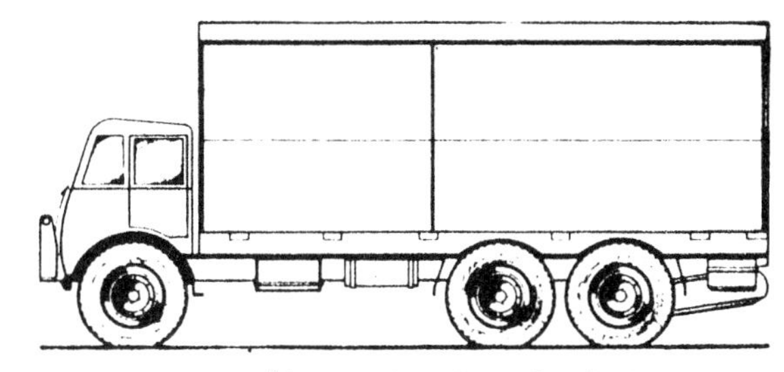

LORRY 10 TON 6 x 4 PRINTING

ON FODEN D.G. 6/12 CHASSIS

BODY. An expandable body; the sides open out forming a completely covered and self-contained machine room. Double doors in rear bulkhead. Detachable metal wheelhouses, four jacks are supplied for levelling lorry and machine.

EQUIPMENT. Demy Offset Printing Machine, Machine Motor etc.

DIMENSIONS

	OVERALL	INSIDE BODY
Length	26' 8½"	19' 11"
Width	(8' 6" (when closed) (15' 0" (over flaps (when open)	(8' 2" (when closed) (14' 9" (when open)
Height	11' 9"	7' 4¾"

WEIGHTS

	UNLADEN	LADEN
F.A.	4 tons 11 cwt.	5 tons 1 cwt.
R.B.	9 tons 16½ cwt.	10 tons 6½ cwt.
Gross	14 tons 7½ cwt.	15 tons 7½ cwt.

LORRY 3 TON 6 x 4 PRINTING

(Early Type)

Early Printing Lorries have a similar but smaller body mounted on Leyland Retriever 3 ton 6 x 4 chassis.
OVERALL DIMENSIONS: Length 22' 6". Width 7' 6". Ht. 12' 0".
Laden Weights: F. Axle 4 tons. R. Bogie 6 tons 12 cwt.

LORRY 10 TON 6 x 4
PHOTO MECHANICAL

ON FODEN D.G. 6/12 CHASSIS

BODY: Shell generally same as Printing body.

EQUIPMENT: A Proving press to take standard demy plates.
A whirler for washing and coating demy plates.
Hot box for drying plates.
Set of three sinks.
Two fixed face-up frames with manual evacuation.
Fixed single point arc.
Tables for developing and examination. Water storage tanks.

DIMENSIONS

	OVERALL		INSIDE BODY
Length	27' 6½"		19' 11"
Width	8' 6"	(when closed)	8' 2"
	15' 0"	(over flaps when open)	(when closed) 7' 4½"
Height	11' 9"		

WEIGHTS

	UNLADEN	LADEN
F. Axle	4 tons 7½ cwt.	4 tons 18 cwt.
R. Bogie	9 tons 9 cwt.	10 tons 8 cwt.
Gross	13 tons 6¾ cwt.	15 tons 4 cwt.

LORRY 3 TON 6 x 4
PHOTO MECHANICAL
(Early type)

LORRY 3 TON 6 x 4 PHOTO MECHANICAL

Early Photo Mechanical lorries had a similar but smaller body mounted on Leyland Retriever chassis.

OVERALL DIMENSIONS: Length 22' 6". Width 7' 6". Height 12' 0".

LADEN WEIGHTS: F. Axle 3¼ tons. R. Bogie 6 tons tons.

LORRY 3 TON 6 x 4
CAMERA C.V.2

ON LEYLAND RETRIEVER CHASSIS

This vehicle is designed to couple with the Dark Room (or Processing Lorry).

BODY. House type. Door on nearside and off-side. Telescopic connection to Process body.

Two water tanks front tank approx. 57 gals. rear tank 49 gals.

DIMENSIONS

	OVERALL	INSIDE BODY
Length	22' 0"	14' 0½"
Width	8' 0"	7' 4½"
Height	12' 0"	7' 6"

LADEN WEIGHTS

F.A.	2 tons 11½ cwt.
R. Bogie	5 tons 12 cwt.
Gross	8 tons 3½ cwt.

LORRY 3 TON 6 x 4
DARK ROOM P.V.2 (OR PROCESSING)

ON LEYLAND CHASSIS

BODY. This vehicle is designed to be coupled to the Camera Lorry.
House type. Body divided into two compartments by a partition, with a central sliding door. Access to body by door in centre of rear. Door on offside in front of partition. Bottom hinged flap in doorway forming walkway when Camera and Process bodies are connected. Fitted with sinks, and water tanks.

DIMENSIONS

	OVERALL	INSIDE BODY
Length	22' 0"	10' 0" front compartment
		3' 10½" rear "
Width	8' 0"	7' 4½"
Height	12' 0"	7' 6" at centre

LADEN WEIGHTS: F.Axle 2 tons 19½ cwt. R. Bogie 6 tons 8½ cwt. Gross 9 tons 7 cwt.

LORRY 3 TON 6 x 4 PLOTTER

ON LEYLAND RETRIEVER CHASSIS

BODY. Special House Type. Rear door. Hatch either side of body to take a canvas chute connecting Recorder Vehicle (see below).

Four air ventilators fitted in roof connecting with rear air louvre. Water tank capacity 43 gals. on nearside rear of body and 32 gals. on offside running board.

CAB. Open type. Canvas canopy on folding superstructure.

DIMENSIONS

	OVERALL	INSIDE BODY
Length	22' 3"	13' 8"
Width	7' 6"	6' 0"
Height	10' 5"	6' 9"

WEIGHTS

	UNLADEN	LADEN
F.A.W.	2 tons 9¼ cwt.	2 tons 19½ cwt.
R.A.W.	4 tons 14¼ cwt.	5 tons 12 cwt.
Gross	7 tons 3½ cwt.	8 tons 11½ cwt.

LORRY 3 TON 6 x 4 RECORDER

ON LEYLAND RETRIEVER CHASSIS

Similar from the general vehicular aspect to the Plotter vehicle described above. Dimensions as Plotter Lorry.

F.W.D., H.A.R. 4 TON 4 x 4

(Payload = 4 short tons or 3 long tons 11 cwt.)
4 x 4 Normal Control Wheelbase 156".

ENGINE: Make, Waukesha BZ. 6 cyl. Petrol. Bore 4". Stroke 4¼". Capacity 320 cu. ins. 4.24 litres. Max. B.H.P. 95 @ 2800 r.p.m. Max. torque 2688 lbs. ins. @ 1200 r.p.m. R.A.C. rating 38.4 h.p.

CLUTCH: Lipe. 13" dia. Single dry plate.

GEARBOX: Five forward and 1 reverse. Ratios 7.53 to 1, 4.3 to 1, 2.52 to 1, 1.42 to 1, 1.00 to 1, r. 7.37 to 1.
TRANSFER CASE: Chain drive incorporating centre differential which can be locked. Ratio 1.7 to 1.

F. AXLE: Full floating, single reduction spiral bevel gear. Ratio 5.29 to 1.

R. AXLE: Full floating, single reduction spiral bevel gear. Ratio 5.29 to 1.

SPRINGS: Semi-elliptic front and rear.

BRAKES: Foot, Hydraulic vacuum servo assisted.
Hand, transmission brake.

STEERING: Left hand drive.

WHEELS AND TYRES: Wheels 20 x 8 steel disc type. Tyres 9.00-20, single front and twin rear.

ELECTRICAL EQUIPMENT: 6 volt.

TURNING CIRCLE: 73' 6".

TRACK: F. 5' 8". R. 5' 10".

GROUND CLEARANCE: 11". Belly clearance 16¾".

CAPACITIES: Fuel 33 gals. Cooling 4½ gals.

PERFORMANCE FIGURES
B.H.P. per ton 11.4.
Tractive effort per ton (100%) 1,120 lbs./ton.
Maximum speed 35 m.p.h.
M.P.G. 6.0 Radius 200 miles.

BODY TYPES FITTED TO F.W.D., H.A.R. 4 ton 4 x 4

There are two body types on F.W.D., H.A.R. models.

(A) Models ex. U.S.A. on S/M.2330 have a U.S. Army type 12 foot cargo body, with flat floor, folding troop carrier seats and canvas tilt on detachable bows. Winch is mounted at the front of the chassis.

(B) Models ex. Canada on S/M.2828 have a wheelarch body, 10' 0" long x 6' 9" wide. Winch on these models mounted between cab and body.

CAB:- Normal Control. Folding top type "A". All enclosed type "B". Pintle hook fitted both types, height 36".

Following Weights and Dimensions apply to Type (A) ex. U.S.A. (Type B. is slightly smaller and lighter).

DIMENSIONS

	OVERALL	INSIDE BODY
Length	22' 9"	11' 9"
Width	8' 2"	7' 4"
Height	9' 6"	5' 2"

WEIGHTS

	UNLADEN	LADEN
F. Axle	2 tons 11 cwt.	2 tons 19½ cwt.
R. Axle	2 tons 8 cwt.	5 tons 12½ cwt.
Gross	4 tons 19 cwt.	8 tons 12 cwt.

MACK EHID. 5 SHORT TON 4 x 2

Max. Payload 10,000 lbs. = 4½ long tons

Normal Control. Wheelbase 14' 2"

(NOTE: Following details cover standard type Mack E.H. ex. S/M 2429. For details of other Mack E.H. series models see page 654.)

ENGINE: Make, Mack EN-354. 6-cyl. petrol. Bore 3 7/8". Stroke 5". Capacity 354 cu. ins. 5.8 litres. R.A.C. rating 36 h.p. B.H.P. 110 @ governed speed of 2820 r.p.m. Max. torque 3060 lbs. ins. @ 1200 r.p.m.

CLUTCH: Single dry plate. 14" dia.

GEARBOX: Mack T.R.51. Five speed and 1 reverse. Ratios, 8.05 to 1, 4.35 to 1, 2.57 to 1, 1.45 to 1, 1.00 to 1, reverse 8.12 to 1.

F. AXLE: Reverse Elliott type.

R. AXLE: Mack RA-44. Double reduction, spiral bevel and spur. Ratio 8.59 to 1.

SUSPENSION: Semi-elliptic front and rear.

BRAKES: Foot, compressed air on all wheels. Hand, transmission brake.

STEERING: Left Hand Drive.

WHEELS & TYRES:- Budd disc wheels 20" x 8". Tyres 9.00-20. 10 ply, dual rear.

ELECTRICAL EQUIPMENT: 6 volt.

TURNING CIRCLE: 65'

CAPACITIES: Fuel 41 gals. Cooling 6½ gals.

PERFORMANCE FIGURES

B.H.P. per Ton. 12.0
Tractive Effort per ton (100%) 1210 lbs./ton
Max. Speed 35 m.p.h.
M.P.G. 7.0. Radius 280 miles

MACK EHID. 5 TON 4 x 2
ex. S/M. 2429

(Note: This is the standard type Mack E.H. model, for details of other types of Mack E.H. series, see page 655.)

BODY: Standard U.S. Army 12 foot cargo body with flat floor, folding troop carrier seats along sides and detachable bows and canvas tilt.

CAB: Normal control, seats, two, folding canvas top. Pintle hook at rear, height from ground 2' 6".

DIMENSIONS

	OVERALL	INSIDE BODY
Length	22' 6"	11' 9"
Width	8' 0"	6' 9"
Height	8' 8"	5' 4"

WEIGHTS

	UNLADEN	LADEN (4½ long tons)
F. Axle	2 tons 1 cwt.	2 tons 15 cwt.
R. Axle	2 tons 11 cwt.	6 tons 9 cwt.
Gross	4 tons 12 cwt.	9 tons 4 cwt.

ex. S/M. 2765

S/M 2765 covers a mixed batch of Mack Lorries and Semi Trailers.
There are four basic types:-

1. Mack EH Normal Control Lorry
This is rated as a 3 ton 4 x 2 Lorry for W.D. service and has similar characteristics to the Mack EHID on page 652. There are 9 different chassis types in this batch of Mack EH models, but some of the variations are only minor.

2. Mack EHU Forward Control Lorry
Generally similar to the EH, but forward control. 2 chassis types. Rated as a 3 ton 4 x 2.

3. Mack EHT. Tractor and Semi Trailer
Rated for W.D. service as a 6 ton 4 x 2-2 semi trailer. Has similar characteristics to the Mack EHT model ex. S/M 2075 shown on page 285. Three chassis types on S/M 2765.

4. Mack EHUT. Tractor and Semi Trailer
Rated for W.D. service as a 6 ton 4 x 2-2. Generally similar to the Mack EHT but is a forward control tractor unit. Two chassis types.

Summary:- All told there are 16 chassis variations within this one batch of vehicles ex. S/M 2765. For instance there are two engine types (some models have the EN 310 and others the EN 354 engine), four rear axles, six gear box variations, 10 variations in wheelbase, and tyre fitments of either 8.25-20 or 9.00-20.

BODY TYPES FITTED TO MACK EH SERIES EX. S/M 2765

There are a number of variations in body size, weights and dimensions, the following are typical of the four basic chassis types shown on facing page.

1. MACK EH 3 TON 4 x 2 G.S.

   DIMENSIONS: Inside body 10' 6" x 7' 0". Overall: 20' 7" x 8' 0" x 9' 6" Ht.

   WTS: Unladen F.A. 1 ton 15 cwt. R.A. 2 tons 8 cwt. Laden: F.A. 2 tons, R.A. 5 tons 11 cwt.

2. MACK EHU 3 TON 4 x 2 G.S.

   DIMENSIONS: Inside body 14' 0" x 7' 0". Overall: 22' 0" x 7' 0" x 9' 9" Ht.

   WTS: Unladen F.A. 2 tons 0 cwt. R.A. 2 tons 10 cwt. Laden: F.A. 2 tons 17 cwt. R.A. 5 tons 1 cwt.

3. MACH EHT 6 TON 4 x 2-2 SEMI TRAILER G.S.

   DIMENSIONS: Inside body 17' 0" x 7' 0". Overall: 31' x 8' 0" x 10' 0".

   WTS: Unladen F.A. 1 ton 15½ cwt. M.A. 2 tons 9¼ cwt. R.A. 1 ton 16¼ cwt Gross 6 tons 1 cwt. Laden: F.A. 1 ton 19¾ cwt. M.A. 4 tons 16¾ cwt. R.A. 5 tons 7½ cwt. Gross 12 tons 4 cwt.

4. MACK EHUT 6 TON 4 x 2-2 SEMI-TRAILER G.S.

   Wts. and Dimensions are generally as Mack EHT unit (item 3 above) except that overall length is slightly less.

TRACTOR 4 x 2, FOR 6 TON SEMI-TRAILER

THE BEDFORD - SCAMMELL

AND

SEMI-TRAILER 6 TON 2 WH. G.S.

Page 126 gives details of the Bedford - Scammell Tractor coupled to a Flat Platform 6 ton semi trailer (now known as Semi-Trailer 6 ton 2 wh. Platform).

The same Bedford-Scammell tractor unit is also used to couple to the Semi-Trailer 6 ton 2 wh. G.S. This semi-trailer (like the flat-platform type) is fitted with the Scammell Automatic coupling device. The G.S. body is 15 ft. long by 7 ft. wide, has fixed sides, drop tailboard and is fitted with detachable, superstructure and tilt.

TYRES: 10.50-16 singles on tractor and semi-trailer.

DIMENSIONS

TRACTOR AND SEMI-TRAILER COUPLED
Length 25' 0". Width 7' 6". Ht. 10' 3".

TRACTOR ONLY
Length 15' 9". Width 6' 7". Ht. 7' 0".

SEMI-TRAILER ONLY
Length 16' 0". Width 7' 6". Ht. 10' 3". Stripped Shipping Ht. 6' 10". Superstructure detached.

WEIGHTS

TRACTOR & G.S. SEMI-TRAILER COUPLED

	UNLADEN	LADEN
F. Axle	1 ton 3¼ cwt.	1 ton 10¾ cwt.
M. Axle	1 ton 6 cwt.	3 ton 14¾ cwt.
R. Axle	19 cwt.	4 ton 12½ cwt.
	3 ton 8¼ cwt.	9 ton 18 cwt.

Tractor Solo: 2 ton 1½ cwt.
G.S. Semi-Trailer Solo: 1 ton 6¾ cwt.

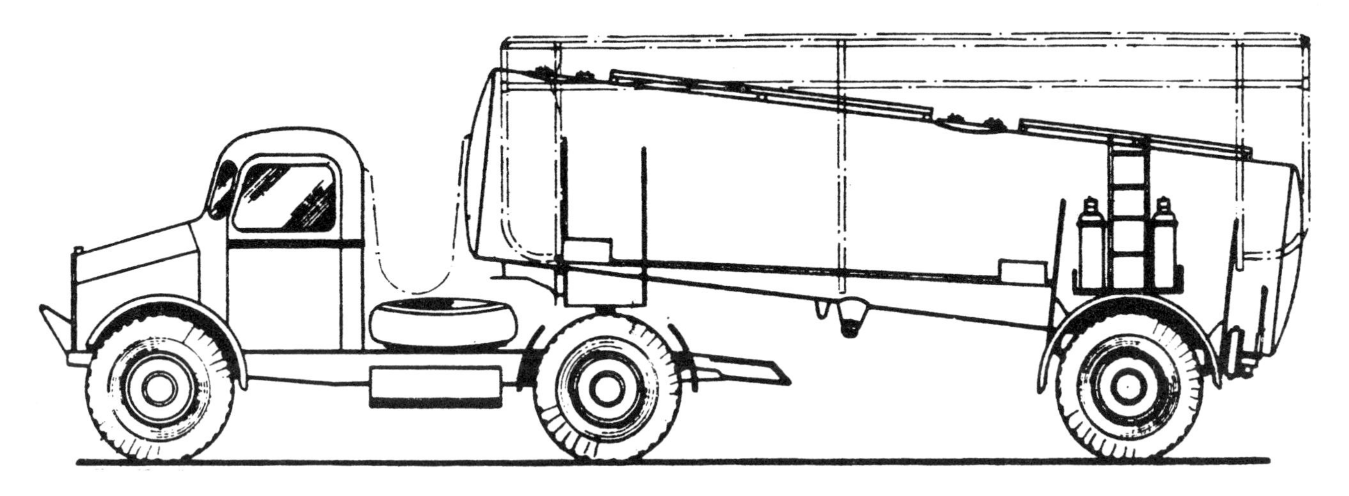

LORRY 6 TON 4 x 2-2, PETROL 1,750 GAL.

BEDFORD SCAMMELL

TRACTOR UNIT

Bedford OXC. fitted with Scammell Automatic coupling device as described page 126, but modified by fitment of fire screen behind cab for this Trailer-Tanker pole.

SEMI-TRAILER

Double compartment Elliptical Tank 16' 9" barrel length x 17' 5" overall, dimensions of ellipse being 5' 9" x 3' 10", constructed throughout of 1/8" thick mild steel plates. 2-10 ft. lengths of 2" and 2-10 ft. lengths of 1½" petrol-resisting hose. Fitted with detachable camouflage cover on superstructure.

TYRES: Tractor and semi-trailer, 10.50-16 singles.

Overall Dimensions

Length 26' 5". Width 7' 4". Height 8' 8". Ht. less superstructure 7' 11".

WEIGHTS

	UNLADEN	LADEN
F. Axle	1 ton 3¼ cwt.	1 ton 4¾ cwt.
M. Axle	1 ton 17 cwt.	4 tons 8 cwt.
R. Axle	1 ton 5¼ cwt.	4 tons 10 cwt.
Gross	4 tons 5½ cwt.	10 tons 2¾ cwt.

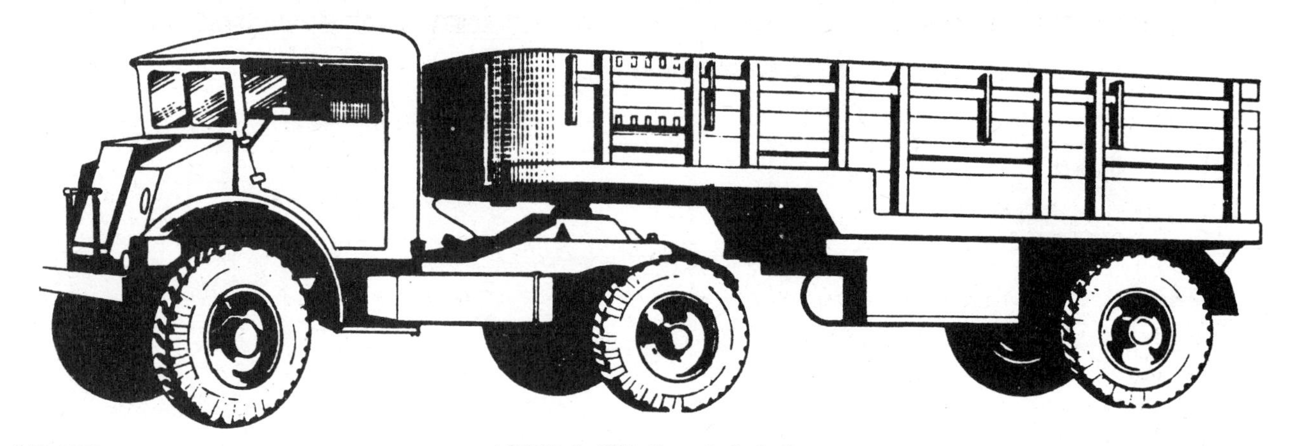

S/M.2857

LORRY 6 TON 4 x 4-2 G.S.

Canadian Form F60T 3 ton 4 x 4 Tractor (Chassis C.395Q) Coupled to 6 ton Canadian Semi-Trailer

GENERAL: This tractor semi-trailer combination has been designed to give a degree of cross-country performance. It is fitted with a ball type coupling (or fifth wheel). The semi-trailer cannot ordinarily be detached from the Tractor. (Can be for shipping.)

TRACTOR UNIT: Similar to standard Ford 3 ton 4 x 4 chassis (shown on page 225) except:-

Wheelbase is 9' 7". Turning Circle is 50' 0".

SEMI-TRAILER: A drop frame design. Inside body length 18' 0", length of rear dropped section 9' 2". Inside width 6' 6". Loading Ht. of rear drop section, 4' 2".

Body has rack sides and tailboard surmounted by detachable canvas tilt and hoops.

Semi-trailer chassis utilises many components common to the Ford Tractor Unit. Tyres are 10.50-20 (as Tractor unit). Brakes are hydraulic (as Tractor) actuated by vacuum servo connection from tractor. Semi-Trailer fitted with W.D. drawbar at rear.

OVERALL DIMENSIONS

	TRACTOR	SEMI TRAILER
Length	14' 2½"	18' 4"
Width	7' 0"	7' 0"
Height	7' 7"	11' 2" (tilt on)
		7' 4" (tilt off)

Length of coupled unit 27' 0"

WEIGHTS

Unladen: Tractor 3 tons 4 cwt. Semi Trailer 2 tons 12 cwt.

	UNLADEN	LADEN
F.A.	2 tons 3 cwt.	2 tons 8½ cwt.
M.A.	2 tons 0 cwt.	4 tons 9¾ cwt.
R.A.	1 ton 13 cwt.	4 tons 17¾ cwt.
Gross	5 tons 16 cwt.	11 tons 16¼ cwt.

LORRY 6 TON 4 x 2 G.S.

DENNIS MAX. Mk. II chassis

This Mk.II edition is similar to the early Dennis Max. shown on page 52, except for the following:-

(1) Fitment of 5 speed Gearbox. Ratios 5.18 to 1, 2.94 to 1, 1.66 to 1, 1.00 to 1, overdrive 0.69 to 1, rev. 6.66 to 1.

(2) Max. speed (in overdrive) 35 m.p.h.

(3) Detail changes to controls and instruments to facilitate water-proofing.

(4) Wheelarch body.

(5) Split cab with A.A. ring in roof.

W.D. Spring Drawbar Gear, Ht. 28"

WEIGHTS AND DIMENSIONS

WEIGHTS

	UNLADEN	LADEN
F. Axle	2 tons 18 cwt.	3 tons 15 cwt.
R. Axle	2 tons 6 cwt.	8 tons 0¼ cwt.
Gross	5 tons 4 cwt.	11 tons 15¼ cwt.

DIMENSIONS

	OVERALL	INSIDE BODY
Length	21' 10"	14' 0"
Width	7' 6"	6' 10"
Height	9' 5"	6' 0"

Stripped Shipping Ht. 7' 0".

LORRY 6 TON 6 x 4 CRANE TURNTABLE

THORNYCROFT "AMAZON" WF/AC6/2

NORMAL CONTROL WHEELBASE 11' 9"

BOGIE CENTRES 4' 6"

ENGINE:- Make, Thornycroft AC6/2. 6-cylinder Petrol. Bore 4 3/8". Stroke 5¼". Capacity 7.76 litres. Max. B.H.P.100. Max. Torque 3660 lbs. ins. @ 1000 r.p.m.

CLUTCH:- Single dry plate.

GEARBOX:- 4-speed forward, 1 reverse. Overall ratios (High) 39.3 to 1, 21.1 to 1, 11.95 to 1, 7.66 to 1, R.58.9 to 1.

AUXILIARY GEARBOX:- Two speed, ratios direct and 2.5 to 1.

FRONT AXLE:- I section

REAR AXLES:- Full floating, overhead worm drive. Ratio 7.66 to 1.

SUSPENSION:- Front, semi-elliptic. Rear, Thornycroft patent rigid beam suspension.

BRAKES:- Foot, Mechanical vacuum servo-assisted on all wheels. Hand, Mechanical on rear wheels. Trailer Brakes, single line vacuum.

WHEELS AND TYRES:- Front 12.75-20 tyres on 8.37 x 20 disc. wheels. Rear 13.50-20 tyres on split type disc wheels.

ELECTRICAL EQUIPMENT:- 12 volt.

TURNING CIRCLE:- 56 feet.

TRACK:- Front 6' 6", Rear 6' 4½".

CAPACITIES:- Fuel 40-gals.

PERFORMANCE FIGURES (Gross wt. 12.6 tons)

B.H.P. per Ton 7.6.
Tractive Effort per Ton (100%) 1,250 lbs./Ton.

LORRY 6 TON 6 x 4
CRANE TURNTABLE

ON THORNYCROFT WF/AC6 CHASSIS

EQUIPMENT:- Coles E.M.A. Mark VII P.E. Crane. powered by 250-v. D.C. generator, driven by Power-Take-Off. Separate 6 h.p. motors for hoisting and derricking and 1 h.p. motor for slewing. Drawbar gear.

DUTY:- 5-tons @ 7' 6" radius
3-tons @ 13' 0" radius
1¼-tons @ 19' 3" radius

Max. Ht. of Lift 27' 0"

DIMENSIONS (JIB HORIZONTAL)

Overall Length 30' 0"
Overall Width 7' 9"
Overall Height 14' 0"

WEIGHT READY FOR SERVICE

GROSS 13 tons 5 cwt.

S/M.2304

INTERNATIONAL KR-8-R
4 x 2 TRACTOR

FOR 7 TON SEMI TRAILER

Wheelbase 12' 5". Normal Control. (Coupled to 7 ton Semi trailer shown opposite)

ENGINE: Make, International FBC-318. Petrol 6 cyl. Bore 3 7/8". Stroke 4½". Capacity 318 cu. ins. (5.2 litres). Max. B.H.P. 85 @ 2600 r.p.m. Max. Torque. 2-160 lbs. ins. @ 1000 r.p.m.

CLUTCH: Rockford. Single dry plate.

GEARBOX: Make, Fuller. Five speed. Ratios 6.5 to 1, 3.7 to 1, 1.9 to 1, 1.00 to 1, overdrive 0.82 to 1, rev. 6.4 to 1.

FRONT AXLE: Make, Eaton I beam.

REAR AXLE: Make, Eaton, full floating, double reduction. Ratio 8.49 to 1.

SUSPENSION: Semi elliptic front and rear.

BRAKES: Hydraulic, vacuum servo assisted. Two line vacuum connections for semi trailer.

STEERING: Left hand drive.

WHEELS AND TYRES: Wheels. 20 x 8. Tyres 9.00-20/36 x 8 12 ply. Twin rear.

TURNING CIRCLE: 54 ft.

ELECTRICAL EQUIPMENT: 6 volt.

TRACK: 5' 10 1/8".

CAPACITIES: Fuel 17½ gals. Cooling 5¼ gals.

WINCH: Garwood 4K1B is fitted behind tractor cab and semi trailer is fitted with necessary sheeves and rollers (S.M.2304 only).

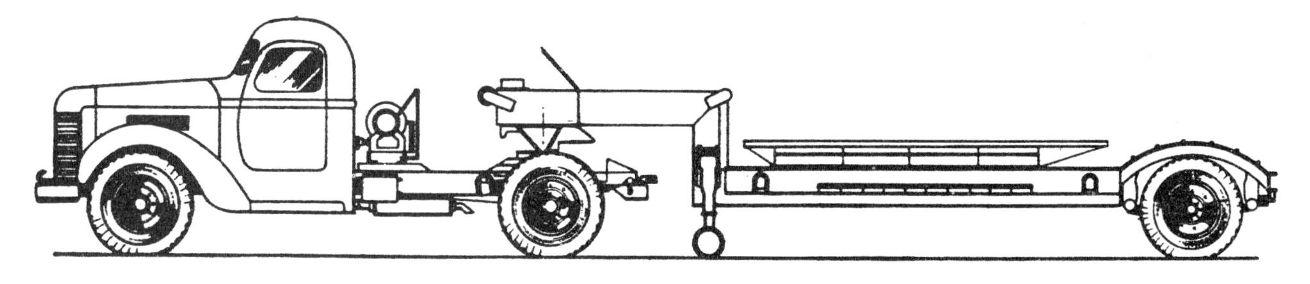

S/M. 2304

LORRY 7-TON 4 x 2-2 LOW LOADING

(Semi-Trailer is made by Hobbs, Lufkin, Steel Products Co., Truck Engineering Corp, American Body and Trailer Co. to a common design.)

Semi-Trailer couples to International Tractor shown opposite

SEMI-TRAILER:- Designed for the transport on roads of tracked tractors and road making equipment.

Tyres:- 9.00-15, 12 ply, twins.

Brakes:- 2 line vacuum operated.

Semi-Trailer is equipped with jockey or landing wheels, can be detached from tractor.

Two 10 ft. loading ramps are provided and loading is effected over the arched rear decking over trailer axle.

Flat load space available is 12' 0" long x 8' 0" wide. Loading Ht. 2' 6".

WEIGHTS

(Tractor & S. Trailer Coupled)

	UNLADEN	LADEN (7 tons)
F. Axle	2 tons 0 cwt.	2 tons 4½ cwt.
M. Axle	3 tons 4 cwt.	5 tons 15 cwt.
R. Axle	2 tons 0 cwt.	6 tons 5¼ cwt.
Gross	7 tons 4 cwt.	14 tons 4¾ cwt.

Semi-Trailer Unladen Wt. 3 tons 0 cwt.

DIMENSIONS:- Semi-Trailer 24' 0" x 8' 0" x 5' 6"
Tractor 18' 11" x 7' 7" x 7' 3"

Length, Tractor and Semi-Trailer coupled, 38' 0".

LORRY 10 TON 4 x 2 - 2 G.S.

Make:- Tractor, International KR.8.

:- Semi-Trailer, Fruehauf Model 220

TRACTOR UNIT is similar to International KR.8 described on page 662 except Wheelbase, 11' 5".

Gearbox, Type 52C, Ratios 8.0 to 1, 4.6 to 1, 2.46 to 1, 1.41 to 1, 1.00 to 1, rev. 8.0 to 1.

No winch fitted.

SEMI-TRAILER:- Straight frame type with slight kick-up over turntable. Semi-trailer has landing wheels and is detachable from tractor unit. Brakes 2 line vacuum operated. Tyres 36 x 8, 14 ply twins.

BODY is G.S. type 20' 0" long x 6' 9" wide.

WEIGHTS (Estimated)

	UNLADEN	LADEN
F. Axle	2 tons 0 cwt.	2 tons 5 cwt.
M. Axle	3 tons 5 cwt.	8 tons 0 cwt.
R. Axle	2 tons 5 cwt.	7 tons 5 cwt.
Gross	7 tons 10 cwt.	17 tons 10 cwt.

DIMENSIONS:- Tractor 18' 11" x 7' 7" x 7' 3".

Semi-Trailer:- 20' 4" x 7' 6" x 7' 9".

Length, Tractor and Semi Trailer coupled, 31' 0".

ON LEYLAND HIPPO MK.II CHASSIS

BODY:- This is a welled type body incorporating wheelarches to give a lower loading height. Weather protection by canvas tilt on detachable hoops. Body can be knocked down for overseas shipment.

CAB:- All-enclosed steel panelled cab. Seats two. Roof of cab split to reduce shipping height.

DIMENSIONS

	OVERALL	INSIDE BODY
Length	27' 3"	19' 6"
Width	8' 1"	7' 3"
Height	10'11"	6' 6"
Cab. Ht.	9' 7½"	

Each wheelarch measures 8' 6" long x 1' 10" wide and 13½" high, and is 3' 10" from rear of body. Ht. of towhook 3' 2".

WEIGHTS

	UNLADEN	LADEN
F. Axle	3 tons 12 cwt.	5 tons 6½ cwt.
R. Bogie	5 tons 4 cwt.	14 tons 1½ cwt.
Gross	8 tons 16 cwt.	19 tons 8 cwt.

NOTES ON LATEST TYPE MACK N.R. AND WHITE 1064

10 TON 6 x 4 G.S.

SUPPLEMENTING EARLIER DATA GIVEN ON

Pages 280-281 for Mack
" 278-279 for White

DUAL TYRE EQUIPMENT

Both Mack and White are now fitted with twin rear tyres size 11.00-24 (i.e. same size as the single front tyres). This change applied commencing chassis No: NR.4D8567 on Mack. Dual tyred vehicles are further identified by new Sply/Mech Nos., viz. S/M 6315 for Macks and S/M 6314 for Whites.

NEW BODY:- Coincident with fitting of dual tyres, a new body was introduced. This is a flat floor body with fixed side sections 14" high surmounted by folding troop seat extension racks giving a total side height of 36½". Weather protection by canvas tilt on detachable hoops. Height floor to hoops 66". Clear inside body dimensions 15' 0" x 7' 0". Loading Height 4' 9".

TRAILER BRAKE CONNECTIONS:- All Whites have 2 line Air Pressure Connections. Mack models have 2 line Air Connections commencing Chassis No.2871 D.

SOFT TOP CABS (Folding canvas top) have been fitted to Macks commencing Chassis No. 2871D.

WHITE AXLE RATIOS. Data on page 278 was published prior to production. Production Axle ratio is 7.33 to 1. Corresponding overall ratios are 45.9 to 1, 25.3 to 1, 12.7 to 1, 7.33 to 1, 4.9 to 1, rev. 59.7 to 1.

TRACTOR 4 x 4
FIELD ARTILLERY

Make: Morris C.8

A new design body has been introduced on the Morris C.8 F.A.T. chassis. The chassis is the Mk.III design as described on page 82 fitted with 10.50-16 tyres. The body varies from the early types shown on page 84 in the following main points:-

1. Roof contour resembles that of G.S. lorry with removable all-over canvas cover on hoops.

2. Body is divided in three main compartments:-

(a) Crew compartment seating 6 including driver, behind which is

(b) Kit locker running full width of body, behind which is

(c) Ammunition compartment with adjustable partitions. Inside Length is adjustable from 40½" to 36¾". Width is 6' 3½". Ht. is 2' 9". At the rear is a drop tailboard. Spare wheel is carried in a compartment underneath.

3. Body has four side doors for quick access, plus A.A. ring, commanders look out in roof.

OVERALL DIMENSIONS: Length 14' 9.5/8". Width 6' 11". Ht. 7' 11". Ht. of towing hook (laden) 2' 4.7/8".

WEIGHTS

	UNLADEN	LADEN (30 cwt.)
F.A.	1 ton 12¾ cwt.	1 ton 16 cwt.
R.A.	1 ton 13 cwt.	2 tons 19 cwt.
Gross	3 tons 5 cwt.	4 tons 15 cwt.

TRACTOR 4 x 4 ANTI-TANK

(Conversions for towing the 17 pdr. gun)

The Tractor 4 x 4 A.T. is based on conversions of the following vehicles:-

(1) Morris C.8 2 pdr.
A.T. Portee..............shown on page 85

(2) Morris C.8.P. S.P.
Predictor........................" " " 137

(3) Chevrolet 8440 2 pdr.
A.T. Portee...................." " " 268

The conversions vary in detail, but all have the following basic features:-

(a) Seating for crew of 8 including driver.

(b) Accommodation for a total of 30 rounds ammo. six of the rounds being "ready for use".

(c) Weather protection by a detachable canvas tilt on superstructure, externally the vehicles being generally similar to that shown on page 85.

(d) Warner electric trailer brake controller has been fitted in the service, but only on those tractors which tow 17 pdr. guns fitted with Warner brakes.

(e) No winch is fitted to these tractors.

WEIGHTS AND DIMENSIONS
Generally applicable to all three conversions.
Overall Length 14' 3". Width 7' 3".
Ht. 9' 0".

Laden Axle Weights, Morris Conversions:-

F.A.	2 tons	6 cwt.
R.A.	2 tons	13 cwt.
Gross	4 tons	19 cwt.

(Note: Chevrolet Conversion approx. 15 cwt. heavier)

S/M. 2487

TRACTOR 4 x 4 LIGHT A.A.

On Canadian Ford F. 30 (C.29Q)
30 cwt. 4 x 4 Chassis

(Similar body mounted on Canadian Chevrolet C.30 Chassis)

THE CHASSIS is similar to the Ford 30 cwt. 4 x 4 (model COIQF) described on page 218. Tyres are 10.50-16 pneumatic.

The vehicle is designed to carry the following crew and equipment.

Seating for driver and mate in the cab; for six of the crew in the two side seats at the rear of the main body; for the brakesman in a special seat at the rear centre of the main body facing rear. The brakesman operates by hand the cable controlled brakes on the bofors gun trailer.

There are eight compartments in the outer sides of the body, four on each side. The upper and lower front compartments carry ammunition cases. The upper rear compartments carry the equipment and supply stores. The lower rear compartment carry tyre chains.

In the front of the main body are two tyre carriers, one for the vehicle spare, the second for the gun spare.

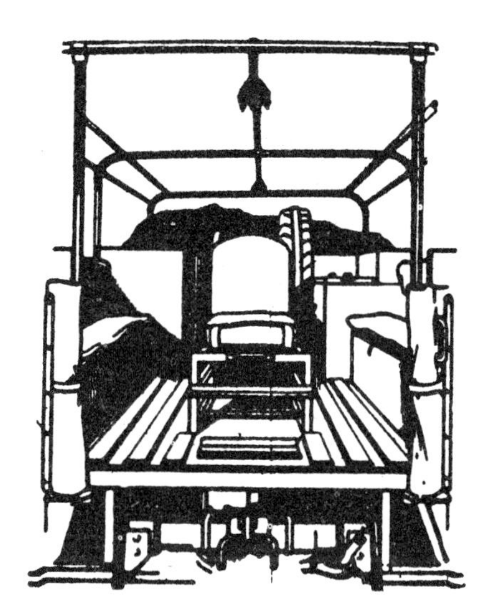

There are two kit lockers for the crew in the cab, and seven open metal trays in the main body for the kits of the crew seated there.

A metal framing runs the whole length of the main body down the centre and this carries two trays, one for the gun sight box, the other for the box of gun spare parts. The brakesman's seat is mounted on the rear of this framework.

OVERALL DIMENSIONS

Length 17' 7". Width 7' 4". Ht. 7' 5"
Stripped Height 6' 6".

WEIGHTS

	UNLADEN	LADEN
F. Axle	2 tons 0½ cwt.	2 tons 16¼ cwt.
R. Axle	2 tons 6½ cwt.	3 tons 18¾ cwt.
Gross	4 tons 7 cwt.	6 tons 14½ cwt.

Ht. of Tow hook 2' 3".

TRACTOR 4 x 4 LIGHT A.A.

on Canadian Ford F.60S 3 ton 4 x 4 chassis

Generally this vehicle is similar to the Tractor 4 x 4 Light A.A. shown on page 671 (overleaf), except that it is mounted on the Canadian Ford 3 ton 4 x 4 134" w.b. chassis which differs in the following main details:-

10.50-20 tyres
six inch constant velocity joints
Vacuum assisted brakes.

The body is as described and illustrated on page 671.

WEIGHTS

	UNLADEN	LADEN
with full tanks and vehicle equipment		
F.A.	2 tons 5¼ cwt.	2 tons 15¾ cwt.
R.A.	2 tons 13¼ cwt.	4 tons 14 cwt.
Gross	4 tons 19 cwt.	7 tons 9¼ cwt.

OVERALL DIMENSIONS

Length 17' 7". Width 7' 4". Ht. 7' 7".
Stripped Ht. 6' 8"

 S.M.2645

CARRIER, SELF-PROPELLED 40 M.M. A.A. (S.P. BOFORS)

on Canadian Ford F.60 B (C.39QB)

3 ton 4 x 4 chassis

GENERAL:- This S.P. Bofors is a Canadian edition of the Morris S.P. shown on page 135.

THE CHASSIS:- Design is based on that of the standard Ford 3 ton 4 x 4 shown on page 225. It differs in the following respects:-

Wheelbase 11' 2".

Track. f. 71.97" rear 70.5"

Ground Clearance, 9.4"

Turning Circles. r.71'; 1.61' 0".

Tyres, 10.50-16 pneumatic. Spare tyre provided but carried in unit transport.

BRAKES, Special hand control to vacuum operating cylinder to lock wheels when firing gun.

Frame, reinforced with additional cross members.

THE CAR is open and seats crew of 4 men. A 2.75 k.v.a. generator is driven by P.T.O. from transfer box. Rear axle spring locking mechanism is fitted in addition to levelling jacks. W.D. type spring drawbar gear at rear.

OVERALL DIMENSIONS:- Length 20' 4". Width 7' 10". Ht. 7' 8".

WEIGHTS

	Unladen	Laden
F. Axle.	2 tons 13 cwt.	3 tons 1 cwt.
R. Axle.	2 tons 19½ cwt.	3 tons 18 cwt.
Gross	5 tons 12½ cwt.	6 tons 19 cwt.

S.M.6242

MACK N.O. 6 7½ TON 6 x 6

Normal Control. Wheelbase 13' 0"
Bogie Centre 4' 10"

ENGINE: Make, Mack EY military engine. Petrol. 6 cyl. Bore 5" Stroke 6". Capacity 707 cu.ins. (11.6 litres) Max. B.H.P. 170 @ governed speed of 2,100 r.p.m. Max. torque 6,600 lbs.ins. (Net 6228 lbs. ins.).

CLUTCH: Lipe 15", single dry plate.

GEARBOX & TRANSFER CASE: Make, Mack TRDXT-36. Gearbox ratios 8.05 to 1, 4.57 to 1, 2.61 to 1, 1.45 to 1, 1.00 to 1, rev. 8.13 to 1. Transfer Ratios direct and 2.50 to 1.

AXLES: Ratio 9.02 to 1. Front axle, triple reduction, single spiral bevel reduction in diff. carrier, double bevel reduction in wheel drive units. Rear Axles, double reduction spiral bevel and spur.

BRAKES: Foot, compressed air. Hand, mechanical.

TRAILER BRACE CONNECTIONS: 2 line Air Pressure and Warner Electric.

SUSPENSION: Semi-elliptic front. Inverted semi-elliptic rear.

STEERING: Left hand drive. Ross cam and lever.

WHEELS & TYRES: Budd disc 24" x 9.10" rims. Tyres 12.00-24, 14 ply, singles front, twin rear.

ELECTRICAL: 6 volt lighting, 12 volt starting.

TRACK: F.6' 6". R. 6'4".

TURNING CIRCLE: 75½' r. lock., 69' l. lock.

CAPACITIES: Fuel 67 gals. Cooling 10½ gals.

PERFORMANCE FIGURES: (load 15,000 lbs.).

B.H.P. per ton: 8.9

Tractive effort per ton (100%) 126 lbs./ton Top.

" " " " " 2550 lbs./ton Bottom.

Max. Speed 31 m.p.h.

M.P.G. 3.5 solo. (2.5 towing)

TRACTOR 6 x 6 HEAVY

MACK N.O. 7½ TON 6 x 6

BODY:- Flat floor G.S. or cargo type. Weather protection by detachable tilt and hoops. Spare wheel carried inside body.

CAB:- All-enclosed with folding windscreen and canvas top. Seats 3.

WINCH:- Make Garwood 5 M.B. mounted at front of tractor.

SPECIAL EQUIPMENT:- This tractor has been specially designed to tow the U.S. 155 mm. gun. It is equipped with a special drawbar incorporating a trail clamp for the 155 mm. M1 gun carriage. A chain hoist is installed at rear of body for use when attaching gun to special drawbar. In addition a standard U.S. Army pintle hook (tow hook) is supplied.

DIMENSIONS

	Overall	Inside Body
Length	24' 9"	11' 8"
Width	8' 6"	7'10"
Height	10' 2"	

WEIGHTS

	Unladen	Laden (15,000 lbs.)
F. Axle	5 tons 10 cwt.	5 tons 10 cwt.
R. Bogie	7 tons 0 cwt.	14 tons 0 cwt.
Gross	12 tons 10 cwt.	19 tons 10 cwt.

Above Laden Weights are for solo use. When towing gun weight is reduced by 2 tons 5 cwt.

ALBION CX 22S - TRACTOR 6 x 4 HEAVY

Normal Control. Wheelbase 14' 8"
Bogie Centres 4' 6"

ENGINE: Make Albion EN.244 C.1., 6 cyl. Bore 4 5/8". Stroke 5½". Capacity 9.06 litres. Max. B.H.P. 100 @ 1750 r.p.m. Max. torque. 4.400 lbs/ins. @ 1000 r.p.m.
GEARBOX: 4 speed and 1 reverse. Ratios 5.79 to 1, 2.9 to 1, 1.63 to 1, 1.00 to 1.
AUXILIARY BOX: 2 speed. Ratios direct and 2.15 to 1.
REAR AXLES: Overhead worm drive. Ratio 8.33 to 1.
SUSPENSION: Front semi-elliptic. Rear, fully articulated bogie, two inverted semi-elliptic springs each side mounted one above the other and pivotting on central trunnion mounting.
BRAKES: Foot, Mechanical air pressure servo assisted on rear wheels. Hand, Mechanical air servo assisted on rear wheels.
TRAILER BRAKE CONNECTIONS: Two line air pressure connections. Gun cylinder for operating cable brakes.
TYRES: 14.00-20 singles front and rear.
TURNING CIRCLE: 62 ft.
TRACK: F. 6' 10". R. 6' 9¼".
GROUND CLEARANCES: 9½" Spring clips, 13½" r. axle.
CAPACITIES: Fuel 75 gals. Cooling 8 gals.
PERFORMANCE FIGURES:-
(Solo, laden weight of 15.6 tons)
B.H.P. per ton 6.4
Tractive Effort per ton (100%).
1260 lbs./ton. Max. Speed 27 m.p.h.
M.P.G. 8 Radius of Action. 600 miles.

TRACTOR 6 x 4 HEAVY

ON ALBION CX32S CHASSIS

BODY:- G.S. type body with welled floor. Weather protection by canvas tilt on detachable superstructure. Crew compartment at front of body, seats 4 and has doors either side. Two additional folding troop seats at rear of body A.A. ring in roof. Loading skids carried in body.

CAB:- All enclosed, seats 3. Commanders traffic look-out in roof.

WINCH:- Tractor equipped with Scammell 8 ton vertical winch. 430 ft. steel cable.

DRAWBAR GEAR:- W.D. type, front and rear, Ht. 3' 1".

WEIGHTS

	Unladen	Laden (7.2 How. Role)
F.A.	3 ton 16 cwts.	3 tons 15½ cwt.*
R. Bogie	6 tons 12¾ cwt.	11 tons 16 cwt.
Gross	10 tons 8¾ cwt.	15 tons 11½ cwt.

* Draft eye load accounts for reduction in laden wt.

DIMENSIONS

	Overall	Inside Body
Length	25' 6"	14' 4¾"
Width	8' 9"	7' 7¼"
Height	10' 4½"	5' 4"

U.S. ARMY HALF TRACKS IN W.D. SERVICE GENERAL NOTES

There are two distinct families of U.S. ½ tracks:-

1. THE STANDARD TYPES, made by White, Autocar and Diamond 'T' to a common chassis design.
   A few of these may be found in W.D. Service.
2. THE SUBSTITUTE STANDARD TYPES made by International. This chassis design is based on that used for the Standard Types, but utilises International engine, axles and other components. The majority of ½ tracks in W.D. Service are these substitute standard types.
   BODY TYPES. A wide variety of body and gun mounts are mounted on these two basic chassis. These are distinguished by M. Nos.
   The main body types are:-
   - Car, Half Track, M.2 made by White, etc.
   - " " " M.9 made by International.
   - Carrier Personnel Half Track, M.3, made by White etc.
   - Carrier Personnel Half Track, M.5, made by International
   - Carriage Motor Multiple Gun, M.13, made by White etc.
   - Carriage Motor Multiple Gun, M.14, made by International.

   It should be noted that M.14 types in W.D. Service have been modified by demounting guns and adding seats to fill a Personnel role, roughly equivalent to that of Car Half Track M.9.

CHASSIS SPECIFICATION OF INTERNATIONAL HALF TRACKS

(Applicable M. 5, M. 9, and M. 14)

ENGINE:- Make, International Rcd. 450-B. Petrol. 6 Cylinders. Bore 4 3/8". Stroke 5". Capacity 451 cu. ins. (7.38 litres). R. A. C. rating 45.9 h.p. Max. B. H. P. 143 @ 2700 r.p.m. Max. torque 4260 lbs. ins. @ 1000 r.p.m.

GEARBOX:- 4 speed and 1 reverse. Ratios: 4.92 to 1 2.60 to 1, 1.74 to 1, 1.00 to 1, rev. 4.37 to 1.

TRANSFER CASE:- Two speed. Ratios Direct and 2.48 to 1.

F. AXLE:- Driven type. F. axle drive optional in either high or low ratios of transfer case. Full floating axle Spiral bevel gears. Ratio 7.16 to 1.

R. AXLE:- Single reduction. Spiral bevel. Ratio 4.22 to 1. Conventional differential. Drive sprocket at forward end of track.

TRACKS:- Endless band type, rubber moulded around steel cable. Adjustable rear idler sprocket.

BRAKES:- Hydraulic vacuum servo assisted operating on front wheels and rear axle drive sprocket. Hand, transmission brake. Warner Electric trailer brake controls.

STEERING:- Orthodox front wheel steering. No assistance from tracks.

FRONT TYRES:- Size 9.00-20, Combat Wheels, Divided type.

DIMENSIONS:- Nominal Wheelbase (F. axle to centre of track) 11' 3½" Length of track on ground 52".
Width of track 12¼".
Width over tracks 76 1/16".
Track (or tread) of front wheels 64½".

PERFORMANCE FIGURES:-

Maximum Speeds:-

Transfer Ratio	Low	High
1st gear	3.4 m.p.h.	8.5 m.p.h.
2nd "	6.5 "	16.0 "
3rd "	9.7 "	24.0 "
4th "	16.9 "	42.0 "
Reverse	3.8 "	9.6 "

Tractive Effort per Ton (100%) 2,210 lbs/Ton bottom gear
" " " " " 180 " " top gear

Petrol Consumption 4½ miles per gallon. (Average condition).

Radius of Action 225 miles.

Capacities Fuel 50 gals. Cooling 6½ gals. (Imperial gallons).

BODY TYPES
TRUCK 15-CWT. HALF TRACKED PERSONNEL

(International M5, M9, and M14)

In W.D. Service all three types are Personnel carriers (The M.14 has additionally been modified to special Command & Load Carrying roles as described overleaf).

SEATING, is provided for a total crew of
13 on the M.5 including one on occasional seat by driver.
10 on the M.9.
10 on the M.14.

ARMOUR is 5/16" homogeneous, except for front windscreen which is 5/8" homogeneous.

WEATHER PROTECTION is by canvas tilt carried on detachable hoopsticks.

DOORS are fitted either side at front with hinged sections above waistline. Rear door is fitted on M.5 and M.9 but not on M.14.

The interior layouts vary chiefly in that large equipment lockers are fitted in the sides behind the driving compartment on the M.9 model. Marked "L" on plan view.

WINCH is fitted to M.14 and may be fitted to certain M.5 and M.9 models. Rated pull 10,000 lbs. When winch not fitted it is replaced by front roller.

WEIGHTS AND DIMENSIONS

Rated Payload 3500 lbs. (31 cwt.)

(Illustration shows M.9 model)

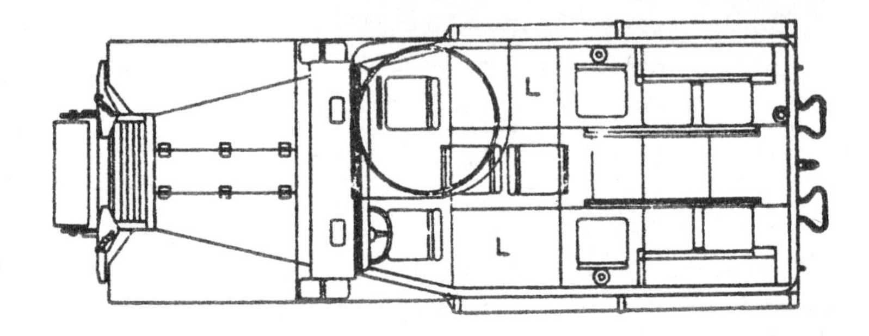

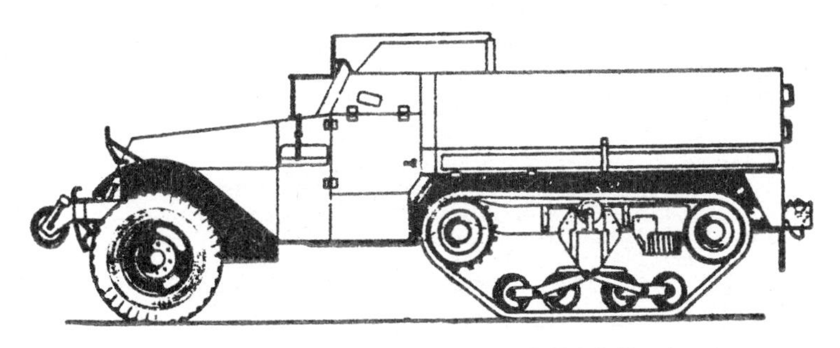

Illustration shows M.9 Half Track

UNLADEN WEIGHTS (Full Fuel)	M.5 and M.9	M.14
F. Axle	2 tons 10 cwt.	2 tons 16 cwt.
Track	4 tons 14½ cwt.	4 tons 7 cwt.
Gross	7 tons 4½ cwt.	7 tons 3 cwt.

LADEN WEIGHTS (Laden 31 cwt.)		(laden 34 cwt.)
F. Axle		2 tons 17 cwt.
Track		5 tons 19¾ cwt.
Gross	8 tons 15½ cwt-	8 tons 16¼ cwt.

OVERALL DIMENSIONS	(less winch)	(with winch)
Length	20' 2"	20' 9"
Width	7' 4"	7' 4"
Height	7' 8"	7' 8"
Ht. (cut down)	7' 3"	6' 2"

INSIDE BODY DIMENSIONS (APPROX.)

M.5. Length behind driver 9' 0" Width between tanks 4'6".

M.9. Length behind bins 6' 0" Width between tanks 4'6".

TOWING: Pintle hook fitted at rear. Height 29½"

M.14. Modified for load Carrier Role. Truck 15 cwt. half tracked G.S. with winch

This is similar to the M.14 in Personnel role except that seats are not fitted in body and rear floor is flat (no wells). Load space is 8' 5" long x 5' 9" wide (except at front where petrol tanks encroach on width). No tilt fitted.

M.14 Modified for Command Role
Truck 15 cwt. half-tracked Command

Modified by fitment of commanders desk and map board at rear body; wireless operation's desk at middle of body with No.19 H.P. and No.19 L.P. sets installed. Fitment of roof lights and canvas cover.

TRUCK 15-CWT 4 x 4
ARMOURED PERSONNEL

Make: G.M. Canada

Model C.15.TA. Chassis Model 8449

ENGINE: Make G.M. 270. 6-cyl. Petrol. Bore 3 25/32". Stroke 4". Capacity 269.5 cu.ins. 4.41 litres, Max. B.H.P. gross 104 @ 3000, net 95 @ 3000 r.p.m. Governed to 2750 r.p.m. Max. torque (net) 2580 lbs. ins @ 1000. R.A.C. rating 34.31 h.p.

CLUTCH: Single dry plate.

GEARBOX: 4 speed. Ratios 5.30 to 1, 2.61 to 1, 1.89 to 1, 1.00 to 1, rev. 5.94 to 1.

TRANSFER BOX: 2 speed. Ratios, direct and 1.87 to 1.

FRONT AXLE: Ratio 6.5 to 1. Same diff. assembly as r. axle. 5" Bendix Weiss joints.

REAR AXLE: Full floating. Ratio 6.5 to 1.

BRAKES: Foot, Hydraulic vacuum servo assisted. Hand, Transmission brake.

STEERING: Right hand drive.

WHEELS AND TYRES: Wheels, divided type 6.00 x 16. Tyres 10.50-16 pneumatic. Spare carried left side of hull.

GROUND CLEARANCES: F. Axle 8".

TRACK: F. 5' 11", R. 5' 10 3/4".

TURNING CIRCLE: 50'.

ELECTRICAL EQUIPMENT: 6 volt.

CAPACITIES: Fuel 40 gls. Cooling 3½ gals.

PERFORMANCE FIGURES

B.H.P. per ton 18.0
Tractive Effort per ton (100%) 940 lbs./ton.
Max. speed 43 m.p.h. @ governed r.p.m.
Petrol Consumption 9 m.p.g. Radius 300 miles.

TRUCK 15-CWT 4 x 4
ARMOURED PERSONNEL

Make G.M. Canada

GENERAL: As a personnel carrier, this vehicle seats 8 men with stowage for their rifles and all their equipment.

As a load carrier, the 6 rear seats can be removed and stowed in the side kit lockers. The rear foot well is covered by a plate. As an ambulance, stretcher support brackets are fitted to take 2 stretchers longitudinally in the body. Stretchers do not encroach in either front seats. Rear seat backs are folded down.

ARMOUR: 14 MM. basis front and 6 MM. basis sides and rear.

VISION: Armoured front ports each carry protectoscopes. Ports are hinged to open up horizontally. Glass windscreens provided for use when ports are open.

WEATHER PROTECTION: Detachable canvas tilt on detachable hoops. When not in use hoops are bunched at front, and tilt is folded back on to unarmoured front roof (or continuation of the front screen).

STOWAGE: Two large stowage compartments above rear wheels either side; each taking camouflage net and 4 large haversacks. Four collapsible basket bins inside body to take total of 8 blankets and 8 small haversacks. Bins can be collapsed and stowed in carrier on rear door. Fittings for 8 rifles. Tool locker in floor. P.O.W. carrier outside rear of body.

DOORS: Rear door 32" wide. Front doors 28" wide with hinged ports.

TOW HOOK: W.D. spring drawbar gear. Ht. 32".

DIMENSIONS

Overall Length 15' 7". Overall Width 7' 7". Overall Height 7' 6½".

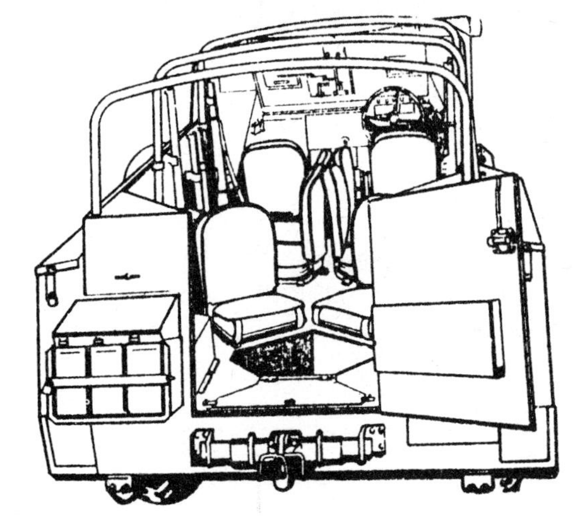

WEIGHTS

UNLADEN		LADEN
Full fuel and spare wheel		(Max. gross rating)
F. Axle.	2 tons 0¼ cwt.	2 tons 3½ cwt.
R. Axle.	2 tons 6¼ cwt.	3 tons 1¾ cwt.
Gross	4 tons 7 cwt.	5 tons 5¼ cwt.

(683)

SM.2424

G.M. LIGHT RECONNAISSANCE CAR

Make: G.M. Canada. Chassis Model 8447

4 x 4. WHEELBASE 8' 5½"

ENGINE:- Make, G.M.270. Petrol 6 cyl. Bore 3.25/32". Stroke 4". Capacity 269.5 cu. ins. Max. B.H.P. 106 @ 3,000 r.p.m. Max. torque 2,580 lbs./ins. @ 1,000 r.p.m. R.A.C. 34.31 h.p.

CLUTCH:- Single dry plate 11½" dia.

GEARBOX:- 4-speed and 1 reverse. Ratios 6.55 to 1, 3.31 to 1, 1.73 to 1, 1.00 to 1. Reverse 7.54 to 1.

TRANSFER BOX:- Single speed with front axle declutching.

AXLES:- FULL floating, spiral bevel drive. Ratio 6.5 to 1.

SUSPENSION:- Front and rear, semi-elliptic. Double acting hydraulic shock absorbers front and rear.

BRAKES:- Foot hydraulic. Hand, mechanical on rear wheels.

STEERING:- Right hand drive. Ball bearing nut and sector.

WHEELS:- Split type, rim 6.00/16.

TYRES:- 9.00/16 Runflat.

ELECTRICAL EQUIPMENT:- 12 volt.

TURNING CIRCLE:- 50' 3". Track F.5'10", r. 5' 10½".

GROUND CLEARANCE:- 8¾".

CAPACITIES:- Fuel 30 gals. Cooling 3 gals.

M.P.G.:- 8. Radius of Action 240 miles.

PERFORMANCE FIGURES:-

B.H.P. per Ton 21.9.
Tractive Effort per Ton (100%) 1,400 lbs./Ton.

(684)

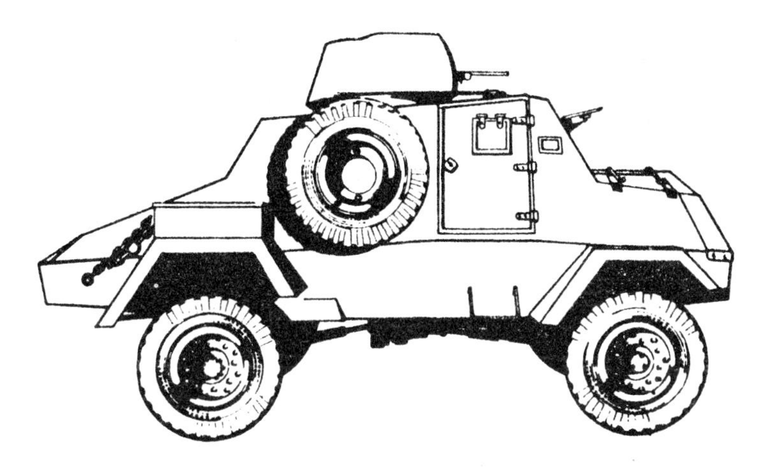

CAR, LIGHT RECONNAISSANCE

CANADIAN G.M.8447 4 x 4

GENERAL DESCRIPTION:- A lightly armed, armoured light reconnaissance car with a machine gun turret. The engine is mounted in the normal forward position and the drive is through all four wheels.

ARMOUR PLATE:- Armour plates are up to 15 m.m. thickness, and Spec. IT 100.

Front 12 m.m. Sides 6 and 8 m.m. Rear 6, 10 and 12 m.m. Top 12 m.m. Engine 6-10 m.m. Turret 8 m.m. Hull 10 m.m.

TURRET:- One man free traverse turret, with spring-loaded (Ground or A.A.) Mtg. for Bren Gun.

ARMAMENT and AMMUNITION:- Bren Gun in turret, and Boys A.T. Rifle with mounting which can be fired from Commander's seat.

Ammunition:- 33 magazines .303 for Bren Gun, 24 magazines for Boys Anti Tank.

Rifle: 1 Box .303 Rifle Grenades. Smoke discharger mounted in centre of front panel. 6 Generators for discharger.

WIRELESS:- No.19 Wireless Set carried on certain vehicles in place of Boys Rifle.

CREW:- 3 men, Driver, Commander and Gunner (Commander on left of Driver, Gunner in turret).

DOORS:- 2 Side entrance doors.

OVERALL DIMENSIONS

Length 14'10½". Width 7' 0". Height 7' 10½".

LADEN WEIGHTS

F.A.	2 tons 7¾ cwt.
R.A.	2 tons 9 cwt.
Gross	4 tons 17 cwt.

MORRIS MK. II 4 x 4 LIGHT RECONNAISSANCE CAR

This is a four wheel drive edition of the Morris Mk.I 4 x 2 L.R.C. shown on pages 356-358.

Apart from the addition of a driving front axle, this Mk.II model varies in the following main details:-

FRONT SUSPENSION:- Orthodox semi-elliptic (instead of independent F.W. coil springs)

TYRES:- 9.00-16 R.F. (instead of 9.25-16 R.F.)

The hull, armour plating and armament of the Mk.II model are similar to the Mk.I shown on page 357.

OVERALL DIMENSIONS:- Length 12' 10". Width 6' 8". Height 6' 4".

WEIGHTS

	UNLADEN	LADEN
F.A.	1 ton 11 cwt.	1 ton 19 cwt.
R.A.	2 tons 2 cwt.	2 tons 7 cwt.
GROSS	3 tons 13 cwt.	4 tons 6 cwt.

CAR 4 x 4 LIGHT RECONNAISSANCE HUMBER MK. III a

Similar to the Humber Mk.III model described on pages 354-5, except:-

Carburettor:- Jet and choke different.

Air Cleaner:- Oil Bath, two stage instead of single.

Front and Rear Axle Rati s:- 4.875-1.

Brakes:- Drum dia., increased to 12".

Suspension:- More leaves. Front springs 12 leaves, Rear 17.

Body:- Equipment Lockers. One each side of body, rear of entrance doors.
Spare Wheel fitted. Bolted to rear of body hull.
P.O.W. Carrier. One on top and one on rear section of each rear mudguard. One on top of nearside front mudguard.

Weights:- Laden. 3 tons 12 cwts. 0 qrs.

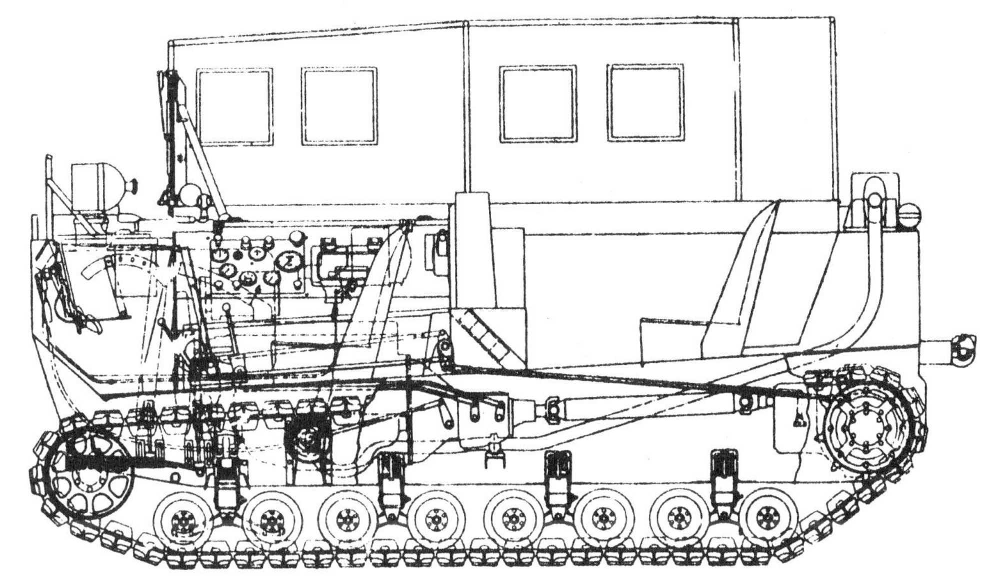

TRUCK 10-CWT TRACKED G.S.

THE WEASEL

BODY: Right hand front is the engine compartment. Left hand front is the driving compartment. Rear left is the passengers compartment. Rear right is the equipment or stowage compartment. Alternatively stowage compartment can be arranged to seat 2 passengers increasing total crew to four.

WEATHER PROTECTION: Windscreen (detachable) incorporating built in defroster and 3 wipers. Detachable bows and canvas cover to body with detachable side curtains. Body heater. Tow hook at rear. Ht. 27$\frac{1}{8}$". Towing eye (or loop) at front.

OVERALL DIMENSIONS

Length 10' 5$\frac{1}{4}$", length less tow hook 9' 11"
Width 5 ft. width (20" tracks) 5' 5".
Ht. 5' 10$\frac{5}{8}$", Stripped Ht. 4' 3".

WEIGHTS

Unladen weight, including full fuel tanks = 3725 lbs. = 1 ton 13$\frac{1}{4}$ cwt.
Crew (of 2) = 340 lbs. = 3 cwt.
Load " = 860 lbs. = 7$\frac{3}{4}$ cwt.
Gross Laden Wt. = 4925 lbs. = 2 tons 4 cwt.

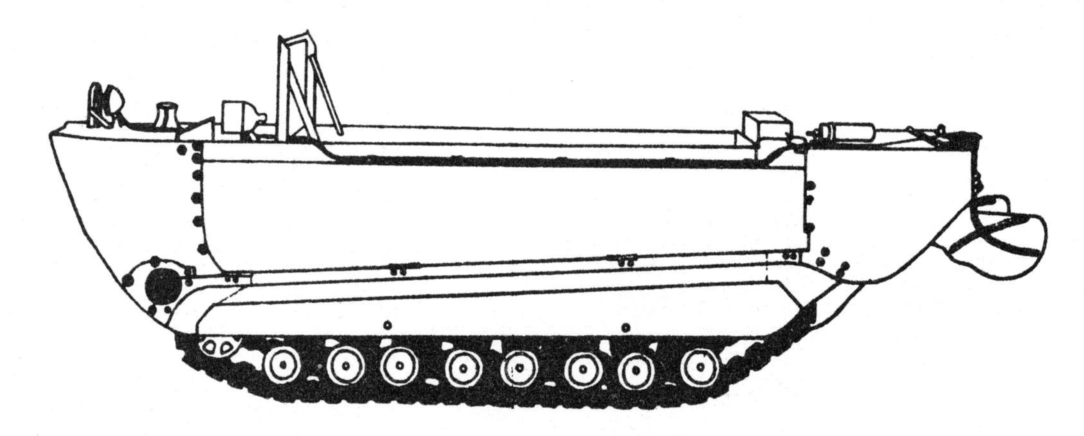

AMPHIBIAN 10-CWT TRACKED G.S.

THE AMPHIBIOUS WEASEL

(U.S. Army Nomenclature is Light Cargo Carrier M.29.C.)

This is a relatively simple conversion of the Truck 10-cwt Tracked G.S.

Bow and stern cells, track side panels, sponson air tanks and cable controlled rudders are added to the standard M.29 model (with 20" tracks).

A capstan is fitted to top of front (or bow) cell, driven by P.T.O. off engine.

Water propulsion is effected by the standard tracks. Water speed approx. 4 m.p.h. Water steering is by two rudders. Track side panels are hinged to provide access to track and suspension.

Tow hook is mounted at stern and tow eye at front.

This amphibian is intended only for operation across slow moving streams or in quiet water. It is not intended for use in surf or highly turbulent waters.

WEIGHTS AND DIMENSIONS
equipped with 20" Tracks

OVERALL DIMENSIONS	
Length over hull	14' 5⅝"
" including rudders	15' 8¾"
Width	5' 7"
Ground Clearance	10½"
Overall Height	5' 10¾"
Free Board (Fully Laden)	
Bow	10½"
Stern	8"
Weights	
Unladen (Full fuel)	4840 lbs. = 2 tons 3¼ cwt.
Crew (2 men)	340 lbs. = 3 cwt.
Payload	860 lbs. = 7¾ cwt.
Gross Laden Wt.	6040 lbs. = 2 tons 14 cwt.
Track Pressure	
Length of Track on ground	78⅛"
Area " " " "	3125 sq. ins.
Ground Pressure	1.93 lbs./sq.ins.

S.M. 6027, 6362

TRUCK 10-CWT. TRACKED G.S.

THE STUDEBAKER "WEASEL"

U.S. Army Nomenclature is "Light Cargo Carrier M.29".

GENERAL: A light fully tracked vehicle designed primarily for snow operation and/or airborne use. Super low ground pressures enable it to traverse snow, mud and swamp conditions impracticable to the ordinary tracked vehicle. In its standard form it is amphibious, but free board is limited. The special amphibious model M.29C. is shown on page 690.

ENGINE: Make, Studebaker. Model 6-170. Petrol. 6 cyl. Bore 3". Stroke 4". Capacity 170 cu. ins. (2.85 litres). Max. B.H.P. Gross 75 @ 3800. Net 55 @ 3600. Net torque 1320 lbs. @ 1400.

CLUTCH: Borg & Beck, 8".

GEARBOX: Synchromesh 3 speed: Ratios 2.66 to 1, 1.56 to 1, 1.00 to 1, r.3.55 to 1.

TRANSFER CASE, AXLE, STEERING: Combined Clark Steering Differential Unit. Transfer Case 2 speed, Ratios .866 to 1 and 2.74 to 1. Axle Ratios 4.87 to 1. Steering effected by two levers mounted in front of driver.

*TRACKS: Rubber belt with four steel cables embedded in each track. Width of track 15". Steel support plates 15" wide and 1" high grousers.

GROUND CONTACT:

Zero Penetration .... 2344 sq. ins.
Pressure .... 2.1 lbs/sq. in.
5" Penetration (Snow) .... 2812 sq. ins.
Pressure .... 1.75 lbs/sq. ins.

SUSPENSION: Four transverse springs bolted under hull. Each carrying four solid rubber tyred bogie wheels at either end.
MAX SPEEDS (in low Range) 4.3, 7.4 and 11.6 m.p.h.
(in High Range) 13.6, 23.2 and 36.4 m.p.h.
TURNING CIRCLE: 24' 0" (lead surface).
GROUND CLEARANCE: 11".
CAPACITIES: Fuel 29 gals. Cooling 9 qts.
PETROL CONSUMPTION: Variable from 0.7 to 7.0 m.p.g. according to load and terrain.
*Note: Later types fitted with 20" wide tracks.

FORD G.P.A. AMPHIBIAN ¼ TON 4 x 4

"THE AMPHIBIOUS JEEP"

GENERAL CHARACTERISTICS:- The design of this amphibian is based on that of the standard land Jeep (i.e. Willys M.B. or Ford G.P.W.) and utilises many components common to the land version. Water propulsion is by a propeller driven by P.T.O. from the transfer case, water steering is by rudder.

ENGINE:- Petrol. 4-cyl. Bore 3⅛" Stroke 4⅜". Capacity 134.2 cu.ins. (2.19 litres)
B.H.P. 60 @ 3600 r.p.m. Max torque 1260 lbs. ins. @ 2000 r.p.m.

GEARBOX:- 3 speed and 1 reverse, as land Jeep.

TRANSFER BOX:- Two speed, incorporating P.T.O. to drive Water propeller.

AXLES:- Front and rear, ratio 4.88 to 1. Differential assemblies interchangeable with land Jeep.

BRAKES:- Foot, hydraulic on all wheels. Hand, transmission brake.

TYRES:- 6.00-16, 6 ply.

ELECTRICAL EQUIPMENT:- 12 volt.

CAPACITY:- Fuel 12½ gallons.

PERFORMANCE FIGURES

Max. speed	land	50 m.p.h.
" "	water	5 m.p.h.
Petrol Consumption,	Land	18 m.p.g.
" "	water	1.5 m.p.g.

TURNING CIRCLE:- Land: 38' 0"
Water: 36' 0"

FORD G.P.A. AMPHIBIAN ¼ TON 4 x 4 (contd.)

"THE AMPHIBIOUS JEEP"

SPECIAL EQUIPMENT

Capstan: Mounted in bows and driven from engine crankshaft.

Bilge Pumps: One power operated pump, capacity 25 g.p.m. plus one hand operated pump.

Tow Hook: At rear. Ht. from ground 28". Lifting and mooring eyes.

DRIVING COMPARTMENT

There are seats for driver and 3 passengers. The rear seat can be folded back on top of the rear deck to provide space for cargo. Space thus available behind front seats measures approx. 60" wide x 22" long.

DIMENSIONS

Wheelbase	84"
Track, f. and R.	49"
Ground Clearance	10" at belly
Overall Length	15' 2"
" width	5' 4"
" Ht.	5' 9"
" Ht. (reducable)	4' 2"

WEIGHTS

	UNLADEN	LADEN (800 lbs.)
F. axle	14½ cwt.	14¾ cwt.
R. axle	15 cwt.	23¼ cwt.
Gross	1 ton 9½ cwt.	1 ton 18 cwt.

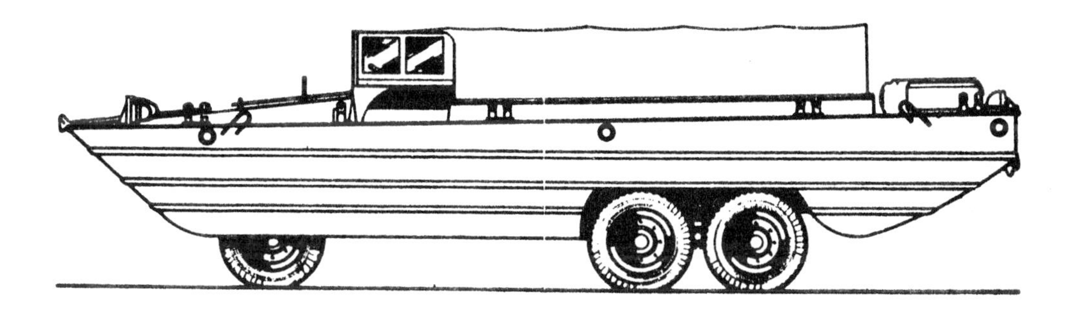

S.M. 2849

G.M.C. D.U.K.W. 353 2½ TON 6 x 6 AMPHIBIAN

Load Capacity 2½ short tons (5000 lbs.)

GENERAL NOTES: This Amphibian has similar automotive characteristics to the conventional 6 x 6 truck. For land operations the vehicle utilises six driving wheels and is steered in the normal way. Water propulsion is by means of a water propeller driven by the engine through the transmission and a special water propellor transfer case. Steering in the water is by a rudder assisted by the action of the front wheels.

ENGINE:- Make G.M.C. model 270. 6-cyl. Petrol. Bore 3 25/32" Stroke 4". Capacity 269.5 cu. ins. 4.41 litres. Max. B.H.P. 95 @ 3000 r.p.m. governed to 3100 r.p.m. Max. torque 2580 lbs. ins. @ 1000 r.p.m.

GEAR-BOX:- Five speeds and 1 reverse. Ratios 6.06 to 1, 3.5 to 1, 1.8 to 1, 1.00 to 1, verdrive .79 to 1. Reverse 6.0 to 1.

G.M.C. D.U.K.W. 353 2½ TON 6 x 6 AMPHIBIAN

(Continued)

Load Capacity 2½ short tons (5000 lbs.)

TRANSFER BOX:- 2 speed. Ratios. 1.16 to 1 and 2.63 to 1. Front axle drive can be disengaged.
AXLE RATIO: 6.6 to 1.
WATER PROP TRANSFER BOX: Ratio .743 to 1. Latest types include an additional overdrive gear primarily for use when reversing.
BRAKES: Foot, Hydraulic vacuum servo assisted on all wheels. Hand transmission brake.
WHEELS AND TYRES: Single disc wheels 18 x 8. Tyres 11.00-18 10 ply, military desert type.
ELECTRICAL EQUIPMENT: 6 volt.
CAPACITY: Fuel 33 gals. Cooling 4¼ gals.

PERFORMANCE FIGURES

Max. speed land	50 m.p.h.	
" " water	6 m.p.h.	(2nd gear)
Petrol Consumption	7 m.p.g.	land (Road circuit)
" "	1 m.p.g.	water (max. speed)
" "	2.5 m.p.g.	water (cruising)

Turning Circles:-
Land, 67' petrol. 70' left. (rear wheel drive)
Water Approx. 45'

SPECIAL EQUIPMENT

Winch: Make. Garwood. Model 42Y4212 driven by P.T.O. from gearbox. Max. pull 10,000 lbs. 300' rope.
Bilge Pumps. Two power driven bilge pumps are fitted, a gear type capacity 50 g.p.m. and a centrifugal type, capacity 130 g.p.m.
TYRE INFLATION. Air cooled tyre pump fitted up to D.U.K.W. 905. Water cooled tyre pump fitted D.U.K.W. 906 to 2005 Automatic tyre inflation system controlled from driver's seat fitted after D.U.K.W. 2005.
TOWHOOK: Fitted at rear. Height from ground approx. 46".
Lifting eyes and mooring eyes fitted both sides.

DRIVING & CARGO COMPARTMENTS

Driving compartment seats 2 and has folding windscreen and folding canvas top.

Cargo Compartment (or Hold) measures 149" long x 82" wide. It is fitted with a detachable canvas cover mounted on detachable bows.

Ht. of hold to top of coaming	28" at front, 23" at rear.
Capacity of hold to top of coaming	196 cu. ft.
Ht. of hold to centre of bows	59" at front, 54" at rear.

Capacity of hold under bows 385 cu. ft.

DIMENSIONS

Wheelbase	164"
Bogie Centres	.44"
Track, f. 63¼", R. 63	5/8"
Ground Clearance, axle,	11¼"
Belly Clearance	17¼"
Loaded Draft, f. wheels	42"; R. wheels 51"
Overall Length	372" (31' 0")
" Width	98" (8' 2")
" Ht. top of bows unladen	106½" (8' 10½")
" Ht. is reducable to	89" (7' 5")

WEIGHTS

	UNLADEN	LADEN (5000 lbs.)
F. Axles	2 tons 2 cwt.	2 tons 6½ cwt.
R. Bogie	4 tons 8½ cwt.	6 tons 12 cwt.
Gross	6 tons 10¾ cwt.	8 tons 18½ cwt.

(Above are weightbridge wts. at W.V.E.E.

Latest type D.U.K.W. from No. 2506 onwards are roughly 3½ cwt. heavier.)

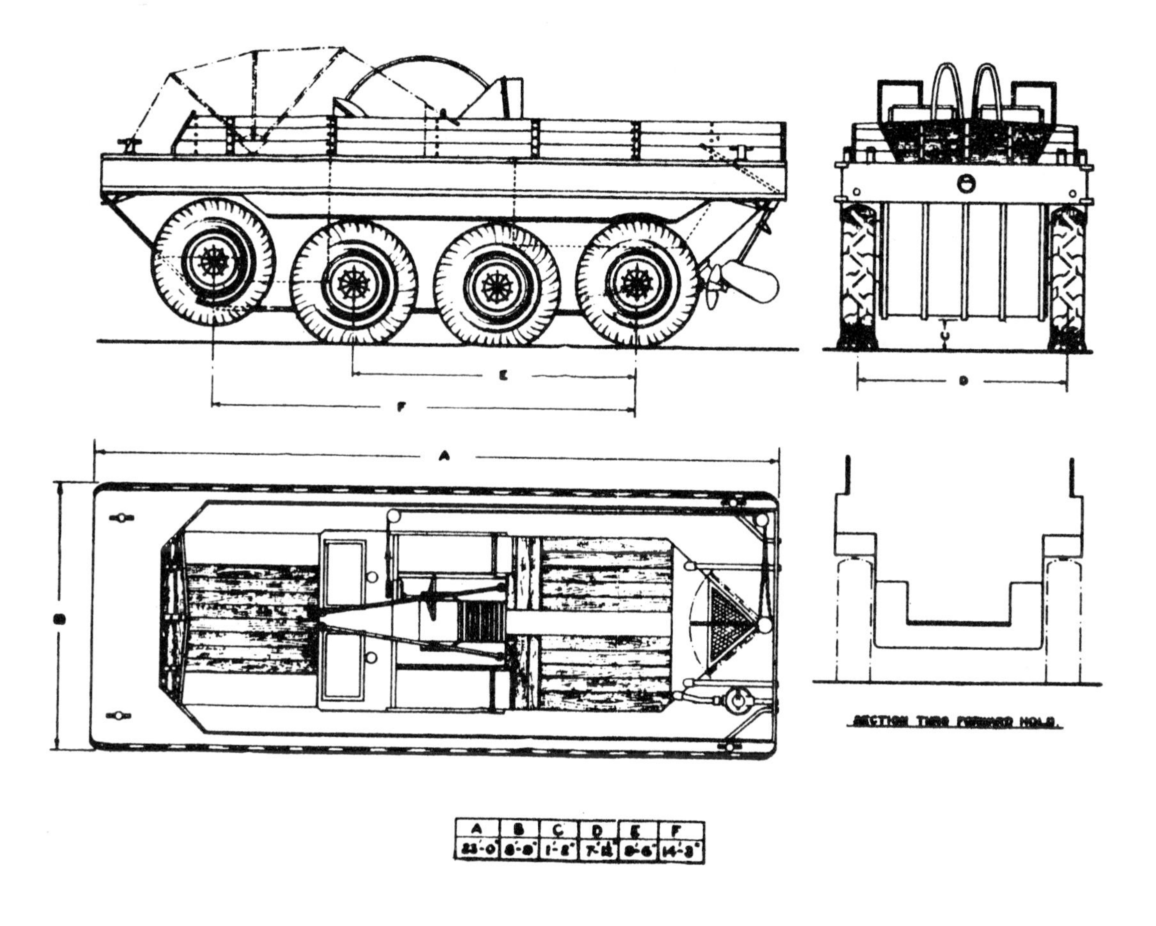

A	B	C	D	E	F
23'-0"	8'-8"	1'-8"	7'-1½"	8'-6"	14'-3"

TERRAPIN MK. I 4 TON 8 x 8 AMPHIBIAN

Made by Morris Commercial

TERRAPIN MK. I 4 TON 8 x 8 AMPHIBIAN

Made by Morris Commercial

GENERAL CHARACTERISTICS

This Amphibian is powered by two engines mounted side by side. The left hand engine drives the four left hand wheels; the right hand engine the four right hand wheels. Land steering is effected by throttling back the appropriate engine. Water propulsion is by two propellers, one driven by the left engine, the other by the right. Water steering is by a rudder and can be assisted by throttling back the appropriate engine.

The special curved axle formation will be noted.

ENGINES: Two. Make Ford. Dagenham. Petrol V.8. Bore 3.06" Stroke 3.75". Capacity 3.62 litres Max. B.H.P. 85 @ 3600. Max Torque 1800 lbs/ins. @ 2000 r.p.m.

CLUTCHES: Two. Make Ford. Operated by a single master clutch pedal.

GEARBOXES: Two. Make Ford. Type 3-speed with Synchro-mesh. Ratios 3.66 to 1, 1.9 to 1, 1.00 to 1, reverse 4.14 to 1. Operated by single master gear change lever.

TRANSFER BOXES: Two. Make Morris. Single speed. Ratio 3.97 to 1. Operated by a single master lever which engages drives to both propellers.

TRANSMISSION: By two lines of worm shafts, each line driving the four wheels on its side of vehicle. Worm reduction 7.75 to 1 (Make Morris).

SUSPENSION: Solid.

TYRES: 12.75-24 Agricultural tractor type with chevron pattern (c. c. tread). Effective radius 23.1".

BRAKES: Mechanical transmission brakes operating on wormshafts. Operated by a single master foot pedal with a single master hand lever for parking. Auxiliary levers provide individual braking on either wormshaft in order to steer the amphibian when being towed.

ELECTRICAL EQUIPMENT: 12 volt.

FUEL TANK: 35 gallon capacity. M.P.G. 2½.

PERFORMANCE FIGURES

Max. Speed, land.	15 m.p.h.
" " water.	5 m.p.h.

Turning Circle 35' land.

SPECIAL EQUIPMENT

BILGE PUMPS: Two constant running power bilge pumps which automatically come into operation when propeller drives engaged. Capacity 50 g.p.m. each. Auxiliary hand operated bilge pump, capacity 8 g.p.m.

NO winch is fitted to Terrapin Mk.I.

DRIVING AND CARGO COMPARTMENTS

Single drivers seat situated roughly centrally.

Two cargo holds.

Front hold measures.

65" long x 85" wide at deck level.
" " x 67" " " sponsons.
" " x 42" " " foot.

Floor of front hold is on two levels.
Front Section 28" long is 49" below deck level.
Rear " 38" " " 53" " " "

TERRAPIN MK. I 4 TON 8 x 8 AMPHIBIAN (contd.)

Total capacity of front hold 112 cubic ft.

Rear hold measures
61" long x 85" wide at deck level.
" " x 67" " " sponsons.

Floor of hold is 38" below deck level.

Total capacity of rear hold 111 cubic feet.

DIMENSIONS

Overall wheelbase 14' 3".

Axle spacing 4' 9", between each axl .

Track: 7' 2".

Overall width across tyres 8' 4".
" " inside " 6' 0".

Min. Ground Clearance 13½" at belly.

Overall Length 23' 0"
" Width 8' 9"
" Height 9' 7"
" " reducible to 8' 3" @ air inlets.
Width at this point 5' 7½".

WEIGHTS

UNLADEN	LADEN
6 tons 16 cwt.	11 tons 8¼ cwt.

Owing to special curved axle arrangement, individual axle weights will vary according to type of road surface. For bridging classification it should be assumed that the entire weight will be evenly disposed on the two middle axles.

DATA BOOK OF WHEELED VEHICLES

PART 7

NEW TRAILERS

This section covers new type trailers produced since the publication of Part 3.

Earlier design trailers are shown in Part 3, pages 317 to 327.

Heavy tank transporter trailers are included with Tank Transporters, pages 501 to 553.

Published by D.G.F.V., Ministry of Supply.

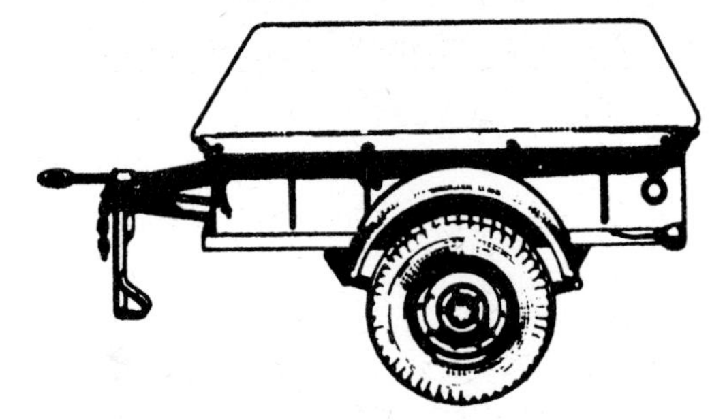

U.S. Army

TRAILER ¼ TON 2 WH.
AMPHIBIAN CARGO

The Jeep Trailer

GENERAL:- This standard U.S. Army trailer has been designed for towing behind the Jeep and utilises as far as possible, standard Jeep components. It is amphibious and has a 6" freeboard when loaded. Payload is 500 lbs.

BRAKES:- Hand Parking only.

TYRES:- 6.00 - 16.

DIMENSIONS

	OVERALL	INSIDE BODY
Length	9' 0½"	6' 0"
Width	4' 8"	* 3' 2"
Height	3' 4"	1' 6"

* Body is 3' 10" wide at top.

WEIGHTS

Unladen 550 lbs. Laden 1,050 lbs.

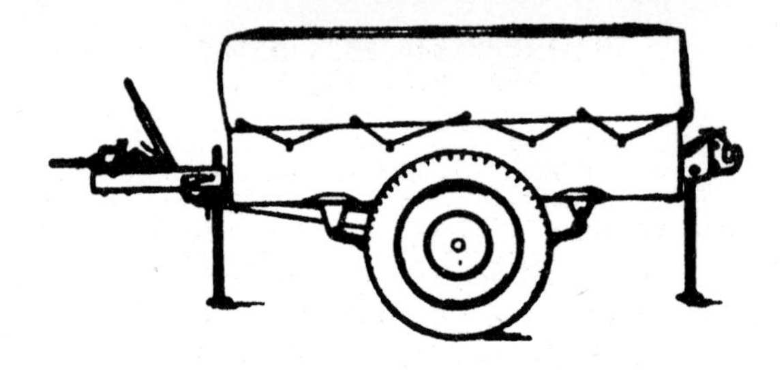

TRAILER 10-CWT. 2 WH.
LIGHTWEIGHT, G.S.

CHARACTERISTICS:- Lightweight trailer, normal towing vehicle is the Jeep.
Trailer is amphibious and floats with a 9" freeboard when carrying a 10-cwt. load.

PAYLOAD:- 10-cwt., but trailer is always known as Lightweight G.S. to avoid confusion with the heavier "Trailer 10-cwt. 2 wheeled G.S." latter shown on Page 706.

BRAKES:- Operated by overrun device, with a separate hand parking lever.

TYRES:- 5.00 - 16.

BODY:- All steel, fixed sided body with canvas cover.
Tubular supporting legs front and rear.
*Tow eye 2" internal diameter. Man handling bars either side of toweye. Tow hook at rear to permit tandem towing.

DIMENSIONS

	OVERALL	INSIDE BODY
Length	9' 1½"	5' 1"
Width	4' 4 3/8"	2' 11"
Height	3' 6 3/8"	1' 8"

WEIGHT:- Unladen 5 cwt. Laden 15 cwt.

* Later types tow eye size increased to 2.13/16" i.d.

TRAILER 2 WH. LIGHTWEIGHT TECHNICAL TYRES

All the following Lightweight technical trailers are based on using components of the Standard Trailer 10-cwt. 2 wh. Lightweight G.S. (shown on page 703), but tyres are 6.00 - 16 in every case. Each trailer has overrun operated brakes with a separate hand parking lever, semi-elliptic springing and standard hubs.

TRAILER 2 WH. LIGHTWEIGHT COMPRESSOR

Overall Length 9' 0". Width 4' 6".
Height 3' 11".
Laden weight 17½ cwt.

TRAILER 2 WH. LIGHTWEIGHT WATER, 100 GALLONS

Virtually one half of the standard 180-gallon water trailer. Capacity 100 gallons instead of 180 and fitted with one hand operated pump and filter unit instead of the two of each fitted to larger model.
Overall Length 8' 7½". Width 4' 6".
Height 3' 10".

WEIGHTS: Unladen 7¼ cwt. Laden 16 wt.

TRAILER 2 WH. LIGHTWEIGHT TYPE Z

Overall Length 9' 4". Width 4' 6".
Height 3' 9½".
Laden Weight 17½ cwt.

TRAILER 2 WH. LIGHTWEIGHT ELECTRICAL REPAIR

Overall Length 9' 4". Width 4' 6".
Height 3' 9½".
Laden Weight 14¾ cwt.

TRAILER 2 WH. LIGHTWEIGHT MACHINERY

Overall Length 9' 4". Width 4' 6".
Height 3' 9½".
Laden Weight 18¼ cwt.

TRAILER 2 WH. LIGHTWEIGHT GENERATOR (5 k.w.)

Overall Length 9' 0". Width 4' 6".
Height 3' 9½".
Laden Weight 19¾ cwt.

TRAILER 2 WH. LIGHTWEIGHT WELDING

Overall Length 9' 0". Width 4' 6".
Height 3' 9½".
Laden Weight 16½ cwt.

TRAILER 2 WH. LIGHTWEIGHT CIRCULAR SAW

Overall Length 9' 7". Width 4' 6".
Height 3' 9½".
Laden Weight 17¾ cwt.

TRAILER 2 WH. LIGHTWEIGHT STORE

Fitted with two sets of bins along the sides. Bins are designed to accept Standard W.T. carton packs.
Central space between bins available for larger M.T. spares.
Overall Length 9' 3". Width 4' 6".
Height 3' 10½".
Unladen Weight 6¾ cwt.

TRAILER 10-CWT. 2 WH. G.S.

(Note: This trailer is not to be confused with the Trailer 2 wh. Lightweight, page 703, which also carries a payload of 10 cwt.)

BRAKES: Operated by overrun device, with a separate hand parking lever.

TYRES: Originally 5.00 - 16.
Later production modified to 5.75 - 16.

BODY: Fixed sides, drop tailboard.
Canvas cover provided which permits loading up to six inches above height of top rave.

Tubular supporting legs front and rear.

Tow eye originally 2" internal Diameter; i.e. designed to take either Jeep hook or 15 cwt. pin and pintle. All types now modified to the standard 2.13/16" size, so enabling trailer to be towed by any prime mover.

DIMENSIONS

	OVERALL	INSIDE BODY
Length	9' 7"	5' 4"
Width	5' 4"	3' 7"
Height	3' 9" (top of raves)	1' 6" (top of raves)

WEIGHT: Unladen 6 cwt. 3 qrs.
Laden 16 cwt. 3 qrs.

NOTE: Above refers to "Trailer 10-cwt. 2 wh. G.S. No. 1". The "Trailer 10-cwt. 2 wh. G.S. No.2" is generally similar but has solid body sides and 6.00-16 tyres.

TRAILER 10 CWT. 2 WH. FOR D.F. SET

GENERAL:- This trailer consists of a lightweight house type body mounted on the standard 10 cwt 2 wh. trailer chassis.
The trailer chassis is as described opposite, except that for the D.F. role the tyres are 5.25 - 16 and that D.F. trailers have always been fitted with the standard 2 21/32" draweye, as the trailer is normally towed by a 3 ton lorry.

THE BODY is fitted with rear door, windows, and seats for two operators together with D.F. equipment. Inside dimensions approx. 5' 6" long x 5' 6" wide x 4' 9" high.

OVERALL DIMENSIONS:- Length 10' 0½", width 6' 6". Height 7' 5".

WEIGHTS:- Unladen 11 cwt. 3 qrs. Laden, equipment only, 15 cwt.

(Note:- Personnel are not carried in trailer on the move.)

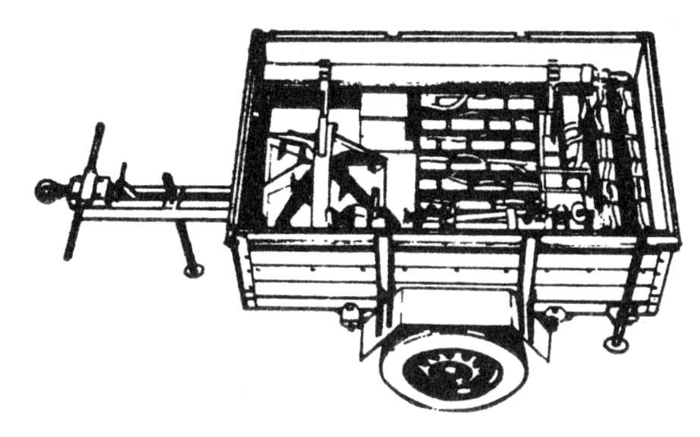

TRAILER 10-CWT. 2 WH. MORTAR

The chassis is generally similar to the standard Trailer 10 cwt. 2 wh. G.S. shown on page 706, except Tyres are 6.00 - 16. This trailer has always been fitted with standard size towing eye to accept any hook. (i.e. 2.13/16" i.d.)

BODY: Open truck type body equipped with fittings to carry either

(a) Mortar Role, 4.2 mortar, load speaker and 24 bombs.

Or (b) Ammunition Role, 44 bombs.

Body shells are identical between both types and are drilled to take either set of fittings.

OVERALL DIMENSIONS

Length	9' 5"
Width	5' 4½"
Height	3' 8"

WEIGHTS

Unladen	7 cwt.
Laden Mortar Role	15 5/8 cwt.
" Ammo "	15 7/8 cwt.

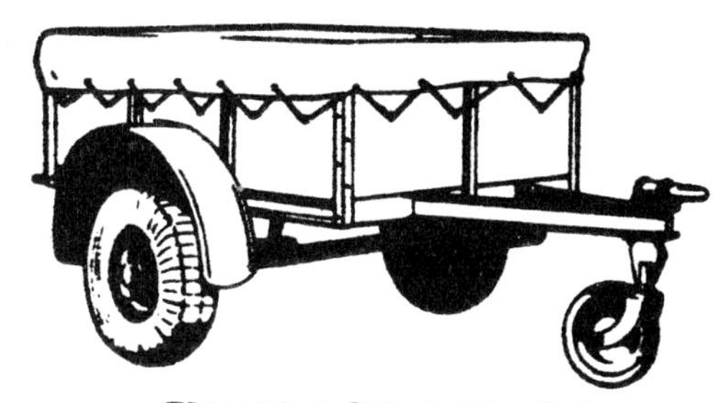

TRAILER 1 TON 2 WH. G.S.
Latest Type with Castor Wheel

(Chassis Plate marked "Trailer Chassis 1 ton 2 wh. No. 5 Mk.I"

This improved 1 ton 2 wh. trailer has now replaced the early type (Shown opposite) in production.

SUSPENSION:- Semi-elliptic.

BRAKES:- Overrun device is not fitted. Simple hand parking lever is only brake.

TYRES:- 9.00 - 13.

HUBS:- Standard H.20LL

SUPPORTING LEGS:- None fitted. Castor wheel is instead fitted to drawbar, and at rear drop tailboard can be lowered and fixed vertically to form loading support.

DIMENSIONS

	OVERALL	INSIDE BODY
Length	10' 9"	6' 8"
Width	6' 5"	4' 6"
Height	4' 3½"	1'10"

Height of Towing eye 2' 7" High Position. 2' 4" Low Position.

WEIGHTS

Unladen 11¼ cwt. Laden 31¼ cwt.

TRAILER 1 TON 2 WH.
WATER 180 GALS.

Similar water tank equipment to 15 cwt. trailer shown on page 320, but mounted on new 1 ton 2 wh. trailer chassis.

Chassis details as page 711, except tyre *size is <u>7.50 - 20</u> and rear support legs are fitted.

<u>DIMENSIONS AND WEIGHTS:</u>- Approx. as page 320, except weight of 1 ton model is 1 ton 17 cwt. Laden.

TRAILER 1 TON 2 WH.
WATER PURIFICATION

Chassis similar to standard 1 ton 2 wh. trailer on page 711, except tyres are 7.50 - 20.

<u>OVERALL DIMENSIONS:</u>- L, 12' 0" W, 6' 8" H, 9' 1".

<u>WEIGHT:</u>- 2 tons 9¾ cwt.

TRAILER 1 TON 2 WH.
STERILIZER

Chassis similar to standard 1 ton 2 wh. trailer on page 710, with 9.00 - 13 tyres as standard fitment.

<u>OVERALL DIMENSIONS:</u>- Length 11' 6". Width 6' 6". Height 6' 9".

<u>WEIGHTS (APPROX.):</u>- Unladen including sterilizer 27 cwt. Laden 30 cwt.

(Note: Similar equipment also fitted to earlier 15 cwt. 2 wh. Trailer.)

---

* First production type 1 ton Water Trailers had 9.00-13 tyres and then 9.25-16 tyres.

TRAILER 1 TON 2 WH.
MACHINERY GAS WELDING

<u>BRAKES:</u> - Overrun operated with hand parking lever.

<u>TYRES:</u> - Early contract 9.00 - 16. Latest production on standard 1 ton 2 wh. trailer chassis (as page 710) with 9.00 - 13 tyres and H.20LL hubs.

<u>SUPPORTING LEGS:</u> - Front and rear.

<u>BODY:</u> - Fixed sided open truck body with drop tailboard. Equipped with fittings to take oxygen and acetylene cylinders, welding outfits etc. Provision for carrying welding bench.

<u>OVERALL DIMENSIONS</u>

Length	11' 6½"
Width	6' 4"
Height	3' 10"

Height of towing eye is variable between 2' 10" and 2' 1".

<u>WEIGHTS</u>

Unladen	15¼ cwt.
Laden	26 cwt.

TRAILER 1 TON 2 WH. MOBILE A.A. COMMAND POST

(Formerly known as S.A. Plotter Trailer)

This drop frame trailer utilises standard 1 ton chassis running gear (see page 710) i.e. 9.00 - 13 tyres, H.20LL Hubs, overrun brakes but three supporting jacks.

It has a specially modified and lengthened drop frame.

The body consists of a flat platform measuring 10' 0" long x 4' 4½" wide which carries a superstructure and canvas canopy. The canopy can be erected to form a tent enveloping the trailer all round.

OVERALL DIMENSIONS (Travelling)
Length 13' 11". Width 6' 4½".
Height 8' 0".

WEIGHTS:- Unladen 1 ton 2 cwt.
Laden 1 ton 10½ cwt.

TRAILER 1 TON 2 WH. PIGEON LOFT (60 Bird)

BRAKES:- Overrun operated with separate hand parking lever.

TYRES:- 7.50 - 20 on H.20LL hubs. (Note, certain early types were fitted with 6.50 - 20 tyres.)

SUPPORTING LEGS:- 1 front, 2 rear.

BODY:- 60 bird pigeon loft with space for stores, equipment, etc.

OVERALL DIMENSIONS:- Length 19' 0".
Width 6' 6". Height 9' 3".

WEIGHTS:- Unladen 1 ton 7¼ cwt.
Laden 1 ton 19 cwt.

15, 17 & 27 K.V.A. - 22 & 24 K.W.
Page 323 gives details of the 22 k.w. generating set mounted on the 2 ton 2 wh. unsprung chassis. The same unsprung chassis was also used to mount the 15 k.v.a. set, details as 22 k.w. on page 323 except weight slightly lower.

THE LATEST GENERATOR TRAILERS are generally similar to the type illustrated on page 323, but differ in the following points:-

(1) They have semi-elliptic suspension.

(2) 10.50 - 13 tyres on standard H.45LL Hubs.

(3) Castor wheel and man handling bars fitted.

Brakes are overrun operated as before.

OVERALL DIMENSIONS:- Length 14' 6". Width 7' 6". Height 9' 3".

The platform of this new trailer is designed to take any one of the following sets.

15 k.v.a.	Gross	Laden	Wt.	3 tons 7¼ cwt.
17 k.v.a.	"	"	"	3 tons 8 cwt. (Approx.)
22 k.w.	"	"	"	3 tons 10¼ cwt.
24 k.v.a.	"	"	"	2 (two) tons 10¾ cwt.
27 k.v.a.	"	"	"	3 tons 11 cwt.

Trailer 2 ton 2 wh. Bakery Generator 22 k.w. 220 volt D.C.

Trailer is similar to 22 k.w. 110 volt D.C.; difference is in generating set.

FOR DETAILS OF SMALLER GENERATOR TRAILERS SEE FOOT OF FACING PAGE

TRAILER 2 TON 2 WH. RADAR A.A. NO.3 MK. III

(Commonly known as the "Babie Maggie")

BRAKES:- Overrun operated, separate hand parking lever.

TYRES:- 9.00 - 16 on standard H.30 hubs.

SUPPORTING LEGS:- 3 special jacks, with castor wheel on drawbar.

BODY:- Special structure to house equipment.

OVERALL DIMENSIONS:-

Length 15' 1". Width 6' 11". Height approx. 9' 0". Track 6' 0".

WEIGHTS:- 1 ton 2½ cwt. unladen.
2 tons 9¾ cwt. laden.

THE SMALLER GENERATOR TRAILERS

5 K.W. GENERATOR TRAILER. Details of this equipment mounted on the Lightweight 10-cwt. 2 wheel trailer are shown on page 705.

6 K.W. GENERATOR TRAILER. This is mounted on the standard 10-cwt. 2-wh. trailer chassis; trailer details as page 706, weights within G.S. loading.

9 K.W. & 9 K.V.A. GENERATOR TRAILERS. These sets are mounted on the 1 ton 2 wh. trailer. Weights within G.S. loadings.

* TRAILER 2 TON 4 WH. G.S.
(Standard U.K. Type)

STEERING:- Turntable.

BRAKES:- Overrun with separate hand parking lever.

TYRES:- 9.00 - 13 on standard H.20LL hubs.

BODY:- Detachable dropsides and a drop tailboard with tarpaulin.

DIMENSIONS

Wheelbase 6' 4". Track 5' 7 1/8".

	OVERALL	INSIDE BODY
Length	15' 0"	9' 10½"
" less tow bar	10' 4"	-
Width	6' 10"	6' 6"
Height	4' 10"	1' 6"
Loading Height	3' 4½"	
Height of draw eye	2' 2"	

* Revised Official Nomenclature is "Trailer, 2 ton 4 wh. G.S. Dropside".

*TRAILER 2 TON 4 WH. G.S.
Canadian Type

Make:- Dominion Co. of Canada

Note:- Although ordered to fill a 2 ton requirement, this trailer has an actual pay-load rating of 6,000 lbs. or roughly 2¾ long tons.

STEERING:- Turntable.

TYRES:- 9.00 - 16 mounted on W.D. type wheels.

BRAKES:- Hydraulic, actuated by overrun device. Hand parking brake on rear wheels.

BODY:- All steel. Fixed sides drop tail-board. Tarpaulin cover.

DIMENSIONS

	OVERALL	INSIDE BODY
Length	16' 4"	12' 0"
Width	6' 8"	6' 3"
Height	6' 5¼"	2' 6"
Loading height	3' 11"	

Wheelbase 7' 9". Track 5' 10½".

WEIGHTS

	UNLADEN	LADEN (2¾ tons)
F. Axle	1 ton 2 cwt.	2 tons 9 cwt.
R. Axle	1 ton 0½ cwt.	2 tons 8 cwt.
Gross	2 tons 2½ cwt.	4 tons 17 cwt.

* Revised Official Nomenclature is "Trailer 2 ton 4 wh. G.S. Open".

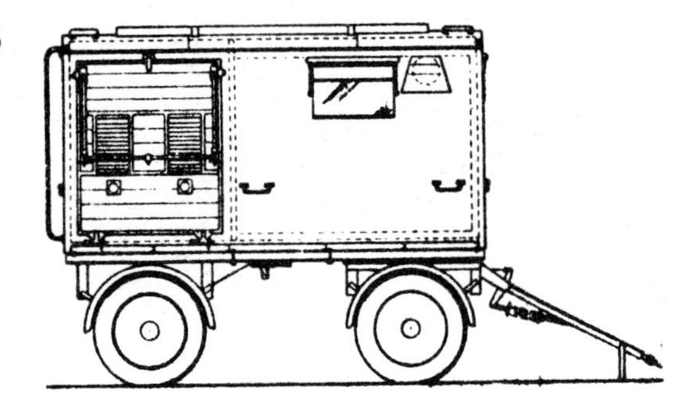

TRAILER 2 TON 4 WH. BEAM WIRELESS

STEERING:- Turntable.

SUSPENSION:- Semi-elliptic.

BRAKES:- Overrun with separate hand parking lever.

TYRES:- 32 x 6 on standard H.15 hubs.

SUPPORTING JACKS:- Four.

BODY:- Special House type.

DIMENSIONS:- Wheelbase 6' 3½". Tracks 5' 0". Overall Length 15' 6". Width 6' 6". Height 8' 9".

WEIGHTS:-

Front Axle	1 ton	4½ cwt.
Rear Axle	1 ton	18¾ cwt.
Gross	3 tons	3¼ cwt.

TRAILER 2 TON 4 WH. PROJECTOR

Page 325 gives details of the 2 ton Projector Trailer fitted with early Projector equipment. Basically these details still apply, but the laden weights have increased appreciably.

The weight of the trailer carrying Projector A.A. 150 cm. Mk.IV(F.S.) including I.F.F. Mk. III is now:-

Front Axle	1 ton 9 cwt.
Rear Axle	2 tons 6½ cwt.

THE NEW 3 TON 4 WH. PROJECTOR TRAILER IS SHOWN ON PAGE 740.

TRAILER 2 TON 4 WH. S.L.C. LOCATOR MK.I*

(Mounted on 2 ton 4 wh. Sound locator Trailer)

SUSPENSION:- Front, single helical. Rear, semi-elliptic.

BRAKES:- Overrun on front wheels only.

TYRES:- 9.00 - 16.

SUPPORTING LEGS:- Two front, one rear.

DIMENSIONS:- Overall length 18' 1". Width 7' 9". Height 11' 0".

Laden Weights:- Front axle 1 ton 2½ cwt.
Rear axle 1 ton 18¼ cwt.

TRAILER 2 TON 4 WH. PREDICTOR A.A. NO.7 MK.I

(Formerly known as I.R. Predictor Trailer)

SUSPENSION:- 3 point. Front, single front spring connected to a central pivotting screw jack on the front axle. Rear, orthodox semi-elliptic.

BRAKES:- Overrun operated on front axle only. Separate hand parking lever.

TYRES:- 9.00 - 16 on standard H.30 hubs.

SUPPORTING JACKS:- Four Bofors type screw jacks. Rear of frame lowered by means of hand-operated winch.

DIMENSIONS:- Wheelbase 13' 6". Track front 5' 0". Rear 6' 4".

OVERALL:- Length 22' 3". Length less tow bar 18' 7". Width 7' 6". Height 8' 3".

LADEN WEIGHTS:-

Front Axle	1 ton 15¼ cwt.
Rear Axle	2 tons 9 cwt.
Gross	4 tons 4¼ cwt.

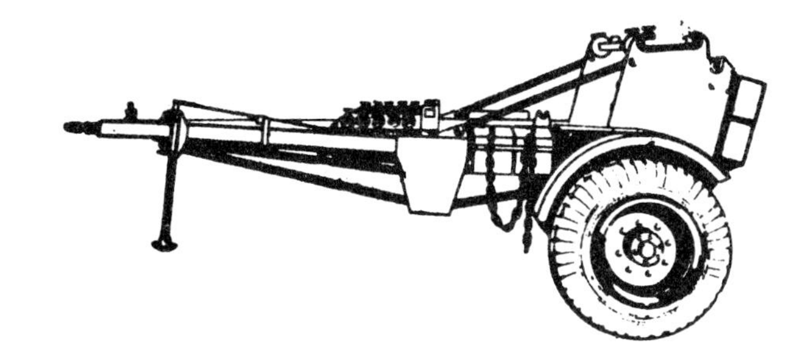

TRAILER 3 TON 2 WH. CABLE DRUM

BRAKES:- Overrun. Hand parking lever.

SUSPENSION.- Quarter elliptic.

TYRES:- 36 x 8 singles. Hubs standard H.45.

JACKS:- Hinged supporting leg at front.

DIMENSIONS:-

Centre tow eye to axle	12' 0"
Overall Length with 8' 0" drum	16' 6"
" Height " " "	9' 2"
" Width	7' 4½"

WEIGHTS:-

Unladen	1 ton 4¼ cwt.
Laden 880 yds. cable	3 tons 7¼ cwt.
" 1000 " "	3 tons 14¼ cwt.

There are three types of trailers:-

1. Trailer 4 ton 4 wh. Bake y. Machinery.

2. Trailer 4 ton 4 wh. Bake y. Dough Trough.

3. Trailer 4 ton 4 wh. Bakery. Oven.

All three trailers are similar in general design. They have

Turntable steering.

Semi-elliptic springing.

Brakes operated by overrun on front axle.

9.00 - 16 tyres.

The bodies consist of a flat platform surmounted by superstructure carrying canvas cover in the case of the Machinery and Dough Trailers. The oven trailer is fitted with a special superstructure to carry the oven.

DIMENSIONS OF DOUGH TRAILER (The largest)
Overall Length 15' 0". Width 7' 6". Height 10' 0".

WEIGHTS OF OVEN TRAILER (The heaviest)
5 tons 5 cwt.

Note:- For details of special Bakery Generator Trailer see under Generator Trailers, page 716.

TRAILER 4 TON 4 WH. A.O.D.
Type 1: WITH BINS
TYPE 2: OPEN WITH DAVIT

These two trailers are similar in chassis design; they vary only in body and fittings.

STEERING:- Turntable.

SUSPENSION:- Semi-elliptic.

BRAKES:- Overrun.

TYRES:- 10.50 - 13 on standard H.45 LL hubs.

BODY:-
A.O.D. with Bins has G.S. type body with fixed sides and drop tailboard surmounted by superstructure and canvas tilt.

A.O.D. OPEN WITH DAVIT has divided drop sides with fixed headboard and drop tailboard. At rear is swivelling davit (or jib). Maximum lift 15 cwt. Height of lift is 5' 6" off floor of body.

DIMENSIONS:- (Applicable to both types except where otherwise indicated.)

	OVERALL:	INSIDE BODY
Length	24' 0"	18' 10"
Width	7' 6"	7' 1"
Height Davit Body	7' 7"	2' 1"
Height Bins Body	10' 0"	6' 1½"

UNLADEN WEIGHTS:-
Trailer A.O.D. with Davit 2 tons 16 cwt.

Trailer A.O.D. with Bins 2 tons 12¼ cwt.

TRAILER 5 TON 4 WH. G.S.

There are two types:-

(1) U.K. Source Trailer 5 ton 4 wh. Platform with Stanchions & Chains.

and

(2) U.S. Source Trailer 5/6 ton 4 wh. G.S. Open.

1. The U.K. Type Trailer 5 ton 4 wh. G.S.

STEERING:- Turntable.

SPRINGS:- Semi-elliptic.

BRAKES:- Two separate connections are provided:-
(a) Single line vacuum system.
(b) Single line pressure system.

Both vacuum and air pressure systems operate the same linkage on the rear wheels. Separate hand parking brake.

TYRES:- 10.50 - 13 on standard H.45 LL hubs.

BODY:- Flat platform with fixed front board, chock rails along the sides and a detachable chock rail at the rear. Detachable stanchions fitted along the sides with chains.

DIMENSIONS

	OVERALL	INSIDE BODY
Length less drawbar	12' 8"	12' 2"
Width	7' 6"	7' 0"
Height	7' 2"	-

WEIGHTS

	UNLADEN	LADEN
FRONT AXLE	1 ton 5¼ cwt.	3 tons 12 cwt.
Rear Axle	19½ cwt.	3 tons 13 cwt.
Gross	2 tons 4¾ cwt.	7 tons 5 cwt.

(2) Trailer 5/6 ton 4 wh. Cargo, Make Hobbs

This is a U.S. type supplied under S/M.2664. (Similar type has been supplied on S/M.2754)

STEERING:- Turntable.

SUSPENSION:- Semi-elliptic.

BRAKES:- Warner Electric. Hand parking brake.

TYRES:- 7.50 - 20 8 ply, twins all round.

(Note: Trailers are shipped complete with Warner Electric Hand Controller for fitting to tractor so that tractors can be modified as required.)

BODY:- Solid sides on stakes (which can be detached). Drop tailboard. Weather protection by tarpaulin.

DIMENSIONS:-

	OVERALL	INSIDE BODY
Length	14' 8" (and 6' 0" drawbar)	14' 0"
Width	7' 9"	7' 3"
Height	7' 0"	3' 0"

WEIGHTS:-

	UNLADEN	LADEN (4½ long tons)*
F.A.	1 ton 9½ cwt.	3 tons 15 cwt.
R.A.	1 ton 5 cwt.	3 tons 9 cwt.
Gross	2 tons 14½ cwt.	7 tons 4 cwt.

* Based on U.S. rating of 5/6 short tons = 11,000 lbs. Actual payload is roughly 5 long tons.

TRAILER 5 TON 4 WH.
SMOKE GENERATORS

There are four types of Smoke Generator (or Hasler) Trailer.

1. Source U.K., Make, Eagle.
2. " " , Make, Brockhouse.
3. " " , Make, Dyson.
4. " U.S.A., Make, Brill (Esso Smoke Generator).

U.K. Types
The following details apply specifically to the Eagle Trailer, and generally to the Brockhouse and Dyson Tractors.

STEERING:- Turntable.

BRAKES:- Single line vacuum system, with separate hand parking lever.

TYRES:- 10.50 - 13 singles (Note:- Dyson trailer fitted 34 x 7 tyres, singles all round).

OVERALL DIMENSIONS:- Length 17' 6" less drawbar. Width 7' 4". Height 9' 0".

WEIGHTS:- Unladen 1 ton 15 cwt.
Laden full smoke generator equipment 6 tons 2 cwt.

The U.S.A. type

Trailer, 2 tons 4 c. w/Smoke Generator

Make, Brill Model M.7.
This is a very compact close-coupled four wheel trailer.
Tyres are 7.50 - 20 8 ply singles.
Brakes are Warner Electric.
Trailer is fitted with four screw jacks and a castor wheel.
Overall Dimensions:- Length 16' 1".
Width 7' 10½". Height 8' 2".
Laden Weight:- 4 tons 7 cwt.

LAUNDRY TRAILERS

1. Trailer 5 ton 4 wh. Washer.
2. " " " Boiler.
3. " " " Dryer

GENERAL:- Early Mk.I Washer, Boiler and Dryer equipment were mounted on four different trailer chassis.

Latest Mk.II Washer, Boiler and Dryer equipments are mounted on a standard Brockhouse 5 ton trailer, as follows:-

STEERING:- Turntable.

SUSPENSION:- Semi-elliptic.

BRAKES:- Overrun operating on front axle with separate hand parking lever.

TYRES:- 10.50 - 13 singles on standard H.45 LL hubs.

BODY:- Platform equipped with superstructure and tile.

DIMENSIONS AND WEIGHTS:-
Overall Length 19' 0" (less drawbar)
" Width 7' 6"
" Height 10' 9" (11' 6" Boiler Trailer)

LADEN WEIGHTS:-
Washer 9 tons
Dryer 6¾ tons
Boiler 6¾ tons

TRAILER 5 TON 4 WH. LAUNDRY C.C.S.

This C.C.S. Laundry is a self contained unit mounted on chassis generally similar to that described above. Body is house type.

OVERALL DIMENSIONS:- Length less drawbar 19' 0". Width 7' 6". Height 12' 6".
Laden Weight 8 tons 15 cwt.

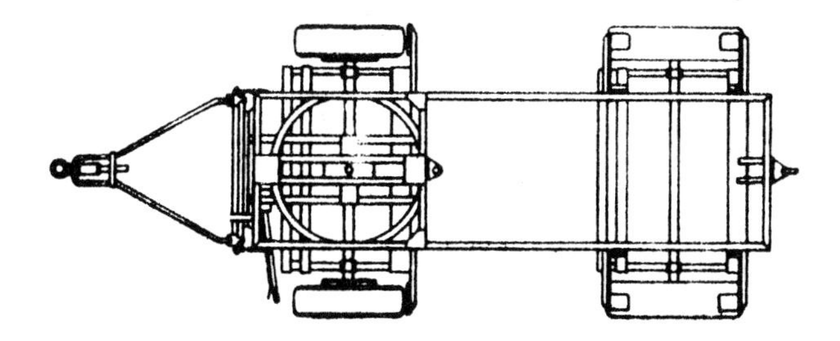

TRAILER 5 TON 4 WH. RADAR A.A. NO.1 MK.1 TRANSMITTER

(Similar Trailer used to carry Radar A.A. No.1)
Mk.I* Transmitter

GENERAL:- An orthodox design flat platform trailer.

STEERING:- Turntable.

SUSPENSION:- Semi-elliptic front and rear.

TYRES:- 9.00 - 16.

BRAKES:- Overrun operated.

WEIGHTS:- Laden Radar A.A. No.1 Mk.1 (TX)

Front Axle	2 tons 3¼ cwt.
Rear Axle	1 ton 19¾ cwt.
Gross	4 tons 3 cwt.

TRAILER DIMENSIONS (With Equipment)
Overall Length 19' 6". Width 7' 6".
Wheelbase 8' 6½". Track front and rear 6' 8½".
Travelling Height 11' 4".

TRAILER 5 TON 4 WH. RADAR A.A. NO.1, MK.1 RECEIVER

A flat platform trailer generally similar to type shown opposite except wheelbase is longer.

STEERING:- Turntable.

SUSPENSION:- Semi-elliptic front and rear.

TYRES:- 9.00 - 16.

BRAKES:- Overrun operated.

WEIGHTS:-

Laden Radar A.A. No.1. Mk.I Equipment

Front Axle	2 tons 3½ cwt.
Rear Axle	2 tons 4¾ cwt.
Gross	4 tons 7 cwt.

TRAILER DIMENSIONS:- Overall Length 21' 9". Width 7' 6½". Wheelbase 11' 9". Track, front and rear 6' 8".

Travelling Height 11' 2".

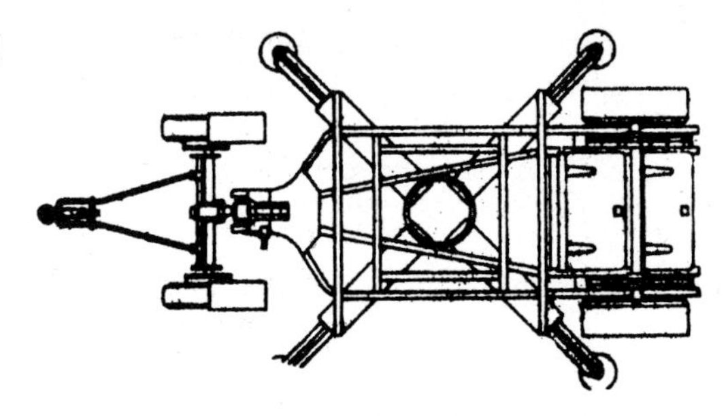

TRAILER 5 TON 4 WH. RADAR A.A. NO.1, MK.1 *

STEERING:- Pivotting front axle.

SUSPENSION:- Front, single quarter elliptic inverted spring. Rear, semi-elliptic springs.

BRAKES:- Overrun operated.

TYRES:- 9.00 - 16.

SUPPORTING JACKS:- Four.

WEIGHTS:- Laden A.A. No.1, Mk.I* (RX)

Front Axle	2 tons 2½ cwt.
Rear Axle	2 tons 8¾ cwt.
Gross	4 tons 12¼ cwt.

Laden A.A. No.1, Mk.II (RX)

Front Axle	2 tons 9¾ cwt.
Rear Axle	3 tons 1¼ cwt.
Gross	5 tons 11 cwt.

TRAILER DIMENSIONS (With Equipment)
Overall Length 20' 9". Width 7' 6".
Wheelbase 13' 6". Track front, 5' 4".
rear, 6' 9".

TRAVELLING HEIGHT Mk.I, 10' 8".

TRAILER RADAR A.A. NO. 1 MK.II TRANSMITTER

(Scammell Lorries Design)

Chassis generally similar to type shown opposite, but jacking and frame details different.

SUSPENSION:- Single front suspension by ¼ elliptic spring. Semi-elliptic rear springs.

STEERING:- Pivotting front axle.

TYRES:- 10-50-16.

JACKING:- Two extending jacks along sides at rear. Single front jack.

BRAKES:- Overrun on front axle.

WEIGHTS

Laden Radar A.A. No.1 Mk.II (TX)

Front Axle	3 tons 6 cwt.
Rear Axle	3 tons 8 cwt.
Gross	6 tons 14 cwt.

TRAILER DIMENSIONS (with equipment)

Wheelbase 15' 9"
Track f. 5' 9 5/8" r. 6' 5 5/16"

Overall length	24' 0"
" width	7' 6"
Travelling Height	11' 4"

Note: Same equipment mounted on a different chassis design, see overleaf.

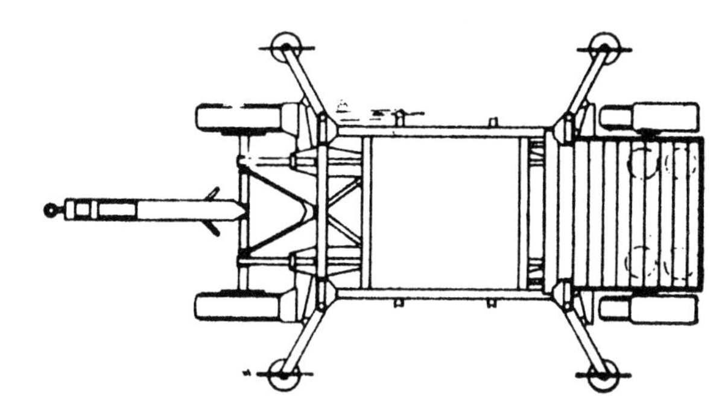

TRAILER 5 TON 4 WH. RADAR
A.A. NO.I MK.II TRANSMITTER

SUSPENSION:- Four point, quarter elliptic front and rear.

STEERING:- Ackermann. Actuated by drawbar pivotting on front axle.

TYRES:- 9.00-20 front and rear.

BRAKES:- Overrun operated.

SUPPORTING JACKS:- Four.

WEIGHTS:- Laden Radar A.A. No.1 Mk.II (TX).

F. Axle	3 tons 5 cwt.
R. Axle	3 tons 7¾ cwt.
Gross.	6 tons 12¾ cwt.

TRAILER DIMENSIONS:- (With equipment)

Overall Length 22' 7". Width 7' 4".
Wheelbase 14' 1". Track f. 6' 7". r. 6' 6".
Travelling Height 11' 2".

(Same equipment mounted on a different Chassis, see page 733).

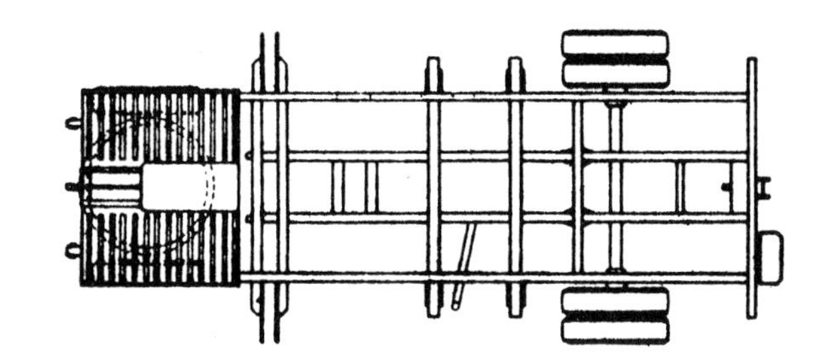

TRAILER 5 TON 4 WH. RADAR
A.A. NO.3 MK.II
RECEIVER & TRANSMITTER

(Trailer also known as the TR or G.L. Mk.IV Trailer)

This is the largest and heaviest of the 5 ton 4 whld. Radar Trailers. It has superseded in production the earlier 5 ton 4 whld. Radar trailers shown on pages 730-734.

STEERING:- Turntable.

TYRES:- 36 x 8, singles front and twins rear.

BRAKES:- Early types, cable operated from tractor. Latest types, single pipe line air pressure system. Early types in service are being modified to air system.

SUPPORTING JACKS:- Three.

WEIGHTS:- Laden with Radar A.A. No.3 Mk.II Equipment.

F. Axle	3 tons 2 cwt.
R. Axle	6 tons 3 cwt.
Gross	9 tons 5 cwt.

OVERALL DIMENSIONS

Length less tow bar 21' 9"
Width 9' 4" over wings; 9' 2" over tyres
Travelling Height 11' 9¾"
Wheelbase 14' 2". Track F. 5' 0½". R. 7' 5 3/8".

TRAILER 5 TON 4 WHEEL
PONTOON 50/60 CLASS

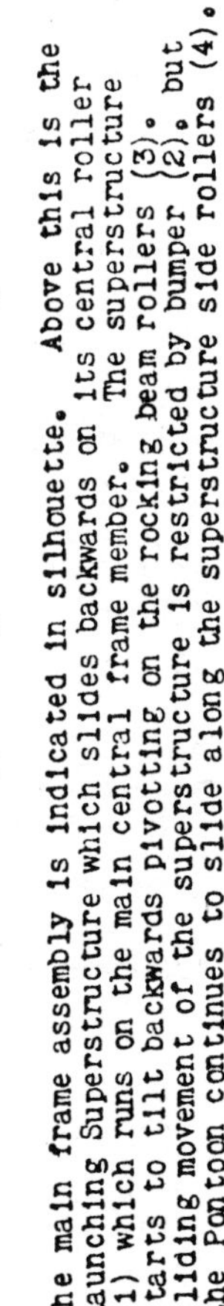

The main frame assembly is indicated in silhouette. Above this is the Launching Superstructure which slides backwards on its central roller (1) which runs on the main central frame member. The superstructure starts to tilt backwards pivotting on the rocking beam rollers (3). Sliding movement of the superstructure is restricted by bumper (2). but the Pontoon continues to slide along the superstructure side rollers (4).

TRAILER 5 TON 4 WH.
50/60 CLASS PONTOON

GENERAL:- Function of trailer is to transport launch and recover the 50/60 class pontoon. Main Frame Assembly consists of drawbar, front axle and turntable which is connected by single central frame member to rear axle. Above this is the Launching Superstructure which carries the pontoon. See Diagram opposite.
Hand winch is fitted at front of main frame assembly.

BRAKES:- Cable operated from tractor.

TYRES:- 9.00-20 twins on standard H.80 hubs.

DIMENSIONS:- Wheelbase, when carrying pontoon, 18' 0". Turning Circle: 68' 0".

When travelling without pontoon, rear axle can be retracted to reduce wheelbase to 12' 6" and turning circle to 48' 0".

Track, front and rear 6' 9½".

OVERALL DIMENSIONS

Length, (inc. drawbar)	32' 8½"
" loaded pontoon	39' 8½"
Width, loaded pontoon	8' 6"
" over detachable bollards used for locating pontoon in the water	10' 0"
Height, solo	8' 6"
" loaded pontoon	8' 3"
" of pontoon platform	4' 5½"

WEIGHTS

	UNLADEN	LADEN
F. Axle	3 tons 6 cwt.	4 tons 14 cwt.
R. Axle	2 tons 17¾ cwt.	5 tons 19¼ cwt.
Gross	6 tons 3¾ cwt.	10 tons 13¼ cwt.

TRAILER 5 TON 4 WHLD.
RADAR A/A NO.3 MK.I

(Also known as A.P.F. Trailer)

Canadian Supply. S/M 2828.

A 4 wheeled drop frame trailer with turntable steering and Warner Electric brakes. Tyres are 7.50-20 (34X7) Singles front and twin rear.

WEIGHTS

F. Axle	3 tons	18 cwt.
R. Axle	5 tons	7 cwt.
Gross	9 tons	5 cwt.

TRAILER 5 TON 4 WHLD.
RADAR A/A NO.4 MK.I

(Also known as Z.P.I. Trailer)

Canadian Supply: S.M. 2828.

A 4 wheeled drop frame trailer with turntable steering and Warner Electric brakes. Tyres are 9.00-20 twins all round.

WEIGHTS:- Laden 6 tons 19 cwt.

PRIME MOVERS FOR ABOVE TRAILERS

See page 650 for details of special F.W.D., H.A.R. 4 x 4 Lorries.

TRAILER 3 TON 4 WH.
RADAR A.A. NO.3 MK.IV

(Chassis generally similar to new 3 ton 4 wh. Projector trailer shown on page 740, except frame modified to accommodate water proofed cabin.)

SUSPENSION:- Three point suspension with pivotted front axle steering.

BRAKES:- Overrun operated on front axle. Separate hand parking lever.

TYRES:- 10.50-16 singles on standard H.30 hubs.

JACKING:- Two Bofors type jacks at rear with special design centrally mounted front jack.

CABIN:- Heat, cold and blast proof. Insulated design with refrigating and air conditioning equipment. Cabin floor welded to frame. Designed to withstand immersion in water to depth of 3' 9" without additional water proofing.

DIMENSIONS

Wheelbase 10' 5".

Track, f. 4' 11½" r. 6' 5½"

Draw eye to front axle 4' 9". Ht. of draw eye 2' 2½".

Overall, including radar equipment. Length less drawbar 14' 7½". Width 7' 6". Ht. stowed for travelling 10' 10½".

ESTIMATED LADEN WEIGHTS

Front Axle.	1 ton 18¾ cwt.
Rear Axle.	2 tons 18 cwt.
Gross.	4 tons 16¾ cwt.

TRAILER 3 TON 4 WH.
PROJECTOR

This is the latest type Projector Trailer designed to take the 150 cm Projector with Mk.VIII control panel.

Earlier 2 ton Projector trailers are shown on pages 721 and 325.

The new Projector Trailer (identified by the chassis nomenclature "Trailer Chassis 3 ton 4 whd. No. 4") has been designed for full cross country performance.

SUSPENSION:- Three point suspension with pivotted front axle steering.

BRAKES:- Overrun operated on front axle. Separate hand parking lever.

TYRES:- 10-50-16 singles on standard H.30 hubs.

JACKING:- Two Bofors type jacks at rear with special design centrally mounted front jack.

DIMENSIONS

Wheelbase: 10' 5"
Track:- F.4' 11", R.6' 5½"

Overall Length:- 14' 10" Less Drawbar
" Width:- 7' 6"

Ht. of frame, 2' 0¼", Ht. projector sub frame, 2' 3¼".

WEIGHTS

	Unladen	Laden (3 tons 8 cwt.)
F. Axle	15¾ cwt.	2 tons 1½ cwt.
R. Axle	1 ton 1¼ cwt.	3 tons 3¼ cwt.
Gross	1 ton 17 cwt.	5 tons 4¾ cwt.

TRAILER 6 TON 4 WH.
CARRIAGE OF TRACKS

There are two types as follows:-

(1) U.S. Army type. U.S. Nomenclature is "Trailer, Full, Flat Bed, 8 ton".

Make, Freuhauf CPT-8 Special. Also made by Winter Weiss.
A low loading, flat platform trailer with a goose neck or kick up section over the turntable.

STEERING:- Turntable.

TYRES:- 9.00-20 10 ply twins all round. Total of 8 running tyres.

BRAKES:- 2 line air pressure. Separate hand parking lever.

BODY:- Flat platform body extending from rear of gooseneck to just behind rear axle.
Rear portion slopes down to meet loading ramps.
Rear portion of platform cut away to permit articulation of rear wheels.
Therefore available load space 7' 11" long by 8' 6" wide.

OVERALL DIMENSIONS:- Length with drawbar 25' 0".
Length less drawbar 19' 9". Width 8' 6".
Ht. to top of gooseneck 4' 10".
Ht. of platform unladen 3' 4".

WEIGHTS:- Unladen 4 tons 9 cwt.

LADEN

Front Axle	5 tons.
Rear Axle	5 tons 10 cwt.
Gross	10 tons 10 cwt.

(2) U.S. Commercial type
Make, JAHN, Model LKS 408

A limited number of these Jahn trailers were imported directly by C.O.H.Q. Basically this Jahn trailer is similar to the Freuhauf trailer described above. It has same size tyres and is generally similar in wts. and dimensions except that width over platform is only 8 feet.

GENERAL NOTES

W.D. TRAILERS IN THE 18-20 TON CLASS

1. Trailer 18 ton 4 twin wheeled low loading.

   Make: Buston Bucyrus.
   Function: To carry 19 R.B. Excavator.
   Further Details: See page 744.

2. Trailer 18 ton 4 twin wheeled low loading.

   Make: Ransome, Sims & Jefferies.
   Function: To carry 19 R.B. Excavator.
   A later design than item (1).
   Further Details: See page 744.

3. Trailer 18 ton 8 wheeled low loading.

   Make: Carrimore. Similar types also made by Eagle & Hands.
   Function: To carry 19 R.B. Excavator.
   Further Details: See page 744.

4. Trailer 20 ton 8 wheeled low loading.

   Make: Crane.
   Function: To carry Excavators.
   Further Details: See page 526.
   Note: Not to be confused with item 6, also made by Crane which is an 8 twin wheeled trailer of 20 ton capacity.

5. Trailer 20 ton 8-twin wheeled low loading.

   Make: Multiwheeler and S.M.T.
   Function: To carry tracked tractors and heavy road making equipment.
   Further Details: See page 524.

6. Trailer 20 ton 8-twin wheeled Transporter.
   Make: Crane.
   Function: To carry tracked tractors and heavy road making equipment and has superseded item 5 in production.
   Further Details: See page 745.

7. Trailer 20 ton 6-twin wheeled low loading.
   Make: Rogers (U.S. type ex. S.M. 2090).
   Function: To carry tracked tractors and heavy road making equipment.
   Further Details: See page 289.

8. Trailer 20 ton 6-twin wheeled low loading.
   Make: Trailmobile (U.S. type ex. S.M. 6193)
   Function: To carry R.E. equipment.
   Further Details: This is generally similar to the Trailmobile semi-trailer shown coupled to a Federal Tractor on page 535, except that it has been converted to a full trailer by addition of a dolly.

9. Trailer 12 twin wheeled Carriage of Rails (30 ton) or Cargo Carrier (20 ton)
   Make: Modified British Mk.I 40 ton Tank Transporter Trailer. (See page 551 for basic trailer details).
   Function: To carry 30 tons of Rails or 20 tons of general cargo.

Details: Modification consists of fitment of heavy timber baulks across body (for rail carrier role) plus fitment of flat floor (for load carrier role).

TRAILER 18 TON 4 TWIN WH. LOW LOADING

Make: Ruston Bucyrus

Designed to carry 19 R.B. Excavator.
Floor surmounted by wooden runners along the sides, 20" wide and 4" higher than floor. Excavator tracks rest on these wooden runners. Excavator is loaded by retracting axle.
BRAKES:- Originally manual from brakeman's seat on trailer. Modified in service to 2 line air pressure operated from tractor.
TYRES:- Originally 8.75-20 twins, modified in service to 10-50-20 twins.
OVERALL DIMENSIONS:- Length 20' 4", Length including tow bar 24' 7". Width 8' 0". Height, floor to top of wooden runners, 2' 11".
UNLADEN WEIGHT:- 5 tons.

TRAILER 18 TON 4 TWIN WH. LOW LOADING

Make: Ransome, Sims & Jefferies

Generally similar to Ruston Bucyrus type, as described above, except that Ransome, Sims & Jefferies trailers have, from the start of production, been equipped with:-

2 line air pressure braking and 10.50-20 tyres.

TRAILER 18 TON 8 WH. LOW LOADING

Original types made by Carrimore are fully described on page 526.
Later types, made by Carrimore, Hands and Eagle to a common design, differ only in clearances to permit fitment of larger tyres.

Tyres 12.00-20 instead of 10.50-20.

This change increases overall length by 2" and the laden height by 1".

TRAILER 20 TON 8 TWIN WHEELED TRANSPORTER

MAKE: CRANES

GENERAL:- This is a flat platform trailer designed for similar role as 20 ton 16 whld. trailer shown on page 524. It differs in that it has a straight (instead of drop) frame and hence platform height is higher. (4' 2")
The trailer uses many similar components to the 40 ton Transporter Trailers (MK.I & MK.II) shown on pages 550-552.

STEERING:- Single turntable as 40 ton trailer.

FRONT AXLE:- Eight wheels spring mounted as two pairs of duals. (as 40 ton trailer)

REAR AXLE:- Eight wheels spring mounted as two pairs of duals (as front axle).

BRAKES:- Westinghouse 2 pipe line air pressure system operating on all wheels. Plus hand wheel for parking (as 40 ton Trailer).

TYRES:- 36 x 8 all round on standard H.80 hubs. Total 16 running tyres and 2 spares.

LOADING RAMPS, carried in sections under frame and placed in position manually. Ramps 2' 0" wide x 15' 1" long.

Jacks provided to support rear of trailer when loading.
"D" type shackles for securing load. Winch roller guide.

DIMENSIONS:- Wheelbase 15' 4". Track as page 550.
Overall:- Length with towbar 26' 9", less tow bar 29' 10". Width 9' 6". Ht. 4' 10".
Platform Space:- 20' 0" x 9' 6" wide. Clear loading space behind raised front section of platform is 16' 6" long.

WEIGHTS

	UNLADEN	LADEN
F. Axle	4 tons 11 cwt.	14 tons 17 cwt.
R. Axle	4 tons 4 cwt.	13 tons 18 cwt.
Gross	8 tons 15 cwt.	28 tons 15 cwt.

TOWING SUPPLEMENT

This section contains the following data:-

(A) Types of Trailer Brake Connections fitted to vehicles in W.D. Service.

(B) List of W.D. vehicles equipped with tow hooks and heights of heoks from ground. Equivalent Trailer Draw Eye Heights.

(C) Recommended Maximum Towed loads for W.D. vehicles.

TYPES OF BRAKING FITTED TO W.D. TRAILERS AND GUNS

These are the types of braking systems fitted to W.D. trailers.

1. Overrun Operated (or, as the U.S. call them, "Impact Brakes"). These are fitted to the majority of the lighter W.D. trailers up to 5 tons capacity. Any tractor of suitable size fitted with a tow hook can tow trailers equipped with overrun operated brakes.

Trailers fitted with the following braking systems can only be towed by tractors equipped with appropriate trailer brake connections.

2. Single Line Vacuum System. Fitted to a number of W.D. medium weight trailers (and many R.A.F. types).

3. Single Line Air Pressure System. Fitted to a few early type heavy trailers and certain specialised medium trailers. Trailers fitted with Single Line Air Pressure brakes can be towed by Tractors equipped with Two Line Air pressure Connections.

4. Two Line Air Pressure System. This is the standard system used on heavy W.D. and U.S. Army trailers and certain U.S. guns.

5. Warner Electric System. Fitted to U.S. Army medium weight trailers and certain guns.

6. Cable Operated Brakes. Fitted to a few specialised trailers and certain guns. These brakes are operated by Tractors equipped with Gun Cylinder or Gun Servo.

(Note: Certain U.S. type semi-trailers in W.D. Service are equipped with Two Line Vacuum System operated brakes. But as they are all semi-trailers, the question of tractor interchangeability does not arise.)

TRACTORS FITTED WITH TRAILER BRAKE CONNECTIONS

The following Symbols are used under the column headed "Applying To":-

(P) to indicate Standard on all Production.

(GM) " " a general modification applying to all of the particular type of tractor and Standard on current production.

(X) " " a special modification to certain tractors only for a particular role.

* Indicates the particular Tractor is fitted with more than one type of trailer brake connections.

WARNER ELECTRIC

	MAKE		APPLYING TO
TRACTOR 4 x 4 A.T.	Morris C.8	(X)	When towing those 17 pdr. guns fitted with Warner brakes.
" "	" C.	(X)	
" "	Chevrolet	(X)	
" "	Ford W.O.T.8	(X)	
*TRACTOR 4 x 4 MEDIUM ARTILLERY	A.E.C.	(P)	
" "	F.W.D. S.U.	(P)	
*TRACTOR 4 x 4 HEAVY A.A.	A.E.C.	(P)	
*TRACTOR 6 x 6 HEAVY	MACK N.M.5	(P)	
TRACTOR 6 x 6 EXTRA HEAVY	" N.O.	(P)	
*LORRY 6 TON 6 x 6 G.S.	WHITE 666	(P)	
LORRY 10 TON 6 x 4 G.S.	Mack N.R. White 1064	(X)	When towing trailer 5 ton 4-wheel, Hobbs.
LORRY 3 TON 4 x 4 G.S.	F.W.D., HAR.	(X)	Only to tractor supplied for towing Radar Trailers. S.M.2828
Truck 15 cwt. Half Track	International M5, M9, M14	(P)	

TRACTORS FITTED WITH TRAILER BRAKE CONNECTIONS (Continued)

* Indicates the particular Tractor is fitted with more than one type of trailer brake connection.

GUN CYLINDER OR GUN SERVO	MAKE	APPLYING TO:-
*TRACTOR 4 x 4 Medium Artillery	A.E.C.	(P)
* " "	F.W.D. S.U.	(P)
*TRACTOR 6 x 4 Heavy Artillery	Albion	(P)
* " "	Scammell	(P)
*TRACTOR 6 x 6 Heavy	Mack N.M.5	(G.M.) - Early types shipped less Gun Servo - now standard in production.
*TRACTOR 6 x 4 Heavy Breakdown	Scammell	(G.M.)

SINGLE LINE VACUUM SYSTEM

TYPE	MAKE	APPLYING TO
LORRY 3 TON 4 x 2 G.S.	Bedford OY.	(X) for towing Haslar Trailers.
LORRY 3 TON 4 x 4 G.S.	Chevrolet	(X) for towing Snow Plow Trailers.
LORRY 3 TON 6 x 4 BREAKDOWN	Austin	(P)
	Dodge V.K.60	(G.M.) for towing Light
	Leyland	(G.M.) Recovery Trailer
LORRY 10 TON 6 x 4 G.S.	Leyland Hippo Mk.II	(P)

(Note: Single Line Vacuum Trailer Brake Connections have been fitted to a number of the following vehicles in the service:- Austin, Albion, Bedford, Chevrolet, Ford, 3 ton 4 x 4 G.S.
Dennis Max, Maudslay, 6 ton 4 x 2 G.S.
Albion, Leyland Mk.I 10 ton 6 x 4 G.S.)

DOUBLE LINE VACUUM SYSTEM

APPLIES TO SEMI TRAILER TYPES.

TRACTORS FITTED WITH TRAILER BRAKE CONNECTIONS (Continued)

SINGLE LINE AIR PRESSURE

Certain early types of A.E.C. Medium & Scammell Heavy Artillery Tractors were fitted with Single Pipe Line Connections. But General Modification has now been carried out to two line system which can operate either single pipe line or two pipe-line equipped trailers.

TWO LINE AIR PRESSURE

TYPE	MAKE	APPLYING TO
LORRY 10 Ton 6 x 4 G.S.	Mack N.R.9	Models N.R.9 and onwards.
	White 1064	All types
*TRACTOR 4 x 4 Medium Artillery	A.E.C.	(G.M.)
* " "	F.W.D. S.U.	(G.M.)
*TRACTOR 6 x 4 Heavy Artillery	Albion	(P)
* " "	Scammell	(G.M.)
*TRACTOR 6 x 6 Heavy	Mack N.M. 5	(P)
* " " Extra Heavy	" N.O.	(P)
TRACTOR 6 x 4 * Breakdown Heavy	Scammell	(G.M.)
TRACTOR 6 x 6 Breakdown Heavy	Diamond T.969	(P)
" "	Ward La France	(P) Latest types only
LORRY 4 Ton 6 x 6 G.S.	Diamond T.968	(P)
*LORRY 6 Ton 6 x 6 G.S.	White 666	(P)
TRACTOR 6 x 4 for 30 Ton Rail Transporter	Federal	(X) - Tractor normally coupled to semi trailer and unsuitable as prime mover unless fitted ballast box as for this role
TRACTOR 6 x 4 for 40 Ton Trailer	Diamond T.980 & 981	(P)
LORRY 5/6 Ton 4 x 4, G.S.	F.W.D. S.U.	(P)

TOWING HOOKS

VEHICLES IN W.D. SERVICE EQUIPPED WITH TOWING HOOKS SHOWING HEIGHT FROM GROUND TO CENTRE OF HOOK WHEN VEHICLE IS UNLADEN

(For Laden Heights subtract 3" from all 3 tonners and Heavier types and 2" from 15-cwt. and lighter types).

TYPE OF HOOK IS INDICATED BY FOLLOWING SYMBOLS

- N.F.S. = Hook NOT fitted as Standard
- W.D. = Standard W.D. hook on spring drawbar gear
- P.P. = Pin and Pintle fitted to early U.K. 15-cwt. trucks
- U.S. = Standard U.S. Army type pintle hook
- J.P. = The smaller type hook used on Jeeps.

See also list of Trailer Draw Eye Heights which show the equivalent heights from the Trailer Draw Eye standpoint Page 812.

Vehicle	Type	Ht.
Br. A.E.C. "Marshall" 3 ton 6 x 4	W.D.	2' 9"
Br. A.E.C. "Matador" M.A.T. 4 x 4	W.D.	3' 2½"
Br. Albion F.T.11 3 ton 4 x 4	W.D.	3' 1¼"
Br. Albion B.Y.1 3 ton 6 x 4	W.D.	2' 9"
Br. Albion B.Y.3 3 ton 6 x 4	W.D.	2' 9"
Br. Albion B.Y.5 3 ton 6 x 4	W.D.	2' 9"
Br. Albion CX6N 10 ton 6 x 4	W.D.	2' 11¼"
Br. Albion CX23N 10 ton 6 x 4	W.D.	3' 2"
Br. Albion CX22S 6 x 4 H.A.T.	W.D.	3' 3¼"
Br. Austin 8 h.p. 4 x 2	None	-
Br. Austin 10 h.p. 4 x 2	None	-
Br. Austin K.30 30-cwt. 4 x 2	N.F.S.	-
Br. Austin K.2 4 x 2 Ambulance	None	-
Br. Austin K.3 3 ton 4 x 2	W.D.	2' 3"
Br. Austin K.5 3 ton 4 x 4	W.D.	3' 0"
Br. Austin K.6 3 ton 6 x 4	W.D.	2' 8½"
U. Bantam 5 cwt. 4 x 4	J.P.	1' 9¼"
Br. Bedford M.W. 15 cwt. 4 x 2 (P.P. (Early	(W.D. (Later	2' 3½"
Br. Bedford O.X. 30 cwt. 4 x 2	N.F.S.	-
Br. Bedford O.Y. 3 ton 4 x 2	W.D.	2' 3½"
Br. Bedford Q.L. 3 ton 4 x 4	W.D.	2' 9½"
Br. Bedford QLT. 3 ton 4 x 4 Troop Carrier	None	-
Br. Bedford QLR. 3 ton 4 x 4 Wireless	W.D.	2' 9½"
C. Chevrolet C. 8A 4 x 4 H. Utility	None	-

TOWING HOOKS (Continued)

		Type	Ht.
C.	Chevrolet C.8 8 cwt. 4 x 2	None	-
C.	Chevrolet C.15 15 cwt. 4 x 2	W.D.	1' 10"
C.	Chevrolet C.15A 15 cwt. 4 x 4	W.D.	2' 4"
C.	Chevrolet C.30 30 cwt. 4 x 4	W.D.	2' 3"
C.	Chevrolet CC60L 1543 2 3 ton 4 x 2	W.D.	2' 3½"
C.	Chevrolet C.60L 8443 3 ton 4 x 4	W.D.	2' 3" early 2' 10" later
C.	Chevrolet C.60S 3 ton 4 x 4	W.D.	2' 3" early 2' 10" later
C.	Chevrolet CGT. 8440 F.A.T. 4 x 4	W.D.	2' 9"
U.	Chevrolet BG. 4 str. 4 x 2	None	-
U.	Chevrolet 4412 3 ton 4 x 2	None	-
U.	Chevrolet 1½ ton 4 x 4	U.S.	2' 6"
U.	Chevrolet Thornton 3 ton 6 x 4	U.S.	2' 10"
Br.	Commer Q2 30 cwt. 4 x 2	N.F.S.	-
Br.	Commer Q4 3 ton 4 x 2	W.D.	2' 8"
Br.	Crossley I.G.L.8 3 ton 6 x 4	W.D.	2' 10"
Br.	Dennis (Tipper)	None	-
Br.	Dennis Max.6 ton 4 x 2	W.D.	2' 8½"
Br.	Dennis Max. Mk.II 6 ton 4 x 2	W.D.	2' 4"
U.	Diamond T. 969 4 ton 6 x 6	U.S.	2' 9"
U.	Diamond T. 980 and 981 6 x 4 Tractor	U.S.	2' 7¼"
C.	Dodge T. 212 8 cwt. 4 x 4	None	-
C.	Dodge D.15 T.222 15 cwt. 4 x 2	W.D.	2' 2"
C.	D. 60L T.110L 3 ton 4 x 2	W.D.	2' 3"
C.	Dodge D.60L T110L6 3 ton 4 x 2	W.D.	2' 3"
U.	Dodge T.213 ½ ton 4 x 4	U.S.	1' 10"
U.	Dodge T. 214 ¾ ton 4 x 4	U.S.	2' 0"
U.	Dodge T.203 30 cwt. 4 x 4	U.S.	2' 10"
U.	Dodge W.F.32 3 ton 4 x 2	None	-
U.	Dodge V.K.62 3 ton 4 x 2	U.S.	2' 9½"
U.	Dodge V.K.60 (W.K.60) 3 ton 6 x 4	U.S.	2' 11"
Br.	E.R.F. 2 C.1.4 6 ton 4 x 2	W.D.	2' 9"
Br.	Foden DG4/6 6 ton 4 x 2	W.D.	2' 10"
Br.	Foden DG6/10 10 ton 6 x 4	W.D.	3' 2"
Br.	Foden DG6/12 10 ton 6 x 4	W.D.	3' 1"
Br.	Ford W.O.A.1. 2 str. 4 x 2	None	-

TOWING HOOKS (Continued)

		Type	Ht.
Br.	Ford W.O.A.2. H. Utility 4 x 2	None	-
Br.	Ford W.O.C.1. 8 cwt. 4 x 2	None	-
Br.	Ford W.O.T.2. 15 cwt. 4 x 2	P.P. Early W.D. Later	2' 1"
Br.	Ford W.O.T.3. 30 cwt. 4 x 2	N.F.S.	-
Br.	Ford W.O.T.8. 30 cwt. 4 x 4	W.D.	3' 1¼"
Br.	Ford W.O.T.6. 3 ton 4 x 4	W.D.	3' 1¼"
C.	Ford C11ADF H Utility 4 x 2 (C29ADF)	None	-
C.	Ford F8 8 cwt. 4 x 2	None	-
C.	Ford F.15 15 cwt. 4 x 2	W.D.	1' 10"
C.	Ford F.15A 15 cwt. 4 x 4	W.D.	2' 4"
C.	Ford F.30 30 cwt. 4 x 4	W.D.	2' 3"
C.	Ford F.6C2L 3 ton 4 x 2	W.D.	2' 1"
C.	Ford F.C. 60L 3 ton 4 x 2 (C298TFS)	W.D.	2' 4½"
C.	Ford F. 60L 3 ton 4 x 4 (C.298QF) 2' 3" Early	W.D. Later	2' 9"
C.	Ford F.G.T. F.A.T. 4 x 4 (C.291QF)	W.D.	2' 9"
C.	Ford F. 60B. S.P. Bofors	W.D.	2' 7"
U.	Ford G.P. 5 cwt. 4 x 4	J.P.	1' 9" Laden
U.	Ford G.P.W. 5 cwt. 4 x 4	J.P.	1' 9" Laden
U.	Ford GPA. Amphibian 5 cwt. 4 x 4	J.P.	2' 4"
U.	Ford 6-cyl. 3 ton 4 x 2	None	-
U.	F.W.D. H.A.R. 4 ton 4 x 4	U.S.	2' 7"
U.	F.W.D. S.U. 5/6 ton G.S.	U.S.	2' 9"
U.	F.W.D. S.U. M.A.T. 4 x 4	U.S.	2' 11½"
U.	G.M.C. DUKW. Amphibian 2½ ton 6 x 6	U.S.	3' 10"
U.	G.M.C. C.C.K.W.352 2½ ton 6 x 6	U.S.	2' 8"
U.	G.M.C. C.C.KW353 2½ ton 6 x 6	U.S.	2' 8"
U.	G.M.C. CCW353 5 ton 6 x 4	U.S.	2' 8"
U.	G.M.C. ACK353 30 cwt. 4 x 4	U.S.	2' 10"
U.	G.M.C. ACKW353 3 ton 6 x 6	U.S.	2' 6"
U.	G.M.C. AC504 3 ton 4 x 2	U.S.	2' 10"
U.	G.M.C. AFWX254 3 ton 6 x 4	U.S.	2' 7"
C.	G.M. Canada 15-cwt. 4 x 4 Armoured Truck	W.D.	2' 7"
C.	G.M. Canada 4 x 4 L.R. Car	None	-

	Type	Ht.
Br. Guy Ant 15 cwt. 4 x 2	P.P.	2' 2"
Br. Guy Quad Ant 15 cwt. 4 x 4	W.D.	2' 5"
Br. Guy Quad Ant F.A.T. 4 x 4	W.D.	2' 8½"
Br. Guy FBAX 3 ton 6 x 4	W.D.	2' 9"
Br. Hillman 10 h.p. Utility 4 x 2	None	-
Br. Humber 8 cwt. 4 x 2	None	-
Br. Humber 8 cwt. 4 x 4	None	-
Br. Humber Snipe 4 str. 4 x 2	None	-
Br. Humber H. Utility 4 x 2	None	-
Br. Humber H. Utility 4 x 4	None	-
Br. Humber Mk. II L.R.C. 4 x 2	None	-
Br. Humber Mk. III L.R.C. 4 x 4	None	-
Br. Humber Mk. IIIA L.R.C. 4 x 4	None	-
U. International HALF TRACKS, M.5, M.9, M.14	U.S.	2' 7"
Br. Karrier K.6 3 ton 4 x 4	W.D.	2' 9"
Br. Karrier C.K.6. 3 ton 6 x 4	W.D.	2' 10"
Br. Leyland Lynx 3 ton 4 x 2	W.D.	2' 2"
Br. Leyland Retriever 3 ton 6 x 4	W.D.	2' 9½"
Br. " " Breakdown	W.D.	2' 2"
Br. Leyland Hippo Mk. II 10 ton 6 x 4	W.D.	3' 3"
Br. Leyland Hippo 10 ton 6 x 4	W.D.	2' 11"
U. Mack EHID 5 ton 4 x 2	U.S.	2' 6"
U. Mack NM 5 6 ton 6 x 6	U.S.	2' 6"
U. Mack N.O. 7½ ton 6 x 6	U.S.	3' 6 5/8"
U. Mack NR 10 ton 6 x 4	U.S.	2' 11"
U. Mack LMSW 6 x 4 Heavy Breakdown	U.S.	3' 0"
U. Mack EXBX 18 ton 6 x 4 T.T.	U.S.	2' 10"
Br. Maudslay Militant 6 ton 4 x 2	W.D.	3' 0"
Br. Morris Com. TERRAPIN Amphibian 4 ton 8 x 8	None	-
Br. Morris Com. PU. 8 cwt. 4 x 2	None	-
Br. Morris Com. PU. 8/4 8 cwt. 4 x 4	None	-
Br. Morris Com. CDF 30 cwt. 6 x 4	W.D.	2' 5"
Br. Morris Com. C.D.S.W. 30 cwt. 6 x 4	W.D.	2' 6"
Br. Morris Com. CS8 15 cwt. 4 x 2	P.P.	2' 0"
Br. Morris Com. C.4 15-cwt. 4 x 2	P.P.	2' 0"
Br. Morris Com. C.8/G.S. 15 cwt. 4 x 4	W.D.	2' 5"

	Type	Ht.
Br. Morris Com. C.8 F.A.T. 4 x 4	W.D.	2' 6" 10' 50-16 tyres
	W.D.	2' 8" 10' 50-20 tyres
Br. Morris Com. C.8/P 4 x 4	W.D.	2' 6"
Br. Morris Com. C.9/B. 4 x 4	W.D.	2' 6"
Br. Morris Motors 10 h.p. Utility 4 x 2	None	-
Br. Morris Motors Mk.I.L., C. 4 x 2	None	-
Br. Morris Motors Mk.II L.R.C. 4 x 4	None	-
Br. Scammell Heavy Gun Tractor 6 x 4	W.D.	3' 0"
Br. Scammell Heavy Breakdown 6 x 4	W.D.	3' 0"
Br. Standard 12 h.p. L Utility 4 x 2	None	-
U. Studebaker WEASEL 10-cwt. Tracked	U.S.	2' 3"
U. Studebaker Amphibious Weasel 10-cwt. Tracked	U.S.	2' 3"
U. Studebaker US6 2½ ton 6 x 4	U.S.	2' 8"
U. Studebaker US 64 5 ton 6 x 4	U.S.	2' 8"
Br. Thornycroft ZS/TC4 3 ton 4 x 2	W.D.	2' 6"
Br. Thornycroft Sturdy WZ/TC4 3 ton 4 x 2	N.F.S.	-
Br. Thornycroft WOF/AC4 3 ton 6 x 4	W.D.	2' 9"
Br. Thornycroft WOF/DC4 3 ton 6 x 4	W.D.	2' 9"
Br. Thornycroft Amazon 6 ton 6 x 4	None	-
Br. Tilling Stevens TS 20/2 3 ton 4 x 2	W.D.	2' 7"
U. Ward La France 4 ton 6 x 6	U.S.	2' 9"
U. White M3A1 15 cwt. 4 x 4	U.S.	2' 4"
U. White 666 6 ton 6 x 6	U.S.	2' 6"
U. White 1064 10 ton 6 x 4	U.S.	2' 11"
U. White 920 18 ton 6 x 4 T.T.	None	-
U. White 922 18 ton 6 x 4 T.T.	U.S.	3' 0"
U. Willys 5 cwt. 4 x 4	J.P.	1' 9" Laden

TRAILER DRAW EYE HEIGHTS

These are the corresponding figures to the Tractor Tow Hook Heights given on page 808.

* Indicates that early production trailers of this type were fitted with the small 2" internal diameter Draw eye which is designed to fit the Jeep Hook and 15 cwt. Truck Pin and Pintle, but will not fit the standard W.D. hook. Later production trailers of this type are now fitted with the 2 13/16" i.d. Draw Eye which will accept any hook. Early types have been modified as required in the service.

U.	Trailer 5-cwt. 2 wh. Amphibious G.S. (The Jeep Trailer)	1' 9 3/8"	
Br.	Trailer 10-cwt. 2 wh. Lightweight G.S.	1' 8¾"	*
Br.	Trailer, 2 wh. Lightweight Technical Types as follows:-		
Br.	Circular Saw	2' 0"	*
Br.	Compressor	2' 0"	*
Br.	Electrical Repair	1' 11"	*
Br.	Generator 5 k.w.	2' 0"	*
Br.	Machinery	1' 11"	*
Br.	Store	1' 11"	*
Br.	Type "Z"	1' 11"	*
Br.	Water, 100-gall.	1' 9"	*
Br.	Welding	2' 0"	*
Br.	Trailer, 10-cwt. 2 wh. G.S.	2' 1½"	*
Br.	Trailer, 10-cwt. 2 wh. D.F.	2' 1½"	
Br.	Trailer, 10-cwt. 2 wh. Generator, 6 k.w.	2' 1½"	*
Br.	Trailer, 10-cwt. 2 wh. Mortar	2' 1¼"	
Br.	Trailer, 15-cwt. 2 wh. G.S.	2' 7"	
Br.	Trailer, 15-cwt. 2 wh. Dental	2' 3"	
Br.	Trailer, 15-cwt. 2 wh. Pole Carrying	2' 7"	
Br.	Trailer, 15-cwt. 2 wh. Water, 180 gall. (See also latest 1 ton Water Trailer)	2' 7"	
Br.	Trailer, 15-cwt. 2 wh. Workshop Servicing	2' 1 1/8"	

Br.	Trailer, 1 ton 2 wh. G.S.	
	Original Mk.I type	2' 6¼"
	Latest Mk.II type (Has two position hook)	2' 7" High Ht. 2' 4" Low Ht.
Br.	Trailer, 1 ton 2 wh. A/A. Command Post	2' 5"
Br.	Trailer, 1 ton 2 wh. Generator 9 k.v.a.	2' 6¼"
Br.	Trailer, 1 ton 2 wh. Generator 9 k.w.	2' 6¼"
Br.	Trailer, 1 ton 2 wh. Machinery Gas Welding (Has four alternative positions)	2' 10" High Ht. 2' 1" Low Ht.
Br.	Trailer, 1 ton 2 wh. Pigeon Loft, 60 Bird	2' 7"
Br.	Trailer, 1 ton 2 wh. Sterilizer	2' 6½"
Br.	Trailer, 1 ton 2 wh. Water 180 gall.	2' 7"
Br.	Trailer, 1 ton 2 wh. Water Purification	2' 8½"
Br.	Trailer 2 ton 2 wh. Generating Set	
	Early Unsprung Design	2' 8"
	Latest Sprung Design	2' 9"
Br.	Trailer 2 ton 2 wh. Radar A/A No.3 Mk.III	2' 8"
Br.	Trailer, 2 ton 4 wh. G.S. Dropside	2' 2"
C.	Trailer, 2 ton 4 wh. G.S. Open	2' 4"
Br.	Trailer, 2 ton 4 wh. Motor Boat	2' 0"
Br.	Trailer, 2 ton 4 wh. Beam Wireless	2' 3¼"
Br.	Trailer, 2 ton 4 wh. Predictor A/A No.7 Mk.I	2' 1"
Br.	Trailer, 2 ton 4 wh. Projector	1' 3"
U.	Trailer, 2 ton 4 wh. Smoke Generator (Esso Smoke Generator)	2' 6" (approx.)
Br.	Trailer, 2 ton 4 wh. Workshop Servicing	1' 3"
Br.	Trailer, 3 ton 2 wh. Cable Drum	2' 5"
Br.	Trailer, 3 ton 4 wh. Projector	1' 7"
Br.	Trailer, 3 ton 4 wh. Radar A/A No.3 Mk.IV	1' 7"
Br.	Trailer, 4 ton 4 wh. A.O.D. W/BINS	2' 2"
Br.	Trailer, 4 ton 4 wh. A.O.D. W/DAVIT	2' 2"
Br.	Trailer, 4 ton 4 wh. Bakery Types	2' 4"
Br.	Trailer, 4/5 ton 4 wh. G.S.	2' 6"
	Trailer 5 ton 4 wh. G.S. (British type)	2' 8"

TRAILER DRAW EYE HEIGHTS (contd.)

U.S. Trailer 5 ton 4 wh. G.S. "Hobbs"	3' 0"
Br. Trailer 5 ton 4 wh. Laundry Types	2' 2½"
Br. Trailer 5 ton 4 wh. Pontoon Type 50/60	2' 8"
Br. Trailer 5 ton 4 wh. Radar A/A No.1 Mk.I Receiver	1' 5"
Br. Trailer 5 ton 4 wh. Radar A/A No.1 Mk.I Transmitter	1' 5"
Br. Trailer 5 ton 4 wh. Radar A/A No.1 Mk.I	1' 5"
Br. Trailer 5 ton 4 wh. Radar A/A No.1 Mk.II Transmitter	1' 5"
Br. Trailer 5 ton 4 wh. Radar A/A No.1 Mk.II Transmitter	1' 5"
Br. Trailer 5 ton 4 wh. Radar A/A No.3 Mk.II (Also known as TR type)	2' 6"
Br. Trailer 5 ton 4 wh. Smoke Generator (or Haslar Trailer)	2' 0"
U. Trailer 6 ton 4 wh. Carriage of Tracks (Formerly known as JAHN Trailer)	3' 0"
Br. Trailer, 7½ ton 6 wh. Light Recovery	2' 3"
U. Trailer, 10 ton 6 wh. Low loading	2' 8"
Br. Trailer 18 ton 4 twin wh. Carriage of Excavator (Ruston Bucyrus or Ransome Sim)	2' 6"
Br. Trailer 18 ton 6 wh. Carriage of Excavator or Tractor (Carrimore Low loading)	2' 5"
Trailer 20 ton 8 twin wh. Low loading (Multiwheeler, S.M.T.)	3' 3"
Trailer 20 ton 8 wh. Carriage of Tractor or Excavator	2' 11"
Trailer 20 ton 8 twin wh. Transporter (Latest Crane design)	2' 8½"
Trailer 12 twin wheel 30 ton Rail Carrier or 20 ton cargo Carrier	2' 8½"
Trailer 40 ton 12 twin wh. Transporter	
British Mk.I Design (Crane)	2' 8½"
British Mk.II Design (Dyson)	2' 8½"
U.S. design made by Rogers, Winter-Weiss & Fruehauf	2' 6"
Trailer 45 ton Tracked Tank Recovery	2' 7"

RECOMMENDED MAXIMUM TRAILED WEIGHT
TO BE TOWED BEHIND ARTILLERY AND BREAKDOWN TRACTORS

It is assumed in all the undermentioned trailed weight figures that the tractors are laden as shown in Weight Sheets of this Data Book. When travelling light, extra ballast will have to be added in order to maintain figures as shown below.

MAKE AND TYPE	Drive	Maximum Recommended Trailed Weight in Tons		
		Normal Roads	Hilly Roads	Cross Country
Morris C8. Field Artillery Tractor	4 x 4	4	3	3
Guy Quad Ant. " " "	4 x 4	NR	NR	NR
Chevrolet CGT. " " "	4 x 4	4	3	3
Ford FGT. " " "	4 x 4	4	3	3
A.E.C. Matador Medium Artillery Tractor	4 x 4	15	10	6.5
F.W.D. S.U.	4 x 4	15	10	6.5
Albion CX22S Heavy Artillery Tractor	6 x 4	15	10	6.5
Scammell Heavy Artillery Tractor	6 x 4	20	15	10
Mack NM5 6 ton 6 x 6	6 x 6	20	15	10
White 666 6 ton 6 x 6	6 x 6	20	15	10
Mack NO 7½ ton 6 x 6	6 x 6	20	15	10
*Diamond T.980 Tractor for 40-ton Trailer	6 x 4	50	50 (X)	N.R.
Morris CDSW. Light A.A. Tractor (Bofors)	6 x 4	4	2	2
Bedford QLD " " " "	4 x 4	6	4	3
Chevrolet Light A.A. Tractor (Bofors)	4 x 4	6	4	3
Ford Light A.A. Tractor (Bofors)	4 x 4	6	4	3
Morris C8. 2 pdr. A.T. Portee	4 x 4	4	3	3
Chevrolet 8440 " " "	4 x 4	4	3	3
Austin 6 pdr. Portee and Fire	4 x 4	6	4	3
Bedford " " " "	4 x 4	6	4	3
Chevrolet Tractor A/T 17 pdr.	4 x 4	4	3	3
Morris Tractor A/T 17 pdr.	4 x 4	3	2	2
Morris C.9.B. S.P. Bofors	4 x 4	3	2	N.R.
Scammell Heavy Breakdown Tractor	6 x 4	20	15	10
Mack LMSW " " "	6 x 4	15	10	8
Ward la France Heavy Breakdown Tractor	6 x 6	15	12	10
Diamond T (4-ton) " " "	6 x 6	15	12	10
Austin 3 ton 6 x 4 Breakdown Gantry	6 x 4	8	6	3

(815)

* Weights quoted are for 7 tons ballast. N.R. = not recommended.
(X) against hilly roads indicates extra ballast may be needed.

RECOMMENDED MAXIMUM GROSS LADEN WEIGHT OF TRAILERS WHEN TOWED BY LOAD CARRIERS WITH FULL LOAD

MAKE AND TYPE	LOAD RATING	DRIVE	Maximum Recommended Trailed weight in Tons		
			Normal Roads	Hilly Roads	Cross Country
*Bantam Car 5 cwts.	5 cwt.	4 x 4	1	1	1/2
*Ford G.P. Car 5 cwts.	5 cwt.	4 x 4	1	1	1/2
*Ford G.P.W. Car 5 cwts.	5 cwt.	4 x 4	1	1	1/2
*Willys M.B. Car 5 cwts.	5 cwt.	4 x 4	1	1	1/2
Dodge ½ Ton Weapons Carrier	10 cwt.	4 x 4	3	2	1
Bedford M.W.	15 cwt.	4 x 2	3	2	1
Ford W.O.T.2	15 cwt.	4 x 2	3	2	1
Guy Ant	15 cwt.	4 x 2	3	2	1
Morris C.S.8	15 cwt.	4 x 2	3	2	1
Chevrolet C.15	15 cwt.	4 x 2	3	2	1
Ford F.15	15 cwt.	4 x 2	3	2	1
Dodge, D.15	15 cwt.	4 x 2	3	2	1
Ford F.15 A	15 cwt.	4 x 4	3	3	1
Guy Quad Ant	15 cwt.	4 x 4	N.R.	N.R.	N.R.
Morris C.8/G.S.	15 cwt.	4 x 4	4	3	2
Chevrolet C.15 A	15 cwt.	4 x 4	4	3	1
Dodge 3/4 Ton Weapons carrier	15 cwt.	4 x 4	3	2	1
White M.3.A.1 15 cwt. Armoured Truck	15 cwt.	4 x 4	3	2	1
G.M. Canada 15 cwt. Armoured Truck	15 cwt.	4 x 4	3	2	1
International Half Tracks, M.5, M.9, M.14	15 cwt.	2 x Tracks	8	6	3
Austin	30 cwt.	4 x 2	4	2	N.R.
Bedford OX	30 cwt.	4 x 2	4	2	N.R.
Commer Q.2	30 cwt.	4 x 2	4	2	N.R.
Ford W.O.T.3	30 cwt.	4 x 2	4	2	N.R.
Morris C.D.F.	30 cwt.	6 x 4	4	2	1
Morris CD/SW	30 cwt.	6 x 4	4	2	1
Ford W.O.T.8	30 cwt.	4 x 4	6	4	3
Chevrolet (C-30)	30 cwt.	4 x 4	6	4	3
Ford (F-30)	30 cwt.	4 x 4	6	4	3
Dodge T. 203	30 cwt.	4 x 4	6	4	N.R.
G.M.C. A.C.K. 353	30 cwt.	4 x 4	6	4	N.R.

* Can be used towing 25 cwt. across country or on roads if limited life not objected to.
N.R. = not recommended.

RECOMMENDED MAXIMUM GROSS LADEN WEIGHTS OF TRAILERS WHEN TOWED BY LOAD CARRIERS WITH FULL LOAD (contd.)

MAKE AND TYPE	LOAD	DRIVE	Maximum Recommended Trailed weight in Tons		
			Normal Roads	Hilly Roads	Cross Country
Austin	3 ton	4 x 2	5	2½	NR
Bedford OY	3 ton	4 x 2	5	2½	NR
Commer Q.4	3 ton	4 x 2	5	2½	NR
Leyland Lynx	3 ton	4 x 2	6	3	NR
Thornycroft ZS/TC4	3 ton	4 x 2	6	3	NR
Tilling Stevens TS20/2	3 ton	4 x 2	6	3	NR
Chevrolet CC60L(1543 x 2)	3 ton	4 x 2	6	3	NR
Ford FC60L(EC098T)	3 ton	4 x 2	6	3	NR
Dodge V.K. 62	3 ton	4 x 2	6	3	NR
G.M.C. A.C. 504	3 ton	4 x 2	6	3	NR
Dodge D.60L (T110L)	3 ton	4 x 2	5	2½	NR
Mack. E.H.	5 ton	4 x 2	5	2½	NR
Albion F.T.11	3 ton	4 x 4	8	6	3
Austin	3 ton	4 x 4	6	4	3
Bedford Q.L.	3 ton	4 x 4	6	4	3
Ford W.O.T.6	3 ton	4 x 4	6	4	3
Karrier K.6	3 ton	4 x 4	8	6	3
Thornycroft	3 ton	4 x 4	8	6	3
Chevrolet C.60L	3 ton	4 x 4	6	4	3
Ford F.60L	3 ton	4 x 4	8	6	3
Dodge V.K.60	3 ton	6 x 4	8	6	3
G.M.C. A.C.K.W.	3 ton	6 x 6	8	6	NR
G.M.C. A.F.W.	3 ton	6 x 4	8	6	NR
A.E.C. Albion, Austin, Crossley, Guy, Leyland, Thornycroft	3 ton	6 x 4	8	6	3
F.W.D., H.A.R.	4 ton	4 x 4	6	4	3
F.W.D., S.U.	5/6 ton	4 x 4	15	10	5
Studebaker US 6	2½ ton	6 x 6	8	6	3
G.M.C., C.C.K.W. 352 and 353	2½ ton	6 x 6	8	6	3
Studebaker US 64	5 ton	6 x 4	4	2	NR
G.M.C. C.C.W. 353	5 ton	6 x 4	4	2	NR

N.R. = not recommended.

RECOMMENDED MAXIMUM GROSS LADEN WEIGHT OF TRAILERS WHEN TOWED BY LOAD CARRIERS WITH FULL LOAD (contd.)

MAKE AND TYPE	LOAD	DRIVE	Maximum Recommended Trailed weight in Tons		
			Normal Roads	Hilly Roads	Cross Country
Diamond T	4 ton	6 x 6	15	12	10
Dennis Max.	6 ton	4 x 2	4	2	N.R.
E.R.F. 2C14	6 ton	4 x 2	4	2	N.R.
Foden	6 ton	4 x 2	6	3	N.R.
Maudslay	6 ton	4 x 2	4	2	N.R.
*Albion CX 23	10 ton	6 x 4	4	N.R.	N.R.
*Albion CX 6	10 ton	6 x 4	4	N.R.	N.R.
*Foden DG/6/10	10 ton	6 x 4	4	N.R.	N.R.
*Leyland Hippo Mk.I	10 ton	6 x 4	4	N.R.	N.R.
Leyland Hippo Mk.II	10 ton	6 x 4	7½	5	N.R.
Mack. NR.	10 ton	6 x 4	"	5	N.R.
White 1064	10 ton	6 x 4	"	5	N.R.

N.R. = not recommended.

* Trailed weight can be increased by corresponding reduction of tractor weight but must not exceed ½ total tractor weight.